Advances in the Evolutionary Analysis of Human Behaviour

Series editor

Rebecca Sear, London, UK

More information about this series at http://www.springer.com/series/11457

Alexandra Alvergne · Crispin Jenkinson
Charlotte Faurie
Editors

Evolutionary Thinking in Medicine

From Research to Policy and Practice

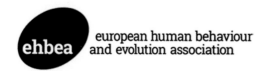

Editors
Alexandra Alvergne
Oxford
UK

Charlotte Faurie
Montpellier
France

Crispin Jenkinson
Oxford
UK

Advances in the Evolutionary Analysis of Human Behaviour
ISBN 978-3-319-29714-9   ISBN 978-3-319-29716-3   (eBook)
DOI 10.1007/978-3-319-29716-3

Library of Congress Control Number: 2016931844

Printed on acid-free paper

This Springer imprint is published by SpringerNature
The registered company is Springer International Publishing AG Switzerland

Preface

Evolution in medicine?! Never heard of it! This quote, in essence, sums up the reaction of a medical doctor who kindly accepted to review the proposal for this book. Far from substantiating the received idea according to which doctors are against any new approach to their field, it shows that health professionals know little about the relevance of evolutionary thinking for medical practice. At first, this may be surprising: the idea that evolution can inform medicine is not new—Erasmus Darwin, Darwin's grandfather and a medical practitioner, hinted at this conceptual breakthrough more than 200 years ago—and evolutionary biologists have pleaded for more evolution into medicine for about two decades. In addition, medicine is repeatedly confronted to evolution: practitioners have to deal with antibiotic resistance, the rapid changes of a virus, or the evolution of tumour cells. Yet, evolution is not part of the medical curriculum of most universities and what is more, most medical students and doctors have just "never heard of it". At second glance, however, this is not surprising.

Until recently, most evolutionary medicine publications did not really target medical practitioners or were published in evolutionary rather than medical journals. Further, most books on the topic are organized into a structure that reflects evolutionary biology sub-fields which are not familiar to medical doctors (e.g. life history theory, host–parasite co-evolution) rather than sub-fields of medicine (cardiology, oncology, obstetrics). Medicine is highly specialized and already requires a considerable amount of knowledge, and one cannot expect its practitioners to teach themselves the basis of evolutionary biology that are required to dive efficiently into the growing literature of evolutionary medicine, a still secondary discipline to medicine. But this is not the whole story. For those medics who have "heard of it", the relevance of an evolutionary framework for the practice of medicine is yet to be demonstrated. Some have argued that in the consultation room, evolutionary thinking may offer little more than a nice story to tell, but will not fix the broken arm. Are they wrong? Arguably the answer is neither yes nor no, but rather that it depends on the field of specialization, the amount of attention it has received from evolutionary scholars and the type of practical implication that is sought (communication with the patient, rethinking the hallmarks of a disease, finding new avenues in cancer therapy, etc.). Still, the question of the impact of evolutionary thinking for practice and policy is one to be asked explicitly.

This book, a collection of 23 chapters, is using a number of unique features to tackle the issues outlined above and in so doing, will enable medical doctors to think evolutionarily: (1) It is organized by medical sub-fields: obstetrics, paediatrics, nutrition, cardiology, oncology, immunity, geriatric, psychiatry and psychology. Those sub-fields that are missing are not omissions from the Editors' part, but rather the expression of a lack of evolutionary studies in some particular medical specialties. For some, it has been either impossible (e.g. toxicology, urology, gastroenterology, dermatology) or extremely difficult (e.g. cardiology) to find contributors. (2) At least half of the contributors are M.D. (medical doctors). The other half is composed of anthropologists, psychologists and population health scientists. This leads to different "cultures" in the manner with which the evolutionary approach is used to address medical issues, and we think it illustrates the richness of the applicability of the evolutionary "toolbox" to serve health and medicine. (3) Each chapter contains a lay-summary and a glossary. A special effort has been made to make the content of this book accessible to a lay reader, whether in evolutionary biology or in medicine. (4) Each chapter contains a section "Implications for policy and practice". The authors have been forcefully instructed to provide an answer to that question, however difficult it might be, pointing out whether or not such implications indeed had already emerged and/or the extent to which they were still speculative at this stage. The result is a collection of sections that contain far-reaching implications for contemporary biomedicine and/or brilliant ideas in waiting to be tested. Indeed, the point of the exercise was not to provide a definitive answer or account of an evolutionary approach to a particular medical topic, but rather to challenge the current state of knowledge and provide the reader with a new lens with which to think about health and medicine.

Although this book is first targeted at health practitioners and medical students, its guiding purpose will serve anyone who is keen on finding out about "evolutionary thinking in medicine", what it means and what it has to offer to mainstream biomedicine, be it patients, students, researchers or the general public.

July 2015   Alexandra Alvergne
                                                                                      Charlotte Faurie

Contents

Editors and Contributors

About the Editors

Alexandra Alvergne is Associate Professor in Biocultural Anthropology at Oxford University and a Fellow of Harris Manchester College in the University of Oxford. She trained as a human behavioural ecologist in France, focusing on the evolutionary and ecological determinants of male reproductive and parental behaviour. She then held a Newton International Fellowship in the Anthropology Department at University College London, where she researched how biological and cultural evolutionary processes intersect in shaping diversity in health decision-making, particularly contraceptive uptake. Now in post at Oxford University, she runs the course "Evolutionary thinking in medicine" for students in Human Sciences (BA), Archeology and Anthropology (BA) and Medical Anthropology (M.Sc.), and she is developing research programmes linking evolution, medicine and anthropology.

Crispin Jenkinson is Professor of Health Services Research, and Director of the Health Services Research Unit (HSRU), at the Nuffield Department of Population Health and a Senior Research Fellow of Harris Manchester College in the University of Oxford. He graduated from Bedford College (University of London) before coming to Oxford where he gained an M.Sc. in Psychology and then undertook research on the psychological impact of long-term illness for a D.Phil. Prior to joining the HSRU in 1992, he was a research fellow at Nuffield College, Oxford. His main research interests include patient reported outcomes and health status measurement, the evaluation of patient experiences of medical care, and methodology. He has extensive experience in developing and validating outcome measures and, in collaboration with others, has conducted randomized controlled trials in which such instruments have been primary end-points.

Charlotte Faurie is a CNRS researcher in Human Evolutionary Biology at the Institute for Evolutionary Sciences in Montpellier University, France. She trained as an evolutionary biologist, focusing on the evolution of the polymorphism of hand preference in human populations. She then held a Marie Curie Post-doctoral Fellowship in the UK, in the Department of Animal and Plant Sciences at Sheffield University, where she investigated the effects of competitive and cooperative interactions among siblings on life-history traits. Back in France, she focused on questions about parental investment, and how sexual selection shapes the evolution of cooperation in humans. She currently leads research programmes on human genetic and behavioural adaptations, and on the medicalization of birth. She teaches evolutionary biology and medicine in several master's programmes in France. She is also a student at the Medical School of Montpellier.

Contributors

Fatima Aboul-Seoud, B.A. Department of Psychology, University of Pennsylvania, Philadelphia, USA

Prof. Steven E. Arnold, M.D. Department of Neurology, Massachusetts General Hospital, Harvard University, Cambridge, USA

Prof. Helen L. Ball, Ph.D. Parent-Infant Sleep Lab, Department of Anthropology, Durham University, Durham, UK

Prof. Gillian R. Bentley, Ph.D. Department of Anthropology, Durham University, Durham, UK

Prof. David F. Bjorklund, Ph.D. Department of Psychology, Florida Atlantic University, Boca Raton, FL, USA

Eleanor Bryant, Ph.D. Division of Psychology, University of Bradford, Bradford, UK

Jorge Correale, M.D. Dr. Raúl Carrea Institute for Neurological Research, FLENI, Buenos Aires, Argentina

Prof. Bernard J. Crespi, Ph.D. Department of Biological Sciences, Simon Fraser University, Burnaby British Columbia, Canada

Robert S. Danziger, M.D. University of Illinois, Chicago, IL, USA

Jimmy Espinoza, M.D., M.Sc., FACOG Division of Maternal Fetal Medicine, Department of Obstetrics and Gynecology, Baylor College of Medicine, Houston, TX, USA; Texas Children's Hospital Pavilion for Women, Houston, TX, USA

Prof. Paul W. Ewald, Ph.D. Department of Biology, University of Louisville, Louisville, KY, USA

Bruno Faivre, Ph.D. Biogéosciences, CNRS UMR 6282, Université Bourgogne Franche-Comté, Dijon, France

Prof. Caleb E. Finch, Ph.D. Department of Neurobiology, Davis School of Gerontology, The Dornsife College, The University of Southern California, Los Angeles, CA, USA

Prof. Claudio Franceschi, Ph.D. Department of Experimental, Diagnostic and Specialty Medicine (DIMES), University of Bologna, Bologna, Italy; Interdepartmental Center "Luigi Galvani" (CIG), University of Bologna, Bologna, Italy; IRCCS Institute of Neurological Sciences, and CNR-ISOF, Bologna, Italy

Zelda Alice Franceschi, Ph.D. Department of History, Cultures and Civilizations (DISCI), University of Bologna, Bologna, Italy

Gordon G. Gallup, Jr., Ph.D. Department of Psychology, University at Albany, State University of New York, Albany, USA

Paolo Garagnani, Ph.D. Department of Experimental, Diagnostic and Specialty Medicine (DIMES), University of Bologna, Bologna, Italy; Interdepartmental Center "Luigi Galvani" (CIG), University of Bologna, Bologna, Italy

Cristina Giuliani, Ph.D. Department of Biological, Geological and Environmental Sciences (BiGeA), Laboratory of Molecular Anthropology and Centre for Genome Biology, University of Bologna, Bologna, Italy

Daniel J. Glass, M.A. Department of Psychology, Suffolk University, Boston, MA, USA

Prof. Neil Greenspan, M.D., Ph.D. Department of Pathology, Case Western Reserve University, Cleveland, OH, USA

Emanuel Guivier, Ph.D. Biogéosciences, CNRS UMR 6282, Université Bourgogne Franche-Comté, Dijon, France

Prof. Alasdair I. Houston, Ph.D. School of Biological Sciences, University of Bristol, Bristol, UK

Prof. Vadym Kavsan, Ph.D. Department of Biosynthesis of Nucleic Acids, Institute of Molecular Biology and Genetics, National Academy of Sciences of Ukraine, Kiev, Ukraine

Cédric Lippens, M.Sc. Biogéosciences, CNRS UMR 6282, Université Bourgogne Franche-Comté, Dijon, France

Karin Machluf, Ph.D. Department of Psychology, Pennsylvania State University, State College, PA, USA

George M. Martin, M.D. Department of Pathology, University of Washington, Seattle, WA, USA

John W. Pepper, Ph.D. Division of Cancer Prevention, Biometry Research Group, National Cancer Institute, Bethesda, MD, USA

Emma Pomeroy, Ph.D. Newnham College, University of Cambridge, Cambridge, UK; Department of Archaeology and Anthropology, University of Cambridge, Cambridge, UK

Ian J. Rickard, Ph.D. Department of Anthropology, Durham University, Durham, UK

Charlotte K. Russell, Ph.D. Parent-Infant Sleep Lab, Department of Anthropology, Durham University, Durham, UK

P. Douglas Sellers II, Ph.D. Department of Psychology, Pennsylvania State University, State College, PA, USA

Artemis P. Simopoulos, M.D. The Center for Genetics, Nutrition and Health, Washington, DC, USA

Gabriele Sorci, Ph.D. Biogéosciences, CNRS UMR 6282, Université Bourgogne Franche-Comté, Dijon, France

Kristina N. Spaulding, Ph.D. Department of Psychology, University at Albany, State University of New York, Albany, USA

Aleksei Stepanenko, Ph.D. Department of Biosynthesis of Nucleic Acids, Institute of Molecular Biology and Genetics, National Academy of Sciences of Ukraine, Kiev, Ukraine

Jay T. Stock, Ph.D. Department of Archaeology and Anthropology, University of Cambridge, Cambridge, UK

Holly A. Swain Ewald, Ph.D. Department of Biology, University of Louisville, Louisville, KY, USA

Prof. emeritus Bernard Swynghedauw, M.D. U942-INSERM, Hôpital Lariboisière, Saint-Rémy-lès-Chevreuse, France

Pete C. Trimmer, Ph.D. School of Biological Sciences, University of Bristol, Bristol, UK

Prof. Stanley Ulijaszek, Ph.D. Unit for Biocultural Variation and Obesity, School of Anthropology, University of Oxford, Oxford, UK

Somogy Varga, Ph.D. Philosophy Department, University of Memphis, Memphis, USA

Lane E. Volpe, Ph.D. The Implementation Group, Colorado, USA; Parent-Infant Sleep Lab, Department of Anthropology, Durham University, Durham, UK

Prof. Jonathan C.K. Wells, Ph.D. Childhood Nutrition Research Centre, UCL Institute of Child Health, London, UK

Chapter 1
Applying Evolutionary Thinking in Medicine: An Introduction

Prof. Gillian R. Bentley, Ph.D.

Lay Summary Evolutionary thinking is beginning to infiltrate medical practice and has the potential to transform how clinicians explain human diseases. Evolutionary medicine takes a long-term view of why humans suffer from various diseases and addresses the reasons behind these. Proponents of this relatively new field argue that clinicians need to understand basic concepts in evolutionary biology and that these should be embedded in the training students receive in medical schools. Historically, in the late nineteenth and early twentieth centuries, medical writings did include evolutionary concepts, but this approach fell out of favour following the excesses of the Second World War. Evolutionary medicine emerged again in the 1990s and has slowly been building momentum around the world with journals, societies, books, and papers expanding in number and visibility. Although biologists and other scientists have been the main proponents, a growing number of physicians and medical students are becoming involved as the field reaches a new maturity.

1.1 A Shift in Perspective in Approaching Medical Issues

What is evolutionary thinking in medicine? In brief, it is the application of basic evolutionary principles derived from the science of biology to understand human susceptibility to disease [1–4]. But it is so much more than this! An evolutionary approach to health and diseases addresses how past and present pathogens, with which we now coexist, behave and change over time [5], is concerned with how individual development within specific environmental contexts can shape suscep-

G.R. Bentley (✉)
Department of Anthropology, Durham University, Durham, UK
e-mail: g.r.bentley@durham.ac.uk

© Springer International Publishing Switzerland 2016
A. Alvergne et al. (eds.), *Evolutionary Thinking in Medicine*,
Advances in the Evolutionary Analysis of Human Behaviour,
DOI 10.1007/978-3-319-29716-3_1

tibilities over the life course [6–8], considers how major changes in human lifestyle such as urbanisation and industrialisation have altered the epidemiological nature of illnesses that can afflict us [9, 10], takes into account phylogenetic history in relation to our vulnerability to specific conditions such as lower back pain [11], and attempts to understand human lifespan and mortality within a broader context of why ageing should exist at all among species [12], among many other topics [3–18]. More recently, evolutionary medicine is reaching out to embrace veterinary practitioners since the two disciplines have much to learn from each other [19, 20].

1.1.1 Novel Questions

So, what are the basic principles that inform an evolutionary approach to medicine? One of the first distinctions between this and clinical medicine is that the former addresses "ultimate" (i.e. evolutionary) questions about health and disease as opposed to "proximate" (i.e. mechanistic) ones [3]. In this respect, Randy Nesse has frequently referred to one of the great biological thinkers, Nikolaas Tinbergen, who contributed to the development of the field of ethology or animal behaviour. Tinbergen developed a set of four questions dealing with mechanism, ontogeny, phylogeny, and function with which to address evolutionary questions about behaviour [3, 21, 22]. These are suggested as highly useful when comparing the kinds of questions asked by medical doctors as opposed to evolutionary biologists, but both ultimate and proximate questions are viewed as complementary. Tinbergen's first question asks about the "proximate" cause of a trait (similar to the kinds of questions a medical doctor might ask), the second addresses immediate developmental issues, the third deals with the development or evolution of a trait on an ultimate level in comparison with other species and over long evolutionary time, while the fourth addresses issues of adaptation in asking about how the trait might affect reproduction and survival.

An example of proximate (clinical) and ultimate (evolutionary) approaches to medicine can illustrate the difference between the two. If presented with a patient suffering from asthma, a medical doctor would presumably take a case history of the patient and would ask about family susceptibility to the condition, the length of time that the patient had experienced symptoms, the degree of severity of the symptoms, and the potential exposures that might trigger the condition. These exposures might relate to the immediate environment at home and elsewhere where the patient was spending their time. The doctor might consider allergens such as dust mites, household cleaners, pollen, and pets. The patient might be referred to a clinic for allergy testing for reactions to specific substances that could then be ruled out as irritants. In contrast, evolutionary medicine approaches illnesses such as allergies and asthma from a more long-term (the "ultimate") perspective. There is an extensive literature arguing that autoimmune disorders such as allergies (including asthma) have become prevalent within contemporary societies since we became removed from ancestral conditions where we coexisted with several pathogens, including intestinal worms called helminths [23–25]. There is evidence to suggest that immunoglobulin E (IgE)—which becomes elevated with allergic conditions—in fact coevolved to respond to our earlier and common coexistence with helminths, which were down-regulating the system for their own benefit. In our cleaner and more hygienic environment, in the absence of helminths, the IgE system is dys-regulated and responds instead to other foreign bodies resulting in autoimmune disorders such as allergies.

Much of this autoimmune topic is related to the "hygiene hypothesis" (see also Chaps. 15 and 17) where recent conditions of extreme cleanliness are thought to have detrimental consequences for the human immune system [25–27]. There is again growing evidence that humans need to "educate" their immune system during development by exposure to a wide variety of bacteria and other organisms (including helminths). The lack of such exposures might trigger a range of autoimmune disorders. Of course, it could be argued that knowledge of human evolutionary history, and the kinds of environments in which we lived in our past, is irrelevant to the way in which doctors treat their patients. But it is precisely this kind of knowledge that is leading to novel treatments of various autoimmune disorders by re-exposing individuals to what are sometimes called "old friends", meaning precisely helminths [28]. Preclinical trials are now evaluating the safety of infecting patients with more benign forms of helminths, such as pinworms, in an effort to achieve remission for a variety of autoimmune disorders including allergies, irritable bowel disease, and multiple sclerosis [29, 30] (see Chap. 17 for the most recent research in this area). It is doubtful whether these kinds of treatments could have arisen without an understanding of human coevolution with other organisms and how this has shaped the human phenotype. Similar understandings are now leading to an appreciation of how contemporary environments are dramatically changing our gut (and other) microbiomes, leading to novel and sometimes serious disorders [31, 32].

1.1.2 Research Areas, Concepts, and Assumptions

The example of allergies falls into one of the themes ("abnormal environments") that George Williams and Randy Nesse originally conceived in their landmark paper in the *Quarterly Review of Biology*, in 1991 [33], in order to create a structure with which to view a variety of human illnesses. They outlined 5 major areas where they felt that an evolutionary approach could make a contribution towards understanding health and illness. These were (1) infectious diseases (see Chaps. 14–17 and 19), (2) host–parasite coevolution (see Chap. 16), (3) injuries, breakdown, and toxins (Chaps. 10, 18 and 23), (4) genetic effects on diseases, and (5) abnormal environments (Chaps. 4–9 and 11). Randy Nesse was to develop this set of themes further in his later articles. "Abnormal environments" became subsumed under the now more common name of "mismatch", popularised in a book by Gluckman and Hanson in 2006 [9] although this was also to develop a more specific meaning in relation to early life development.

Authors supporting evolutionary approaches to medicine have gone on to argue that medical doctors need to understand the concept of adaptation and natural selection [2–4, 34, 35]. First, they have frequently pointed out that natural selection (which is just one of the forces shaping human evolution) works extremely slowly—on the order of thousands of years [36]. This fact has frequently led to the misunderstanding, even among people trained in evolutionary thinking, that humans have stopped evolving. An exemplar of this kind of misunderstanding is the concept of the "environment of evolutionary adaptedness" (EEA) introduced by the psychologist, John Bowlby, in 1969 [37] and later adopted, par excellence, by the Santa Barbara School of evolutionary psychology [38–39]. The EEA theory posits that humans are essentially adapted to the environment in which we spent thousands of years as Palaeolithic foragers, although the precise date and provenience of this environment, as well as the entire concept, have been heavily debated [40–41]. However, the concept of an EEA has tied in nicely with the theme of "mismatch" or "abnormal environments", where humans are frequently seen to be maladapted to contemporary environments where diets rich in fats and sugars and a highly sedentary lifestyle can lead to chronic conditions such as obesity and metabolic disorders [9, 10]. More recent molecular research that is rapidly expanding has, however, pointed out that humans are undergoing constant microevolution in response to both mutations and changing environments [42–45]. The growing field of epigenetics that examines how molecular markers turn genes on and off, or up and down, is also contributing to our growing understanding of how developmental and life course plasticity can alter the human phenotype [e.g. 46–48].

1.2 Contributions to Biomedicine

1.2.1 Rethinking the Optimal Body

Humans and other organisms have evolved features that are basically "jury-rigged compensations for a fundamentally defective architecture" [36: 63]. As our natural environments altered, causing various features to evolve, including bipedalism, evolution had to work on an existing structural design and to modify this where possible to develop features that would be adapted to a changing environment. Nesse's quote above is in relation to the vertebrate eye which is subject to several potential malfunctions as a result of the legacy of "jury-rigged" design compromises, or what has been called elsewhere "historical legacies" [15]. Nesse refers to myopia, detached retinas, and glaucoma as some examples of how the human eye can fail due to its intrinsic and evolved design [36]. Similarly, humans are vulnerable to a series of back problems not experienced by our quadrupedal relatives as a result of our evolution towards an upright posture [15]. Understanding these kinds of evolutionary constraints and the inevitable "trade-offs" in our physiology can be helpful in considering human vulnerabilities and potential treatments for many ubiquitous problems.

Myopia, or short-sightedness, which seems to be a problem caused through gene–culture (design and environment) interaction or coevolution is another excellent example of human developmental vulnerability and "mismatch". Human cultural evolution has led to children spending many hours indoors, away from sunlight, and in close-up work such as reading or electronic screens [49–51]. These practices during childhood, when the eye is still growing, have created a situation where approximately half of the individuals in Europe and the USA are now myopic, while the global development of Southeast Asian countries is leading there to what is described as an epidemic of myopia [50, 51].

The sum of such vulnerabilities forces us to re-evaluate the human body, not as an optimally designed machine, but rather as a series of compromises that indeed has left us vulnerable to a variety of conditions, particularly as we age. In fact, ageing represents the ultimate "trade-off" in evolutionary terms. As so elegantly expressed by George Williams [52, 53], ageing itself has evolved as a by-product of repro- ductive effort earlier in life. The basic point here is that life itself has not evolved to promote personal happiness and longevity and cannot continue without successful reproduction and surviving offspring which are favoured despite the trade-offs. The starkness of this biological statement makes for uncomfortable reading for highly cultural organisms that have developed traits that do indeed (in particularly favourable environments) promote health, longevity, and happiness, sometimes at the expense of individual reproduction [52, 53]. This should not, however, lead us to dispute the clinical significance of vulnerabilities in evolutionary design, and could help in understanding the source of individual disease and decline.

As an exemplar of the ultimate outcome of evolutionary success, as we age, humans and other sexually reproducing organisms suffer from what is termed the "declining force of natural selection" [54]. Our genes are passed on through reproduction, which generally occurs earlier in life. Traits that are maladaptive early in life would tend to be "selected out" because they would be passed on to our offspring who might not survive with these characteristics. However, deleterious traits that are expressed later in life, when successful reproduction is less likely, will not be selected against and can contribute to the ageing process. Furthermore, traits that are beneficial earlier in life and that promote reproductive effort might be associated with negative effects later on in life, a trade-off known as "antagonistic pleiotropy" [52, 53]. Again, this situation promotes the ageing process. Understanding these concepts can be helpful in researching and treating senescent conditions (see Chaps. 18, 19, 21).

Some design features of the human body are more vulnerable than others due to chance or stochastic events in our evolutionary history, as well as the possible action of other evolutionary forces aside from natural selection. These other forces are mutation, genetic drift, and gene flow (otherwise known as migration). Genetic mutations are, in fact, relatively rare and, by their nature, stochastic and are likely to have had a fairly minimum impact on the evolution of specific traits. Exceptions can occur where gene mutations affect control regions. An example of this is where humans evolved the capacity to digest lactose in adulthood—one of the better documented cases of recent human gene–culture coevolution [55–57]. Genetic drift occurs in situations where small populations representing a sample of a wider gene pool remain in relative isolation and develop unique or specific genetic profiles as a result of their small sizes and isolation (this could be through geography or cultural practices). Where such isolated populations expand, a phenomenon known as the "founder effect" can occur, where specific and deleterious traits that occur by chance in the sample population increase in representation. Examples of such founder effects exist among the Amish in the USA where polydactyly is relatively common [58]. Population bottlenecks can also occur at various points in time where populations suffer serious demographic decline. The remaining small populations are likely to experience genetic drift.

1.2.2 Rethinking Medical Practice

Perhaps the least misunderstood and most accepted evolutionary concept in biology and medicine is that of competition with a variety of pathogens with whom we coexist. This has led to one of the most urgent crises in modern medical care, namely the emergence of antibiotic resistance ([59–61], Box 1.1). We are in fact in an "arms race" against rapidly evolving micro-organisms such as bacteria. The crisis is so urgent that the UK Chief Medical Officer, Professor Sally Davies, stated in March 2013 that "the danger posed by growing resistance to antibiotics should be ranked along with terrorism on a list of threats to the nation" [62]. Similarly, in May 2015, Germany's Health Minister, Hermann Gröhe, opined that: "If antibiotics are no longer effective, treatment options could return to those of a pre-Penicillin age" [63]. Understanding the concept of evolved, antibiotic resistance has led to a growing recognition to limit the use of antibiotics in everyday medical settings to cases where it is clear that a patient is in fact suffering from a bacterial (as opposed to a viral) infection, or in cases of less severe infections to allow patients to recover by themselves [64]. A similar adaptive approach is being discussed in relation to cancer chemotherapy in order to limit the potentially damaging effect of emerging resistant tumour cells [65, 66].

> **Box 1.1 Adaptation and Natural Selection: The Example of Antibiotic Resistance**
>
> **Antibiotic Use** Antibiotics were first developed for medical use in the early twentieth century, primarily to treat bacterial infections, although they also work against fungi and protozoa. As the pharmaceutical industry grew, and antibiotics were readily available, they were increasingly prescribed and have also been used prophylactically and extensively in farming to prevent loss of livestock from common infections. Unwittingly, the overuse—and sometimes inappropriate use—of antibiotics has led to the evolution of antibiotic resistance among communities of bacteria. What does this mean?

Fundamentals of Natural Selection As pointed out by Darwin in the "The Origin of Species" [90], natural selection requires three key features: first, organisms within a species must vary in their characteristics or traits. Secondly, traits must be heritable which means they can be passed on to future generations. Thirdly, the variants must have differing reproductive value in a given environment, some being fitter (i.e. reproducing more) than other. Several bacterial traits meet these preconditions, allowing the evolution of different frequencies of variants in their population.

Antibiotic Resistance Resistant bacteria are "selected for" if they have acquired—through inheritance or random mutation—characteristics which enable them to survive being targeted with antibiotics. Individual bacteria from a population might just by chance have properties that confer resistance, meaning they are better adapted. Surviving bacteria can then reproduce and pass the resistant traits to their descendants. The resulting population has evolved to resist the previously effective antibiotic (Fig. 1.1). The chances of resistance arising are related both to variability present in a population and to the rate at which it reproduces, as each round of reproduction will produce new variants into a population pool (either through replication errors or through recombination events). The reason why antibiotic resistance can evolve so quickly is because of the rapid reproductive rate of these tiny organisms, which can take from minutes to about 24 h.

Resistance to Evolutionary Terminology Despite these evolutionary fundamentals, Antonovics et al. [91] have pointed out a discrepancy in evolutionary terminology in academic papers that discuss antibiotic resistance, and they urge its increasing use. Specifically, medical papers only used evolutionary terms 3 % of time (preferring words like "emergence") compared to 68 % for evolutionary biologists. Antonovics and colleagues speculate that resistance to evolutionary terms may have been encouraged to avoid "controversy". In some countries, resistance to evolutionary thinking is linked to strong religious sentiment and may be influencing how antibiotic resistance is described. Note that the website for the Center for Disease Control, the US watchdog for infectious and other diseases around the world, also omits the term evolutionary in describing antibiotic resistance [92]. Garry Trudeau, the well-known creator of the satirical comic strip, Doonesbury, played on this theme by suggesting a creationist patient diagnosed with tuberculosis should be treated with older antibiotics that no longer work against current strains of the disease [93].

A more controversial area is that of "defence", i.e. viewing certain physiological, pathogenic reactions as adaptive in nature rather than as negative consequences to illness that must be treated [15]. An obvious area where this is discussed is how to treat fevers in patients. It is increasingly recognised that a raised temperature is a physiological, adaptive response to an invading pathogen [67]. Raising the body's

Fig. 1.1 Resistance acquired
through selection by
antibiotics

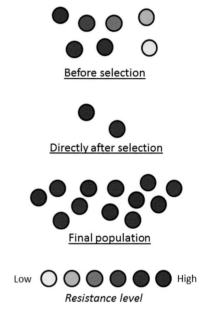

temperature inhibits the ability of a pathogen to overcome the body's defences. Fevers, particularly in children, were also treated in the past in order to prevent febrile convulsions, but it is now recognised that fevers alone do not cause this reaction, as outlined in UK National Institute for Clinical Excellence (NICE) guidelines [68].

A less recognised area of "defence" is developing in emergency medicine. Mervyn Singer, a consultant in intensive care medicine at University College London Hospitals, examined medical records for survivors from major historical battles such as Trafalgar and Waterloo [69–71]. Surprisingly, only a relatively small proportion of men died from trauma directly sustained on the battlefields (as opposed to those that suffered from infections as a sequela to injuries). These statistics led to a re-evaluation of the impact of many medical interventions that have come into use in emergency medicine, such as blood transfusions and ventilation designed to mitigate against blood loss and low oxygen levels in critical care patients. Singer [71: 1] has argued that emergency medicine has, in fact, "overventilated, overfluidised, overfed, overtransfused, and oversedated, and that these all contributed significantly to harm". Instead, emergency medicine needs fewer interventions that interfere with natural physiological adaptations designed to try and keep the individual alive in the face of massive trauma.

Singer [71] has also argued that: "We deferentially followed the seemingly unassailable logic that normal healthy values would provide the optimal milieu for either maintaining organ function or hastening recovery". This leads to a consideration of how doctors approach the concept of "norms". It is reasonable that doctors use concepts of "normal" values for a number of diagnostics to evaluate the health of patients. In contrast, evolutionary biology embraces the concept of variation, since this is essential for the process of natural selection and adaptation. Appreciation of the importance of individual variation is filtering into medicine and forms the foundation for advocates of a more personalised medicine. It is also increasingly applicable to our understanding of tumour evolution and the unique properties of cancer in treating individual patients [72–74].

1.3   Doctors and Evolution: An Evolving Relationship

Given the applicability of evolutionary perspectives to medicine, why are evolutionary concepts unfamiliar to so many doctors? The two areas have not always been divorced. As pointed out by Fabio Zampieri [75], from approximately 1880 to 1940, Darwinian approaches to medicine enjoyed an early heyday referred to by him as "medical Darwinism", where many of the themes which today find resonance and relevance among practitioners were also prominent, including how infectious diseases might evolve, how "civilisation" might create predispositions to specific illnesses, and a concern with cancer. However, a preoccupation with heredity and inherited susceptibility to disease (known as diathesis and later constitutionalism) played into a growing interest in eugenics, a term coined in 1883 by Darwin's cousin, Francis Galton, who was impressed by the principles of selective breeding shown so clearly in relation primarily to plants and pigeons in Darwin's famous book "The Origin of Species", and who sought to bring these principles to bear on "improving" the human condition. The misuse of eugenic principles during the 1940s onwards led to a socially mandated suppression for several decades of any kind of Darwinian approaches to health [75].

Since the publication of the 1991 landmark article "Dawn of Darwinian Medicine" by Williams and Nesse [33], the problem in the revival of evolutionary medicine has been in attempting to demonstrate to clinicians the relevance and utility of evolutionary approaches. Fortunately, the field has developed far enough that considerable progress is already being made, and evolutionary medicine can perhaps be said to be entering into a new maturity. There are a number of relevant, edited books that have been published that are also suitable as teaching texts [13, 17, 18], two primary textbooks [14, 76], and other single-authored volumes [15, 16, 77]. The field has a new International Society, *the International Society for Evolutionary Medicine and Public Health* (ISEMPH) that met for the first time in March 2015 in Tempe, Arizona. There are two new journals that began in 2012 and 2013, respectively, the *Journal of Evolutionary Medicine* published by Ashdin and *Evolutionary Medicine and Public Health* published by Oxford University Press, and there are now a myriad of articles that have been published in other places that are far too numerous to mention.

Earlier, however, it would be fair to say that a few different strands of interests were developing within evolutionary medicine without a great deal of overlap between contributors. Aside from the growing number of articles by Randy Nesse, two volumes edited by Stephen Stearns and Jacob Koella were published in 1999 [78] and 2008 [17] that focused primarily on topics related to host–pathogen coevolution, and genetics and vaccine development, although a couple of articles related to ageing and reproductive health overlapped with some social science topics. Edited volumes also appeared from scholars primarily in anthropology, concentrating on topics derived more from a social science perspective including maternal and infant health, and environmental mismatch [18, 79]. A fourth approach emerged somewhat later and has been spearheaded by physicians whose research represents the field of early life development or foetal programming, stimulated by work in the 1980's by David Barker at the University of Southampton. This parallel group, with strong overlap with some topics relevant to evolutionary medicine, developed into the *International Society for Developmental Origins of Health and Disease* (DOHaD) in 2003. The Society also established the *Journal of Developmental Origins of Health and Disease* in 2009. As stated on its website, its main aim is for "the scientific exploration of early human development in relation to chronic disease in later life" [80] (see Chap. 6), and it clearly has overlapping interests with the theme of "mismatch" from evolutionary medicine. The DOHaD group meets every two years in different locations around the globe, with the 2017 meeting in Cape Town, South Africa. In relation to evolutionary medicine, this developmentally-focused subfield has been represented primarily by Peter Gluckman, Alan Beedle, and Mark Hanson, who also wrote the first textbook in evolutionary medicine [14], albeit with a strong focus representing their particular interests. Finally, Paul Ewald (see Chap. 14) wrote several early influential articles and a book in 1994 (*The Emergence of Infectious Diseases*, [5]) that was to have a profound effect on the field of evolutionary medicine. In fact, Williams' and Nesse's section on infectious diseases in their 1991 article was heavily influenced by an earlier [81] article by Ewald. The latter is also editor in chief of the *Journal of Evolutionary Medicine*. Happily, the original disaggregation of the field into distinct sub-areas is disappearing, as evidenced by the converging of many authors into more recent edited volumes, or as coauthors of papers, perhaps representing a maturation of the field as it develops. A unifying figure across most of these separate strands has been Randy Nesse, who now heads the new *Center for Evolution and Medicine* at the Arizona State University [82], and spearheaded the inaugural international meeting in 2015 of the ISEMPH that was also hosted at Tempe, Arizona [83].

Nesse has also been one of the most vociferous advocates that evolutionary theory should form the foundation for medical education in the future [3, 4, 35, 84]. In the last few years, a number of US and European individuals from biological and medical backgrounds, funded by the National Science Foundation and the National Evolutionary Synthesis Center (NESCent) in Durham, North Carolina, USA, have been developing new questions for the US Medical Exams—the Medical College Admissions Tests (MCATs)—that will require premedical students to have much more knowledge of evolutionary biology than was previously necessary [34]. A number of student societies for evolutionary medicine are springing up on medical school campuses, as exemplified by Michelle Blyth at the Louisiana State

University [85], and a couple of innovative medical schools are conducting Grand Rounds with evolutionary biologists (not just medical doctors) in tow [86]. Durham University began a new MSc programme in evolutionary medicine in 2011 that attracts UK intercalating medical students who have completed their fourth or fifth year of medical school and who will hopefully carry on their careers equipped with the additional tools of evolutionary thinking [87]. Training in evolutionary medicine in mainland Europe can also be obtained at the Zurich Institute for Evolutionary Medicine [88].

All of these developments should give us hope that evolutionary medicine is finally coming into its own and has reached a stage where it can gradually infiltrate into many areas of traditional medical practice. Although the field was criticised in 2012 as having more "breadth than depth" [89: 246], the growing and intimate involvement of medical doctors with the field, as exemplified by the Grand Rounds at UCLA, belies this criticism. It has been almost 25 years since the publication of Williams' and Nesse's [33] article on the "Dawn of Darwinian Medicine" and 65 years since the demise of "Medical Darwinism" [75]. Perhaps this is the dawn of a new era that we might call "embedded evolutionary medicine" when clinicians and their trainers actually embrace evolutionary concepts and join forces with evolutionary scholars interested in health issues.

Glossary

Autoimmune disorders	Occur where the body produces antibodies against its own components (called autoantibodies) and attacks specific cells in the body. The causes of such autoimmune diseases are often unknown. They include conditions such as multiple sclerosis, rheumatoid arthritis, inflammatory bowel disease, and type 1 diabetes
Foetal programming	Refers to the potential for programming for alternative phenotypes during foetal life based on the environment experienced in utero and particularly where nutritional deficits constrain optimal foetal development during gestation
Genotype	Refers to the genetic make-up of an individual
Host–pathogen coevolution	Refers to the arms race that exists between an individual organism (the host) and a variety of other organisms that can cause diseases in that host (pathogens). Both hosts and pathogens will adapt over time to their coexistence, as they are under constant selective pressure for reproduction and survival

Immunoglobulin E	(IgE) is an antibody found in mammals and thought to have evolved as mammals became infested with parasitic worms (helminths) and protozoa (including malarial parasites). This antibody is produced and becomes elevated in allergic conditions such as asthma
Phenotype	Refers to the sum of the genetic make-up of an individual (its genotype) modified by environmental influences experienced during growth, development, and maintenance across the life course
Phylogeny	The evolutionary relationships between species across long time spans
Polydactyly	A genetic condition characterised by an excess number of digits or fingers
Population bottleneck	Occurs where populations of individuals reach sufficiently low numbers and variability in the gene pool that genetic drift is likely to occur

References

1. Eaton SB, Strassman BI, Nesse RM, Neel JV, Ewald PW, Williams GC, Weder AB, Eaton SB 3rd, Lindeberg S, Konner MJ, Mysterud I, Cordain L (2002) Evolutionary health promotion. Prev Med 34:109–118
2. Gluckman PD, Low FM, Buklijas T, Hanson MA, Beedle AS (2011) How evolutionary principles improve the understanding of human health and disease. Evol Appl 4:249–263
3. Nesse RM, Stearns SC (2008) The great opportunity: evolutionary applications to medicine and public health. Evol Appl 1:28–48
4. Nesse RM, Bergstrom CT, Ellison PT, Flier JS, Gluckman P, Govindaraju DR, Niethammer D, Omenn GS, Perlman RL, Schwartz MD, Thomas MG, Stearns SC, Valle D (2010) Evolution in health and medicine. Sackler colloquium: Making evolutionary biology a basic science for medicine. In: Proceedings of the National Academy of Sciences USA, vol 107. Issue no Suppl 1, pp 1800–1807
5. Ewald P (1994) Evolution of infectious disease. Oxford University Press, Oxford
6. Barker DJP (1998) Mothers, babies and health in later life, 2nd edn. Churchill Livingstone, Edinburgh
7. Gluckman PD, Hanson MA (2004) The fetal matrix: evolution, development and disease. Cambridge University Press, Cambridge
8. Gluckman PD, Hanson MA, Pinal C (2005) The developmental origins of adult disease. Matern Child Nutr 1:130–141
9. Gluckman PD, Hanson MA (2006) Mismatch: why our world no longer fits our bodies. Oxford University Press, Oxford
10. Pollard TP (2008) Western diseases: an evolutionary perspective. Cambridge University Press, Cambridge
11. Castillo ER, Lieberman DE (2015) Lower back pain. Evol Med Public Health 2015(1):2–3

12. Partridge L (2007) Aging and evolutionary medicine. In: Nesse R (ed) Evolution and medicine: how new applications advance research and practice. The Biomedical & Life Sciences Collection, Henry Stewart Talks Ltd, London (online at http://hstalks.com/?t= BL0141566)
13. Elton S, O'Higgins P (eds) (2008) Medicine and evolution: current applications and future prospects. Society for the Study of Human Biology Series 48. CRC Press, Boca Raton
14. Gluckman P, Beedle A, Hanson M (2009) Principles of evolutionary medicine. Oxford University Press, Oxford
15. Nesse RM, Williams GC (1996) Why we get sick: the new science of Darwinian medicine. Vintage Books, New York
16. Perlman R (2013) Evolution and medicine. Oxford University Press, Oxford
17. Stearns SC, Koella JC (eds) (2008) Evolution in health and disease, 2nd edn. Oxford University Press, Oxford
18. Trevathan WR, Smith EO, McKenna JJ (eds) (2007) Evolutionary medicine: new perspectives. Oxford University Press, Oxford
19. LeGrand EK, Brown CC (2002) Darwinian medicine: applications of evolutionary biology for veterinarians. Can Vet J 43:556–559
20. Natterson-Horowitz B, Bowers K (2012) Zoobiquity: what animals can teach us about health and the science of healing. Vintage, New York
21. Nesse RM (2011) Ten questions for evolutionary studies of disease vulnerability. Evol Appl 4:264–277
22. Tinbergen N (1963) On aims and methods of ethology. Zeitschrift fur Tierpsychologie 20:410–433
23. Barnes KC, Grant AV, Gao P (2005) A review of the genetic epidemiology of resistance to parasitic disease and atopic asthma: common variants for common phenotypes? Curr Opin Allergy Clin Immunol 5:379–385
24. Hurtado AM, Frey MA, Hurtado I, Hill K, Baker J (2008) The role of helminthes in human evolution: implications for global health in the 21st Century. In: Elton S, O'Higgins P (eds) Medicine and evolution: current applications and future prospects. Society for the Study of Human Biology Series 48. CRC Press, Boca Raton, pp 151–78
25. Versini M, Jeandel PY, Bashi T, Bizzaro G, Blank M, Shoenfeld Y (2015) Unraveling the Hygiene Hypothesis of helminthes and autoimmunity: origins, pathophysiology, and clinical applications. BMC Med 13:81
26. Okada H, Kuhn C, Feillet H, Bach JF (2010) The 'hygiene hypothesis' for autoimmune and allergic diseases: an update. Clin Exp Immunol 160:1–9
27. Strachan DP (1989) Hay fever, hygiene, and household size. B Strachan DP. Hay fever, hygiene, and household size. Br Med J 299:1259–1260
28. Rook GA, Lowry CA, Raison CL (2013) Microbial 'old friends', immunoregulation and stress resilience. Evol Med Public Health 1:46–64
29. Garg SK, Croft AM, Bager P (2014) Helminth therapy (worms) for induction of remission in inflammatory bowel disease. Cochrane Database Syst Rev 1:CD009400
30. Helmby H (2015) Human helminth therapy to treat inflammatory disorders—where do we stand? BMC Immunol 16:12
31. Dethlefsen L, McFall-Ngai M, Relman DA (2007) An ecological and evolutionary perspective on human-microbe mutualism and disease. Nature 449:811–818
32. Parker W, Ollerton J (2013) Evolutionary biology and anthropology suggest biome reconstitution as a necessary approach toward dealing with immune disorders. Evol Med Public Health 2013:89–103. doi:10.1093/emph/eot008
33. Williams GC, Nesse RM (1991) The dawn of Darwinian medicine. Q Rev Biol 66:1–22
34. Hidaka BH, Asghar A, Aktipis CA, Nesse RM, Wolpaw TM, Skursky NK, Bennett KJ, Beyrouty MW, Schwartz MD (2015) The status of evolutionary medicine education in North American medical schools. BMC Med Educ 15:38

35. Nesse RM, Stearns SC, Omenn GS (2006) Medicine needs evolution. Science 311:1071–1073
36. Nesse RM (2005) Maladaptation and natural selection. Q Rev Biol 80:62–70
37. Bowlby J (1969) Attachment and loss, vol 1: attachment. Basic Books, New York
38. Barkow L, Cosmides L, Tooby J (eds) (1992) The adapted mind. Oxford University Press, Oxford
39. Symons D (1979) The evolution of human sexuality. Oxford University Press, Oxford
40. Foley R (1995) The adaptive legacy of human evolution: a search for the environment of evolutionary adaptedness. Evol Anthropol 4:194–203
41. Irons W (1998) Adaptively relevant environments versus the environment of evolutionary adaptedness. Evol Anthropol 6:194–204
42. Bolund E, Hayward A, Pettay JE, Lummaa V (2015) Effects of the demographic transition on the genetic variances and covariances of human life-history traits. Evolution 69:747–755
43. Cochran G, Harpending H (2011) The 10,000 year explosion: how civilization accelerated human evolution. Basic Books, New York
44. Hawks J, Wang ET, Cochran GM, Harpending HC, Moyzis RK (2007) Recent acceleration of human adaptive evolution. Proc Natl Acad Sci USA 104:20753–20758
45. Milot E, Mayer FM, Nussey DH, Boisvert M, Pelletier F, Réale D (2011) Evidence for evolution in response to natural selection in a contemporary human population. Proc Natl Acad Sci USA 108:17040–17045
46. Duncan EJ, Gluckman PD, Dearden PK (2014) Epigenetics, plasticity, and evolution: how do we link epigenetic change to phenotype? J Exp Zool (Mol Dev Evol) 322B:208–220
47. Jablonka E, Raz G (2009) Transgenerational epigenetic inheritance: prevalence, mechanisms, and implications for the study of heredity and evolution. Q Rev Biol 84:131–176
48. Núñez-de la Mora A, Bentley GR (2008) Early life effects on reproductive function. In: Trevathan WR, Smith EO, McKenna JJ (eds) New perspectives on evolutionary medicine. Oxford University Press, New York, pp 149–168
49. Prepas SB (2008) Light, literacy and the absence of ultraviolet radiation in the development of myopia. Med Hypotheses 70:635–637
50. Rose KA, Morgan IG, Ip J, Kifley A, Huynh S, Smith W, Mitchell P (2008) Outdoor activity reduces the prevalence of myopia in children. Ophthalmology 115:1279–1285
51. Dolgin E (2015) The myopia boom. Nature 519(7543):276–278
52. Williams GC (1957) Pleiotropy, natural selection and the evolution of senescence. Evolution 11(4):398–411
53. Mace R (2013) Social science: the cost of children. Nature 499(7456):32–33
54. Medawar PB (1952) An unsolved problem in biology. Printed lecture. HK Lewis and Company, London
55. Beja-Pereira A, Luikart G, England PR, Bradley DG, Jann OC, Bertorelle G, Chamberlain AT, Nunes TP, Metodiev S, Ferrand N, Erhardt G (2003) Gene-culture coevolution between cattle milk protein genes and human lactase genes. Nat Genet 35:311–313
56. Heyer E, Brazier L, Segurel L, Hegay T, Austerlitz F, Quintana-Murci L, Georges M, Pasquet P, Veuille M (2011) Lactase persistence in Central Asia: phenotype, genotype, and evolution. Human Biol 83:379–392
57. Tishkoff SA, Reed FA, Ranciaro A, Voight BF, Babbitt CC, Silverman JS, Powell K, Mortensen HM, Hirbo JB, Osman M, Ibrahim M, Omar SA, Lema G, Nyambo TB, Ghori J, Bumpstead S, Pritchard JK, Wray GA, Deloukas P (2007) Convergent adaptation of human lactase persistence in Africa and Europe. Nat Genet 39:31–40
58. Ruiz-Perez VL, Ide SE, Strom TM, Lorenz B, Wilson D, Woods K, King L, Francomano C, Freisinger P, Spranger S, Marino B, Dallapiccola B, Wright M, Meitinger T, Polymeropoulos MH, Goodship J (2000) Mutations in a new gene in Ellis-van Creveld syndrome and Weyers acrodental dysostosis. Nat Genet 24:283–286

59. Laxminarayan R, Duse A, Wattal C, Zaidi AK, Wertheim HF, Sumpradit N, Vlieghe E, Hara GL, Gould IM, Goossens H, Greko C, So AD, Bigdeli M, Tomson G, Woodhouse W, Ombaka E, Peralta AQ, Qamar FN, Mir F, Kariuki S, Bhutta ZA, Coates A, Bergstrom R, Wright GD, Brown ED, Cars O (2013) Antibiotic resistance-the need for global solutions. Lancet Infect Dis 13:1057–1098
60. Read AF (2014) Woods RJ (2014) Antiobiotic resistance management. Evol Med Public Health 2014(1):147
61. Shallcross LJ, Howard SJ, Fowler T, Davies SC (2015) Tackling the threat of antimicrobial resistance: from policy to sustainable action. Philos Trans R Soc Lond B Biol Sci 370 (1670):20140082
62. http://www.bbc.co.uk/news/health-21737844
63. http://www.euractiv.com/sections/health-consumers/german-government-poised-tackle-growing-antibiotic-resistance-314587
64. https://www.nice.org.uk/guidance/conditions-and-diseases/infections/antibiotic-use
65. Cunningham JJ, Gatenby RA, Brown JS (2011) Evolutionary dynamics in cancer therapy. Mol Pharm 8:2094–2100
66. Gatenby RA, Silva AS, Gillies RJ, Frieden BR (2009) Adaptive therapy. Cancer Res 69:4894–4903
67. Best EV, Schwartz MD (2014) Fever. Evol Med Public Health 2014:92
68. https://www.nice.org.uk/guidance/cg160/chapter/Key-priorities-for-implementation#antipyretic-interventions
69. Singer M, De Santis V, Vitale D, Jeffcoate W (2004) Multiorgan failure is an adaptive, endocrine-mediated, metabolic response to overwhelming systemic inflammation. Lancet 364 (9433):545–548
70. Singer M, Glynne P (2005) Treating critical illness: the importance of first doing no harm. PLoS Med 2(6):e167
71. Singer M (2013) Advancing critical care: time to kiss the right frog. Crit Care 17(Suppl 1):S3
72. Greaves M (2013) Cancer stem cells as 'units of selection'. Evol Appl 6:102–108
73. Merlo LMF, Pepper JW, Reid BJ, Maley CC (2006) Cancer as an evolutionary and ecological process. Nat Rev Cancer 6:924–935
74. Purushotham AD, Sullivan R (2010) Darwin, medicine and cancer. Ann Oncol 21:199–203
75. Zampieri F (2009) Medicine, evolution, and natural selection: an historical overview. Q Rev Biol 84:333–355
76. Stearns SC, Medzhitov R (2016) Evolutionary medicine. Sinauer Associates, Sunderland
77. Lieberman D (2013) The story of the human body: evolution, health and disease. Pantheon Books, New York
78. Stearns SC (ed) (1999) Evolution in health and disease. Oxford University Press, Oxford
79. Trevathan WR, Smith EO, McKenna JJ (eds) (1999) Evolutionary medicine. Oxford University Press, Oxford
80. http://www.mrc-leu.soton.ac.uk/dohad/index.asp?page=2
81. Ewald PW (1980) Evolutionary biology and the treatment of signs and symptoms of infectious disease. J Theor Biol 86:169–176
82. https://sites.google.com/a/asu.edu/cemph/
83. https://www.regonline.com/builder/site/tab3.aspx?EventID=1604576
84. Nesse RM (2008) The importance of evolution for medicine. In: Trevathan WR, Smith EO, McKenna JJ (eds) Evolutionary medicine. Oxford University Press, New York, pp 416–432
85. https://vimeo.com/channels/isemph2015/123341495
86. http://www.evmed.ucla.edu/emm.php
87. https://www.dur.ac.uk/anthropology/postgraduatestudy/taughtprogrammes/evolutionarymedicine/
88. http://www.iem.uzh.ch/index.html
89. Valles SA (2012) Evolutionary medicine at twenty: rethinking adaptationism and disease. Biol Philos 27:241–261
90. Darwin C (2009) The annotated origin: a facsimile of the first edition of on the origin of species. Harvard University Press, Cambridge

91. Antonovics J et al (2007) Evolution by any other name: antibiotic resistance and avoidance of the E-word. PLoS Biol 5:e30
92. http://www.cdc.gov/drugresistance/
93. http://www.gocomics.com/doonesbury/2005/12/18

Part I
Obstetrics

Chapter 2
"Foetal–Maternal Conflicts" and Adverse Outcomes in Human Pregnancies

Jimmy Espinoza, M.D., M.Sc., FACOG

Lay Summary

In most pregnancies, there is a delicate balance between foetal demands and the maternal supply of nutrients; however, in some instances, abnormal foetal–maternal interactions can lead to pregnancy complications. These abnormal interactions can occur at the uteroplacental interface, in the placental vascular system and at the level of foetal–maternal signalling. Some of the consequences of abnormal foetal–maternal interactions include pregnancy complications such as foetal growth restriction, pre-eclampsia, gestational diabetes, preterm parturition and in extreme cases foetal death. We propose that an absolute reduction in the blood flow to the uteroplacental unit may participate in the mechanisms of disease in foetal growth restriction, early-onset pre-eclampsia and maternal thrombophilias, whereas a relative reduction in the supply line due to an excessive foetal demands for nutrients may be more relevant in the mechanism of injury in late-onset pre-eclampsia and gestational diabetes. It is possible that some of these pregnancy complications may have evolved as survival strategies for the foetus or the mother. In this context, interventions aimed at modulating the maternal blood pressure during pregnancy or delaying preterm parturition should be tailored to maximize both maternal and perinatal outcomes.

J. Espinoza (✉)
Department of Obstetrics and Gynecology, Division of Maternal Fetal Medicine,
Baylor College of Medicine, Houston, TX, USA
e-mail: Jimmy.Espinoza@bcm.edu

J. Espinoza
Texas Children's Hospital Pavilion for Women, Houston, TX, USA

© Springer International Publishing Switzerland 2016
A. Alvergne et al. (eds.), *Evolutionary Thinking in Medicine*,
Advances in the Evolutionary Analysis of Human Behaviour,
DOI 10.1007/978-3-319-29716-3_2

19

2.1   Introduction

> The placenta appears to be a ruthless parasitic organ existing solely for the maintenance and
> protection of the fetus, perhaps too often to the disregard of the maternal organism.
> Ernest W. Page AJOG 1939.

Successful pregnancies depend on a balance between increasing foetal demand for
nutrients and a measured maternal investment to safeguard her reproductive future
[1]. Failure of the well-orchestrated maternal foetal interaction may lead to a
conflict of interests between the mother and her foetus and subsequent pregnancy
complications [2]. The term "foetal–maternal conflict" refers to a conceptual
framework whereby foetal growth and development can happen sometimes at the
expense of the maternal well-being [1–5]. This term has being used to describe
clinical situations in which evolutionary adaptations of the mother appear to be in
conflict with those of her foetus [1, 2, 4, 5]. At the histological level, foetal–
maternal conflicts have been implicated in the mechanisms by which abnormal
trophoblast invasion leads to the failure of physiologic transformation of the spiral
arteries [2, 6–9], chronic uteroplacental ischaemia [10] and pregnancy complications including pre-eclampsia [1–13], preterm parturition [14, 15], foetal growth
restriction [11, 13, 16, 17] and foetal death [18]. The conventional obstetrical view
is to compartmentalize and treat pregnancy complications as if problems arising
during pregnancy have either foetal or maternal origin. In contrast, the evolutionary
approach to pregnancy complications is to consider them as a result of abnormal
foetal–maternal interactions. This chapter reviews the evidence supporting the
notion that foetal growth restriction, early- and late-onset pre-eclampsia, preterm
parturition and gestational diabetes may result from inadequate foetal–maternal
interactions.

2.2   Research Findings

2.2.1   Foetal Growth Restriction and Abnormal
Foetal–Maternal Interactions

One of the most common expressions of the "foetal–maternal conflict" is mani-
fested in abnormalities of foetal growth. Paternally derived imprinted genes tend to
maximize foetal growth; in contrast, maternally derived imprinted genes tend to do
the opposite [5] presumably as an evolutionary strategy to protect the maternal
well-being. A remarkable example of this view is the observation that human
triploidic foetuses, for which the extra set of chromosomes (a total of 69 rather than
the normal 46 chromosomes) is derived from the mother (dyginic triploidy), are
associated with early-onset foetal growth restriction even in the first trimester of
pregnancy (Fig. 2.1). These foetuses not only have very small bodies and dispro-
portionably large heads but also very thin and small placentas. Moreover, there is

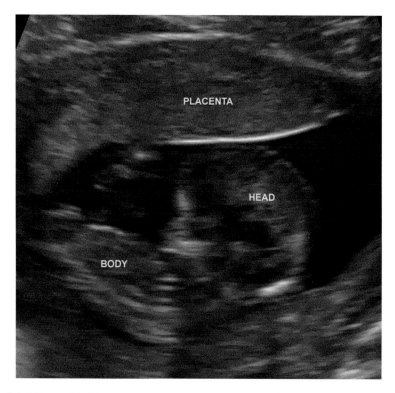

Fig. 2.1 Ultrasonographic findings in a foetus with dyginic triploidy in the first trimester of pregnancy note the significant disproportion between the foetal head and the body

sonographic evidence that dyginic triploidic foetuses show blood redistribution to the foetal brain in the first and early second trimester of pregnancy [19]. Asymmetric early-onset foetal growth restriction in triploidic foetuses may be due to the chromosomal anomaly, but it is possible that blood redistribution starting very early in pregnancy may contribute to the phenotype seen in dyginic triploidy. In contrast, triploidic foetuses for which the extra set of chromosomes is derived from the father (dyandric triploidy) are associated with partial mole pregnancies. These foetuses tend to be of normal size but have large molar placentas and are often associated with early-onset pre-eclampsia before 20 weeks of gestation. It is noteworthy that pregnancy complications that are normally seen in the second or third trimester, including foetal growth restriction and pre-eclampsia, can be seen as early as the first or early second trimester in dyginic or dyandric triploidic pregnancies, respectively.

Genomic imprinting is the process by which one copy of a gene is silenced due to its parental origin. New high-throughput molecular techniques indicate that several hundred genes are imprinted. Experimental studies provide additional evidence that foetal growth is regulated by paternally or maternally derived imprinted genes. The insulin-like growth factor II gene is paternally expressed in the foetus and the placenta; deletion of a transcript of this gene in mice leads to reduced placental growth and foetal growth restriction [4]. Grb10 is an adapter signalling protein that appears to control foetal growth independently from insulin-like growth factor 2 [20]. Grb10 is a potent growth inhibitor, and the majority of its gene expression arises from the maternally derived allele which is located on chromosome 7 [20]. In mice, disruption of this allele results in overgrowth of both the embryo and placenta, such that the mutant mice at birth are 30 % larger than normal [20]. In humans, about 10 % of individuals affected by Silver–Russell syndrome, characterized by severe foetal growth restrictions, inherit both copies of chromosomes 7 from their mother, and it has been proposed that an overexpression of the GRB10 gene accounts for these restrictions in growth [20].

2.2.2 Chronic Ischaemia of the Uteroplacental Unit: Early-Onset Pre-eclampsia and Foetal Growth Restriction

During pregnancy, the uterus and placenta form an anatomic and functional uteroplacental unit. An absolute uteroplacental ischaemia may result from (i) placental bed disorders; (ii) vascular insults to the placenta; or (iii) abnormal foetal placental circulation. Abnormalities in the placental bed and subsequent failure of physiologic transformation of the spiral arteries in the first or early second trimester [6, 7] limit the blood flow to the uteroplacental unit. Indeed, high impedance to blood flow in both uterine arteries, a surrogate marker of chronic reduction of the blood flow to the uteroplacental unit [21, 22], is associated with the failure of the normal physiologic transformation of the spiral arteries in placental bed biopsies from patients with pre-eclampsia [11, 13, 16, 23] and those with foetal growth restriction [11, 13, 16, 17]. However, not all patients with these pregnancy complications have evidence of failure of physiologic transformation of the spiral arteries [11–13, 16, 17, 23]. Moreover, this pathological finding is not limited to patients with pre-eclampsia or foetal growth restriction because it has also been described in a subset of patients with preterm parturition [14, 15] and foetal death [18].

Additional mechanisms leading to absolute uteroplacental ischaemia include insults to the placental vasculature during pregnancy. Recent reports indicate not only that pre-eclampsia is associated with placental vascular lesions consistent with "underperfusion", but also that the earlier the gestational age at which pre-eclampsia develops, the higher the prevalence of lesions consistent with placental ischaemia [24, 25]. Indeed, the frequency of placental histological lesions consistent with "maternal underperfusion" is as high as 75 % in pre-eclampsia that develop between 25 and 27 weeks and as low as 13 % in pre-eclampsia that develop at more than 41 weeks [25]. These observations suggest that there may be a dose response between the magnitude of uteroplacental ischaemia and the timing of onset of pre-eclampsia.

A remarkable example of the latter is the development of pre-eclampsia before 20 weeks of gestation in patients with mole or partial mole. These pregnancy complications are characterized by the presence of "avascular placental villi" or placental villi with capillary remnants [26]. Thus, by definition, mole and partial mole may represent an extreme in the spectrum of ischaemic disease of the trophoblast [27]. The dose response between the magnitude of uteroplacental ischaemia and the timing of the development of pre-eclampsia suggests that there is an absolute or relative "trophoblast ischaemic threshold" beyond which pre-eclampsia develops as a foetal adaptive strategy in an attempt to improve the blood perfusion to the foetal and placental tissues. It is possible that the response to this threshold may be modified by gene–environment interaction [28], the magnitude of angiogenic imbalances [27, 29] and foetal signalling in response to absolute or relative uteroplacental ischaemia [30, 31].

Accumulating evidence indicates that chronic reduction of blood flow to the uterus and placenta is associated with imbalances between circulating angiogenic and anti-angiogenic factors characterized by an excess of the soluble form of vascular endothelial growth factor (VEGF) receptor 1 (sFlt-1) and the soluble endoglin (s-Eng) as well as low circulating maternal concentrations of both VEGF and placental growth factor (PlGF) [27]. Clinical and experimental evidence indicates that angiogenic imbalances are associated with the maternal manifestations of pre-eclampsia, eclampsia and HELLP syndrome [27]. Teleologically, it is difficult to believe that natural selection did not select against pre-eclampsia, which can endanger the survival of both the mother and the foetus. From the evolutionary point of view, it is possible that in preeclamptic patients, the foetus may stimulate the placental release of anti-angiogenic factors to increase the maternal blood pressure in an attempt to increase the blood flow to the placental and foetal tissues. The magnitude of the angiogenic imbalances, gene–environment interaction (Subtle differences in genetic factors that cause some people to possess a low risk for developing a disease through an environmental insult, while others are much more vulnerable) and other factors may determine whether a patient with chronic trophoblast ischaemia will develop pre-eclampsia, foetal growth restriction, both or any of the other intermediate phenotypes including gestational hypertension and gestational proteinuria [27].

Recent reports suggest that among patients with pre-eclampsia, the foetus may use adenosine among other signalling mechanisms in order to increase the maternal blood pressure in an attempt to compensate for limited blood flow to the foetal and placental tissues [27, 30, 32]. In one study [32], the authors compared the foetal plasma concentrations of adenosine from normal pregnancies with those from pre-eclampsia; patients with pre-eclampsia were sub-classified into patients with and without abnormal uterine artery Doppler velocimetry. The results of the study indicated that foetal plasma concentrations of adenosine were significantly higher in patients with pre-eclampsia with abnormal uterine artery Doppler velocimetry than in normal pregnancies. The authors concluded that patients with pre-eclampsia and sonographic evidence of chronic uteroplacental ischaemia have high foetal plasma concentrations of adenosine and proposed that in these patients the foetus may use the adenosine system and/or other signalling mechanisms to increase the maternal blood pressure in an attempt to increase uteroplacental blood flow. An elegant in vitro study provided additional evidence in support of this view [33]. In this study, the authors determined the adenosine concentrations in foetal venous per-fusates using isolated dual-perfused human placental cotyledons. In the latter experimental setting, both the foetal and maternal compartments of the placenta are perfused under controlled conditions. The authors reported that a substantial reduction in the perfusion of the maternal compartment of the placenta was asso-ciated with a significant increase in foetal venous perfusate concentrations of adenosine and a concomitant increase in foetoplacental perfusion pressure. Furthermore, perfusate pressure and the concentration of adenosine in the foetal compartment returned to baseline levels on reperfusion of the "maternal" circuit [33]. A more recent study using cultures of placental cells indicates that the administration of adenosine to the cultures significantly increases the concentration of the anti-angiogenic factor sFlt-1 in the cell culture media under normoxic con-ditions and that the addition of dipyridamole (an adenosine transporter antagonist which increases extracellular adenosine concentration) to cell cultures leads to a significant increase in the concentrations of sFlt-1 in the culture media [34]. Moreover, although hypoxia was associated with a twofold increase in the con-centrations of sFlt-1 in the cell culture media, blockade of adenosine signalling (using a non-specific adenosine receptor antagonist) blunted the hypoxic effect on the concentrations of sFlt-1 and VEGF to a level similar to normoxic conditions [34]. These results indicate that adenosine signalling is important for placental overexpression and release of sFlt-1 under both normoxic and hypoxic conditions. An excess of sFlt-1 is associated with endothelial dysfunction, maternal hyper-tension and the liver and renal injury described in pre-eclampsia. Collectively, this evidence suggests that foetal signalling may play an important role in the devel-opment of pre-eclampsia in the context of chronic reduction of blood flow to the uteroplacental unit.

2.2.3   Maternal Thrombophilias

Histological vascular lesions have been described in the foetal and/or maternal side of the placenta in mothers with inherited and acquired maternal thrombophilias [35–38], but not in foetal thrombophilias [39]. Thus, chronic placental ischaemia may contribute to the increased rate of adverse pregnancy outcomes observed in patients with thrombophilias [38, 40–42]. In the context of the foetal–maternal conflict, it is possible that the evolutionary advantage of preserving thrombophilic genes in a particular population is to favour the maternal well-being over that of her foetus, in addition to reducing the risk of peripartum haemorrhage [43, 44].

2.2.4 Late-Onset Pre-eclampsia

Absolute uteroplacental ischaemia appears to be less relevant in the pathophysiology of late-onset pre-eclampsia, defined as the onset of pre-eclampsia beyond 34 weeks of gestation [45, 46]. Evidence in support of this view includes the recent observation that more than half of patients with late-onset pre-eclampsia do not have placental histological lesions consistent with "maternal underperfusion" [47]. Furthermore, late-onset pre-eclampsia is frequently associated with foetuses that are adequate or large-for-gestational age [45, 48–53]. We proposed that in these cases, an increased foetal demand for substrates that surpass the placental ability to sustain foetal growth may induce foetal signalling for placental overproduction of anti-angiogenic factors and subsequent "compensatory" maternal hypertension [27]. Thus, it is possible that a relative uteroplacental ischaemia due to a mismatch between a limited uteroplacental blood flow and increased foetal demand for nutrients may be central to the development of late-onset pre-eclampsia. It is possible that in both early and late-onset pre-eclampsia, the foetus may signal for the onset of pre-eclampsia. In early-onset pre-eclampsia, real reduction in blood flow appears to be central to the disease; in contrast, in late-onset pre-eclampsia, foetal over-demand may create a state of relative scarcity of nutrients, which in turn would prompt foetal signalling to elevate the maternal blood pressure.

A large metanalysis demonstrated that overzealous attempts to control blood pressure during pregnancy are associated with foetal growth restriction [54]. These observations suggest that pre-eclampsia may have evolved as one of the foetal strategies to compensate for a relative or absolute uteroplacental ischaemia.

2.2.5 Abnormal Foetal–Maternal Interactions and Preterm Parturition

Foetal strategies to cope with chronic uteroplacental ischaemia may include growth restriction, foetal signalling to increase the maternal systemic blood pressure leading into pre-eclampsia [30, 31] or preterm parturition to exit an inadequate intrauterine environment. The observation that the absence of physiological transformation of spiral arteries is also present in a subset of patients with spontaneous preterm delivery [14, 15] suggests that the clinical manifestations of "foetal–maternal conflict" may also include preterm parturition. Smallness at birth may be the result of different insults during pregnancy including chronic reduction of blood flow to the uteroplacental unit. In some growth-restricted foetuses, spontaneous preterm parturition (Delivery before 37 weeks of gestation) may represent a survival strategy to exit an inadequate intrauterine environment [55]; failure of this adaptive strategy may result in foetal or neonatal death. Evidence in support of this view includes the observations that spontaneous preterm parturition is associated with foetal growth abnormalities [56–66]. Of note, the association of smallness at birth (less than 10th percentile for gestational age) and prematurity confers a higher risk of foetal [67] or neonatal death [68] among other adverse perinatal outcomes [68–73]. To the extent that preterm parturition is a survival strategy to exit an inadequate intrauterine environment, the safety of tocolysis (interventions to stop uterine contractions) in the growth-restricted premature foetus should be re-evaluated.

2.2.6   Gestational Diabetes

David Haig in a very insightful article proposed that gestational diabetes mellitus (GDM), among other pregnancy complications, may also be the result of a foetal–maternal conflict [1]. Dr. Haig proposed that a mother and her foetus compete after every meal over the glucose share that each one receives in a way that

> The longer the mother takes to reduce her blood sugar, the greater the share taken by her fetus. [1]

In the last half of pregnancy, there is an increased tissue resistance to the action of insulin; to compensate for this, the mother increases insulin production. According to the foetal–maternal conflict hypothesis, this is caused by foetal signalling using placental allocrine hormones including human placental lactogen (hPL) and human placental growth hormone among others, to guaranty its adequate glucose supply, whereas the increased production of insulin would be a maternal countermeasure [1]. Thus, the nutrient content in the maternal blood may be determined by the balance between foetal signalling using placental-derived hormones and maternal counter-measures. Human experimentation done in the late 1960s provides evidence supporting the notion of the diabetogenic effect of hPL [74, 75]. Indeed, intravenous infusion of physiological amounts of hPL to non-pregnant subjects is associated with glucose intolerance despite increased insulin responses [74]. It is possible that failure of a well-orchestrated maternal–foetal interaction, between foetal signalling increasing the placental production of diabetogenic hormones and maternal coun-termeasures increasing insulin production, may lead to GDM. Thus, gestational diabetes would develop if a woman were unable to increase her insulin production sufficiently to match the increased peripheral insulin resistance.

A large population-based study indicated that GDM is an independent factor for the development of pre-eclampsia after controlling for confounding factors including maternal age, parity, BMI, smoking and chronic hypertension or renal disease (Adjusted Odds-Ratio: 1.61, 95 % CI: 1.39–1.86) [76]. Moreover, a large retrospective study in the USA involving 1813 women with GDM demonstrated that the rates of pre-eclampsia among those with poor glycaemic control were about twice as high as those with better glycaemic control (18 % vs. 9.8 %; OR: 2.56, 95 % CI: 1.5–4.3) [77]. However, there is limited literature in regard to the timing of onset of pre-eclampsia among women with GDM.

In a study involving 45 patients with GDM demonstrated normal placental histology in 80 % of them [76], thus, placental vascular lesions are not common in the foetal or maternal side of the placenta in women with GDM. Since this pregnancy complication is associated with large-for-gestational age neonates, it is possible that in women with GDM who develop pre-eclampsia, an increased foetal demand for substrates that surpass the placental ability to sustain foetal growth may induce foetal signalling for placental overproduction of anti-angiogenic factors and subsequent "compensatory" maternal hypertension. Additional studies are needed to explore the role of angiogenic imbalances in these patients.

2.3   Implications for Policy and Practice

The recognition that some pregnancy complications may be due to abnormal foetal–maternal interactions is important for the clinical management of these pregnancy complications. In the context of a long-lasting reduction of blood flow to the foetal and placental tissues, the foetus may signal the placental release of "pressor substances", which could elevate the maternal blood pressure in an attempt to increase the delivery of nutrients to the foetus. Any medical attempt to "normalize" the blood pressure in the mother could be deleterious to the foetus by preventing the beneficial effect of a compensatory mechanism. The use of medications to lower the blood pressure should be aimed at reducing the blood pressure to a level that will prevent cardiovascular accidents in the mother, but not at "normalizing" the blood pressure. Similarly, the use of medications to stop the uterine contractions in women with preterm labour should be judiciously used in foetuses that are growth-restricted because it is possible that the foetus may have initiated the process of premature labour in order to exit a hostile intrauterine environment.

Glossary

Pre-eclampsia	Hypertensive disorder of pregnancy that typically starts after the 20th week of pregnancy.
Genomic imprinting	The process by which one copy of a gene is silenced due to its parental origin.
Mole pregnancy	Results from a genetic error during the fertilization process that leads to growth of abnormal placenta within the uterus.
Grb10	Growth factor receptor-bound protein 10 also known as insulin receptor-binding protein.
Spiral arteries	Small arteries that are remodelled into highly dilated vessels during pregnancy to increase the blood supply to foetal and placental tissues.

Placental bed disorders	Refers to defective placentation in the human which is associated with pregnancy complications such pre-eclampsia, foetal growth restriction, and foetal death.
Uteroplacental ischaemia	During pregnancy, the uterus and placenta form a functional unit. This term refers to reduced blood flow to this unit.
Angiogenic factors	Promote the viability and growth of endothelial cells. Foetal signalling: proposed pathways used by the foetus to alter the maternal of placental physiology.
HELLP syndrome	A severe form of pre-eclampsia characterized by abnormal liver enzymes, low platelets and destruction of red blood cells.
Adenosine	Compound that plays an important role in energy transfer signal transduction and regulation of blood flow to various organs.
Uterine artery Doppler velocimetry	Ultrasonographic technique to evaluate the characteristics of blood flow in vessels.
VEGF	Vascular endothelial growth factor is a signalling protein involved in the formation and growth of blood vessels.
sFlt-1	Splice variant of VEGF receptor 1an excess of this soluble form in the circulation can reduce the bioavailability of VEGF (anti-angiogenic).
Thrombophilia	Abnormality of blood coagulation that increases the risk of thrombosis.
Tocolysis	Medical interventions to reduce or stop uterine contractions.
Allocrine hormones	Foreign hormones being taken up and eliciting a response in an organism.

References

1. Haig D (1993) Genetic conflicts in human pregnancy. Q Rev Biol 68:495–532
2. Pijnenborg R, Vercruysse L, Hanssens M (2008) Fetal-maternal conflict, trophoblast invasion, preeclampsia, and the red queen. Hypertens Pregnancy 27:183–196
3. Page EW (1939) The relation between hydatid moles, relative ischemia of the gravid uterus, and the placental origin of eclampsia. Am J Obstet Gynecol 37:291–293

4. Constancia M, Hemberger M, Hughes J, Dean W, Ferguson-Smith A, Fundele R (2002) Placental-specific IGF-II is a major modulator of placental and fetal growth. Nature 417:945–948

5. Fowden AL, Coan PM, Angiolini E, Burton GJ, Constancia M (2011) Imprinted genes and the epigenetic regulation of placental phenotype. Prog Biophys Mol Biol 106:281–288

6. Espinoza J, Romero R, Mee KY, Kusanovic JP, Hassan S, Erez O (2006) Normal and abnormal transformation of the spiral arteries during pregnancy. J Perinat Med 34:447–458

7. Pijnenborg R, Vercruysse L, Hanssens M (2006) The uterine spiral arteries in human pregnancy: Facts and controversies. Placenta 27:939–958

8. Burton GJ, Woods AW, Jauniaux E, Kingdom JC (2009) Rheological and physiological consequences of conversion of the maternal spiral arteries for uteroplacental blood flow during human pregnancy. Placenta 30:473–482

9. Burton GJ, Yung HW (2011) Endoplasmic reticulum stress in the pathogenesis of early-onset pre-eclampsia. Pregnancy Hypertens 1:72–78

10. Burton GJ, Jauniaux E (2004) Placental oxidative stress: From miscarriage to preeclampsia. J Soc Gynecol Investig 11:342–352

11. Olofsson P, Laurini RN, Marsal KA (1993) High uterine artery pulsatility index reflects a defective development of placental bed spiral arteries in pregnancies complicated by hypertension and fetal growth retardation. Eur J Obstet Gynecol Reprod Biol 49:161–168

12. Sagol S, Ozkinay E, Oztekin K, Ozdemir N (1999) The comparison of uterine artery doppler velocimetry with the histopathology of the placental bed. Aust N Z J Obstet Gynaecol 39:324–329

13. Aardema MW, Oosterhof H, Timmer A, van RI, Aarnoudse JG (2001) Uterine artery doppler flow and uteroplacental vascular pathology in normal pregnancies and pregnancies complicated by pre-eclampsia and small for gestational age fetuses. Placenta 22:405–411

14. Kim YM, Chaiworapongsa T, Gomez R, Bujold E, Yoon BH, Rotmensch S (2002) Failure of physiologic transformation of the spiral arteries in the placental bed in preterm premature rupture of membranes. Am J Obstet Gynecol 187:1137–1142

15. Kim YM, Bujold E, Chaiworapongsa T, Gomez R, Yoon BH, Thaler HT (2003) Failure of physiologic transformation of the spiral arteries in patients with preterm labor and intact membranes. Am J Obstet Gynecol 189:1063–1069

16. Lin S, Shimizu I, Suehara N, Nakayama M, Aono T (1995) Uterine artery doppler velocimetry in relation to trophoblast migration into the myometrium of the placental bed. Obstet Gynecol 85:760–765

17. Madazli R, Somunkiran A, Calay Z, Ilvan S, Aksu MF (2003) Histomorphology of the placenta and the placental bed of growth restricted foetuses and correlation with the doppler velocimetries of the uterine and umbilical arteries. Placenta 24:510–516

18. Avagliano L, Bulfamante GP, Morabito A, Marconi AM (2011) Abnormal spiral artery remodelling in the decidual segment during pregnancy: From histology to clinical correlation. J Clin Pathol 64:1064–1068

19. Wu RT, Shyu MK, Lee CN, Wu CC, Hwa HL, Lin CJ (1995) Sonographic manifestation and doppler blood flow study in fetal triploidy syndrome: Report of two cases. J Ultrasound Med 14:555–558

20. Charalambous M, Smith FM, Bennett WR, Crew TE, Mackenzie F, Ward A (2003) Disruption of the imprinted Grb10 gene leads to disproportionate overgrowth by an Igf2-independent mechanism. Proc Natl Acad Sci USA 100:8292–8297

21. Kuzmina IY, Hubina-Vakulik GI, Burton GJ (2005) Placental morphometry and doppler flow velocimetry in cases of chronic human fetal hypoxia. Eur J Obstet Gynecol Reprod Biol 120:139–145

22. Prefumo F, Sebire NJ, Thilaganathan B (2004) Decreased endovascular trophoblast invasion in first trimester pregnancies with high-resistance uterine artery doppler indices. Hum Reprod 19:206–209

23. Voigt HJ, Becker V (1992) Doppler flow measurements and histomorphology of the placental bed in uteroplacental insufficiency. J Perinat Med 20:139–147

24. Moldenhauer JS, Stanek J, Warshak C, Khoury J, Sibai B (2003) The frequency and severity of placental findings in women with preeclampsia are gestational age dependent. Am J Obstet Gynecol 189:1173–1177
25. Ogge G, Chaiworapongsa T, Romero R, Hussein Y, Kusanovic JP, Yeo L (2011) Placental lesions associated with maternal underperfusion are more frequent in early-onset than in late-onset preeclampsia. J Perinat Med
26. Benirschke K, Kaufmann P, Baergen R (2006) Pathology of the human placenta. Springer, New York
27. Espinoza J, Uckele JE, Starr RA, Seubert DE, Espinoza AF, Berry SM (2010) Angiogenic imbalances: The obstetric perspective. Am J Obstet Gynecol 203:17–18
28. Sandrim VC, Palei AC, Cavalli RC, Araujo FM, Ramos ES, Duarte G (2009) Vascular endothelial growth factor genotypes and haplotypes are associated with pre-eclampsia but not with gestational hypertension. Mol Hum Reprod 15:115–120
29. Burton GJ, Charnock-Jones DS, Jauniaux E (2009) Regulation of vascular growth and function in the human placenta. Reproduction 138:895–902
30. Espinoza J, Espinoza AF (2011) Pre-eclampsia: A maternal manifestation of a fetal adaptive response? Ultrasound Obstet Gynecol 38:367–370
31. Espinoza J, Espinoza AF, Power GG (2011) High fetal plasma adenosine concentration: A role for the fetus in preeclampsia? Am J Obstet Gynecol
32. Espinoza J, Espinoza AF, Power GG (2011) High fetal plasma adenosine: A role for the fetus in preeclampsia? Am J Obstet Gynecol. doi:10.1016/j.ajog.2011.06.034
33. Slegel P, Kitagawa H, Maguire MH (1988) Determination of adenosine in fetal perfusates of human placental cotyledons using fluorescence derivatization and reversed-phase high-performance liquid chromatography. Anal Biochem 171:124–134
34. George EM, Cockrell K, Adair TH, Granger JP (2010) Regulation of sFlt-1 and VEGF secretion by adenosine under hypoxic conditions in rat placental villous explants. Am J Physiol Regul Integr Comp Physiol 299:R1629–R1633
35. Gogia N, Machin GA (2008) Maternal thrombophilias are associated with specific placental lesions. Pediatr Dev Pathol 11:424–429
36. Sebire NJ, Backos M, Goldin RD, Regan L (2002) Placental massive perivillous fibrin deposition associated with antiphospholipid antibody syndrome. BJOG 109:570–573
37. Redline RW (2006) Thrombophilia and placental pathology. Clin Obstet Gynecol 49:885–894
38. Arias F, Romero R, Joist H, Kraus FT (1998) Thrombophilia: A mechanism of disease in women with adverse pregnancy outcome and thrombotic lesions in the placenta. J Matern Fetal Med 7:277–286
39. Ariel I, Anteby E, Hamani Y, Redline RW (2004) Placental pathology in fetal thrombophilia. Hum Pathol 35:729–733
40. Kupferminc MJ, Many A, Bar-Am A, Lessing JB, Ascher-Landsberg J (2002) Mid-trimester severe intrauterine growth restriction is associated with a high prevalence of thrombophilia. BJOG 109:1373–1376
41. Paidas MJ, Ku DH, Arkel YS (2004) Screening and management of inherited thrombophilias in the setting of adverse pregnancy outcome. Clin Perinatol 31:783–805
42. Kupferminc MJ, Rimon E, Ascher-Landsberg J, Lessing JB, Many A (2004) Perinatal outcome in women with severe pregnancy complications and multiple thrombophilias. J Perinat Med 32:225–227
43. Lindqvist PG, Dahlback B (2008) Carriership of Factor V Leiden and evolutionary selection advantage. Curr Med Chem 15:1541–1544
44. Lindqvist PG, Svensson PJ, Dahlback B, Marsal K (1998) Factor V Q506 mutation (activated protein C resistance) associated with reduced intrapartum blood loss–a possible evolutionary selection mechanism. Thromb Haemost 79:69–73
45. Rasmussen S, Irgens LM (2003) Fetal growth and body proportion in preeclampsia. Obstet Gynecol 101:575–583

46. Aardema MW, Saro MC, Lander M, De Wolf BT, Oosterhof H, Aarnoudse JG (2004) Second trimester doppler ultrasound screening of the uterine arteries differentiates between subsequent normal and poor outcomes of hypertensive pregnancy: Two different pathophysiological entities? Clin Sci (Lond) 106:377–382

47. Soto E, Romero R, Kusanovic JP, Ogge G, Hussein Y, Yeo L (2011) Late-onset preeclampsia is associated with an imbalance of angiogenic and anti-angiogenic factors in patients with and without placental lesions consistent with maternal underperfusion. J Matern Fetal Neonatal Med

48. Odegard RA, Vatten LJ, Nilsen ST, Salvesen KA, Austgulen R (2000) Preeclampsia and fetal growth. Obstet Gynecol 96:950–955

49. Xiong X, Demianczuk NN, Buekens P, Saunders LD (2000) Association of preeclampsia with high birth weight for age. Am J Obstet Gynecol 183:148–155

50. Xiong X, Fraser WD (2004) Impact of pregnancy-induced hypertension on birthweight by gestational age. Paediatr Perinat Epidemiol 18:186–191

51. Rasmussen S, Irgens LM, Espinoza J (2014) Maternal obesity and excess of fetal growth in pre-eclampsia. BJOG 121:1351–1358

52. Rasmussen S, Espinoza J, Lee W, Martin SR, Belfort MA (2014) Re: Customized growth curves for identification of large-for-gestational age neonates in pre-eclamptic women. Ultrasound Obstet Gynecol 43:165–169

53. Espinoza J, Lee W, Martin SR, Belfort MA (2014) Customized growth curves for identification of large-for-gestational age neonates in pre-eclamptic women. Ultrasound Obstet Gynecol 43:165–169

54. von Dadelszen P, Ornstein MP, Bull SB, Logan AG, Koren G, Magee LA (2000) Fall in mean arterial pressure and fetal growth restriction in pregnancy hypertension: A meta-analysis. Lancet 355:87–92

55. Romero R (1996) Prenatal medicine: The child is the father of the man. Prenat Neonatal Med 1:8–11

56. Ott WJ (1988) The diagnosis of altered fetal growth. Obstet Gynecol Clin North Am 15:237–263

57. Deter RL, Rossavik IK, Harrist RB (1988) Development of individual growth curve standards for estimated fetal weight: I Weight estimation procedure. J Clin Ultrasound 16:215–225

58. Ott WJ (1993) Intrauterine growth retardation and preterm delivery. Am J Obstet Gynecol 168:1710–1717

59. Zeitlin J, Ancel PY, Saurel-Cubizolles MJ, Papiernik E (2000) The relationship between intrauterine growth restriction and preterm delivery: An empirical approach using data from a European case-control study. BJOG 107:750–758

60. Goldenberg RL, Nelson KG, Koski JF, Cutter GR (1985) Low birth weight, intrauterine growth retardation, and preterm delivery. Am J Obstet Gynecol 152:980–984

61. Bukowski R, Gahn D, Denning J, Saade G (2001) Impairment of growth in fetuses destined to deliver preterm. Am J Obstet Gynecol 185:463–467

62. Secher NJ, Kern HP, Thomsen BL, Keiding N (1987) Growth retardation in preterm infants. Br J Obstet Gynaecol 94:115–120

63. Morken NH, Kallen K, Jacobsson B (2006) Fetal growth and onset of delivery: A nationwide population-based study of preterm infants. Am J Obstet Gynecol 195:154–161

64. Mercer BM, Merlino AA, Milluzzi CJ, Moore JJ (2008) Small fetal size before 20 weeks' gestation: Associations with maternal tobacco use, early preterm birth, and low birthweight. Am J Obstet Gynecol 198:673–677

65. Hediger ML, Scholl TO, Schall JI, Miller LW, Fischer RL (1995) Fetal growth and the etiology of preterm delivery. Obstet Gynecol 85:175–182

66. MacGregor SN, Sabbagha RE, Tamura RK, Pielet BW, Feigenbaum SL (1988) Differing fetal growth patterns in pregnancies complicated by preterm labor. Obstet Gynecol 72:834–837

67. Odibo AO, Cahill AG, Odibo L, Roehl K, Macones GA (2011) Prediction of intrauterine fetal death in small-forgestational-age fetuses: Impact of including ultrasound biometry in customized models. Ultrasound Obstet Gynecol

68. Garite TJ, Clark R, Thorp JA (2004) Intrauterine growth restriction increases morbidity and mortality among premature neonates. Am J Obstet Gynecol 191:481–487
69. Aucott SW, Donohue PK, Northington FJ (2004) Increased morbidity in severe early intrauterine growth restriction. J Perinatol 24:435–440
70. Bernstein IM, Horbar JD, Badger GJ, Ohlsson A, Golan A (2000) The vermont oxford network. Morbidity and mortality among very-low-birth-weight neonates with intrauterine growth restriction. Am J Obstet Gynecol 182:198–206
71. Hemming K, Hutton JL, Bonellie S (2009) A comparison of customized and population-based birth-weight standards: The influence of gestational age. Eur J Obstet Gynecol Reprod.Biol 146:41–45
72. Zhang X, Platt RW, Cnattingius S, Joseph KS, Kramer MS (2007) The use of customised versus population-based birthweight standards in predicting perinatal mortality. BJOG 114:474–477
73. Engineer N, Kumar S (2010) Perinatal variables and neonatal outcomes in severely growth restricted preterm fetuses. Acta Obstet Gynecol Scand 89:1174–1181
74. Beck P, Daughaday WH (1967) Human placental lactogen: Studies of its acute metabolic effects and disposition in normal man. J Clin Invest 46:103
75. Samaan N, Yen SC, Gonzalez D, Pearson OH (1968) Metabolic effects of placental lactogen (HPL) in man. J Clin Endocrinol Metab 28:485–491
76. Ostlund I, Haglund B, Hanson U (2004) Gestational diabetes and preeclampsia. Eur J Obstet Gynecol Reprod Biol 113:12–16
77. Yogev Y, Xenakis EM, Langer O (2004) The association between preeclampsia and the severity of gestational diabetes: The impact of glycemic control. Am J Obstet Gynecol 191:1655–1660

Chapter 3
Obstructed Labour: The Classic Obstetric Dilemma and Beyond

Emma Pomeroy, Ph.D., Prof. Jonathan C.K. Wells, Ph.D. and Jay T. Stock, Ph.D.

Lay Summary The obstetric dilemma (OD) theory proposes that walking on two feet (bipedalism) favours narrower hips, while the large human brain favours wider hips with a more spacious birth canal through which the baby's head can pass. These competing demands on pelvic size and shape were argued to have resulted in a tight fit between foetal head size and the maternal birth canal, causing a long, painful childbirth and high risks of obstructed labour in humans. While this 'classic' OD has been widely accepted among anthropologists and medical researchers as an explanation for the human pattern of childbirth, obstructed labour may not be an inevitable consequence of the OD. Rather, rates of obstructed labour may vary over space and time in relation to more recent changes in growth, diet, weight and health resulting from alterations to our environment (both natural and caused by humans). Importantly, this suggests that the modern burden of obstructed labour, which accounts for up to 14 % of maternal deaths in low- and middle-income countries, can be reduced by focussing on lifestyle factors. Changing the typical birth posture may also help, and a closer examination of parental and foetal body measurements may aid in identifying the mothers most at risk of labour complications.

E. Pomeroy (✉)
Newnham College, University of Cambridge, Cambridge, UK
e-mail: eep23@cam.ac.uk

E. Pomeroy · J.T. Stock
Department of Archaeology and Anthropology, University of Cambridge,
Pembroke Street, Cambridge CB2 3QG, UK

J.C.K. Wells
Childhood Nutrition Research Centre, UCL Institute of Child Health, London, UK

© Springer International Publishing Switzerland 2016
A. Alvergne et al. (eds.), *Evolutionary Thinking in Medicine*,
Advances in the Evolutionary Analysis of Human Behaviour,
DOI 10.1007/978-3-319-29716-3_3

33

3.1 Introduction

The OD [1] suggests that humans have a uniquely protracted, painful birth with increased risk of obstructed labour (where mechanical problems such as a large foetal head, shoulder dystocia or malpresentation of the foetus prevent its passage through the birth canal [2]) because two key characteristics of our species place antagonistic demands on pelvic form [3–5]. Bipedalism was proposed to favour a narrow pelvis for efficiency [6], while the exceptionally large size of the human brain necessitated a larger pelvis with a more spacious birth canal, resulting in a tight fit between foetal head and maternal bony birth canal. The OD has become widely accepted in anthropology and medicine, and blamed for a variety of birth complications including cephalopelvic disproportion, shoulder dystocia, and rising Caesarean rates [7–9]. However, this classic notion of a persistent OD leading to birth complications among humans is increasingly being challenged. In this chapter, we briefly outline some recent challenges to the OD and then focus on the link between the OD and obstructed labour in contemporary population. Rather than seeing obstructed labour as an inevitable outcome of the OD, we evaluate how rates probably vary temporally and between populations as a result of the influences of lifestyle and environmental conditions on maternal and offspring phenotype. By providing insight into the causes of such problems, an evolutionary view offers novel perspective on appropriate strategies to manage and reduce their impact.

3.2 Research Findings

3.2.1 Pelvic Morphology, Parturition and Locomotion

Proponents of the OD argue that the close fit between neonatal head size and the birth canal led to the evolution of uniquely human prolonged and painful labour, necessitating a rotational birth mechanism, assistance during childbirth, and increasing rates of maternal and/or neonatal mortality due to obstructed labour [4, 10, 11]. However, the uniqueness of these characteristics is open to question, and more fundamentally, evidence for key assumptions underlying the OD has been challenged [12–14], as we review briefly here.

The existence of selective constraints of bipedalism on pelvic morphology is not well supported. The alternative proposal that a broader pelvis offers greater walking efficiency [15] has recently been supported by empirical studies (summarised in [14, 16]). In addition, while shorter individuals generally have smaller pelves, birth canal size is maintained among certain recent small-bodied populations despite an absolutely narrower pelvis, suggesting a wider birth canal is compatible with adequate locomotor efficiency [17, 18]. Furthermore, pelvic breadth is also under climatic selection [19–21]. If the heat-retaining benefits of greater pelvic breadth in cold conditions outweigh hypothetical penalties in locomotor efficiency, it makes

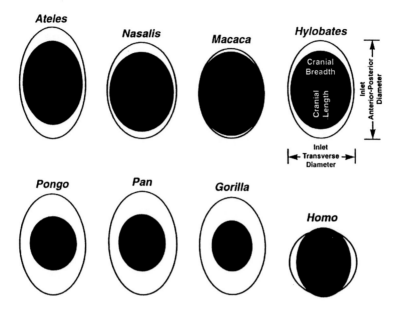

Fig. 3.1 The relationship between maternal pelvic inlet and neonatal head size in various primates, including our closest relatives the great apes (*bottom row*) (as redrawn in [4]). *Ateles:* spider monkey, *Nasalis:* proboscis monkey, *Macaca:* macaque, *Hylobates:* gibbon, *Pongo:* orang-utan, *Pan:* chimpanzee, *Gorilla:* gorilla and *Homo:* modern humans

little sense that pelvic dimensions could not increase to alleviate obstetric problems, which are likely under strong selection [22].

There also appears to be scope for natural selection to have decreased obstetric constraints on pelvic morphology. The relatively close fit between neonatal head size and maternal bony birth canal is not uniquely human [23] (Fig. 3.1). Some mammals have a much greater degree of disparity between neonatal and maternal pelvic canal size than humans, yet solutions to the problem have evolved in those species. For example, in free-tailed bats, the maternal intrapubic ligament stretches to ~ 15 times its usual length during birth [24]. Significant increases in human bony birth canal size could be achieved with relatively small increases in diameter that fall within existing variation in humans, suggesting that the OD could have been relieved relatively easily by natural selection. A mere 3 % increase in maternal pelvic canal diameter would permit a 10 % increase in neonatal brain size and so reduce the OD substantially [25]. There is also evidence to suggest a degree of uncoupling between locomotor and obstetric features of the pelvis in our evolutionary lineage, allowing more independent evolution of these characters than in great apes [26]. Finally, the bony birth canal is as variable as other pelvic or limb bone dimensions [27] and as variable in males as in females [28, 29], which is inconsistent with female pelvic morphology being under particularly high obstetric constraint.

However, it is also worth noting that culture is a key adaptive response to a wide range of factors among humans, and some argue that childbirth assistance from other individuals ('midwives') may have evolved relatively early in the human evolutionary lineage [4, 10]. This could have reduced the selective pressure for biological adaptation to the tight fit between the birth canal and infant head.

3.2.2 Cephalopelvic Disproportion and Obstructed Labour: The Consequences of an Environmentally Induced Obstetric Dilemma?

The focus here is on the relationship of obstructed labour to the OD. Obstructed labour is a significant problem in contemporary agricultural and industrial populations, particularly in low- and middle-income countries where it accounts for 4, 9.5 and 13.5 % of maternal deaths in Africa, Asia and Latin America, respectively (though it is notable that post-partum haemorrhage and infections and hypertension actually account for more maternal deaths [30]). Rather than being an inevitable consequence of the OD, as some have suggested previously, obstructed labour may vary in frequency relative to social and environmental conditions, and understanding the relationship between the environment and obstructed labour is therefore relevant to trying to reduce its burden.

Short maternal stature is among the strongest predictors of obstructed labour and Caesarean delivery [31–33], so the long-term trends in height may also offer insight into the temporal origin of obstructed labour. Smaller women generally have smaller pelves and thus the greater risk of cephalopelvic disproportion (a form of obstructed labour where the infant's head is too large to pass through the mother's bony birth canal, necessitating Caesarean delivery) [32]. While size at birth is influenced by both maternal and paternal genetics [34–36], maternal phenotype (height and weight) is a primary determinant of foetal nutrition so that shorter, thinner mothers typically give birth to smaller, lighter babies [36–38]. In many populations, short female stature results from environmental conditions (poor diet, disease) in the current and preceding generations [39]. Although smaller mothers do have smaller and lighter babies, 'brain-sparing' growth may still increase the risk of cephalopelvic disproportion by preserving head (brain) size at the expense of other organs [40, 41].

The greater plasticity of height and weight compared with head size could mean that altered environmental conditions cause more rapid change in maternal body size than head circumference and thus elevate obstructed labour risk [12]. When small stature evolves over an extended period, the size of the birth canal is preserved [17, 18], but obstructed labour may be more frequent where maternal body size has decreased relatively recently and appropriate adaptations in pelvic morphology are lacking [12]. Modern rates of obstructed labour could therefore have arisen with the adoption of agriculture in the last 10,000 years [12, 42], which was frequently associated with markedly decreased stature compared with preceding hunter-gatherer populations, probably due to increased rates of infection associated with living in settlements, nutritional deficiencies resulting from a less varied diet, and increased famine risk among agriculturalists [43, 44]. Alternatively, current obstructed labour rates could reflect more recent fluctuations in income and food security in low- and middle-income countries [12].

We know little about past rates of obstructed labour, and estimating childbirth-related death rates in the past is difficult [12, 45]. While the archaeological record offers examples of deaths in childbirth, these are rare cases where the infant remains within the birth canal (reviewed in [12]). As humans were hunter-gatherers for over 90 % of our species' history, data from modern hunter-gatherers could offer some insight into past obstructed labour rates. Unfortunately, few relevant data are available and they may be problematic, especially for extrapolating to preagricultural societies [12]. Anecdotally though, birth is quicker and obstructed labour extremely uncommon in hunter-gatherer groups including small-bodied populations ('pygmies') [42, 46], suggesting greater coordination of height and pelvic canal dimensions in these populations.

The problem of mismatched maternal height and offspring head size may be further exacerbated by the current global obesity and diabetes 'epidemics' [12]. Maternal overweight, obesity, excess pregnancy weight gain and gestational or pre-existing diabetes are associated with foetal macrosomia [37, 47, 48] and negative birth outcomes including obstructed labour [48–50]. As obesity rates rise, foetal macrosomia and associated problems may become more common unless female pelvic size also increases. The relationships between height, overweight, obesity and obstetric dimensions are not well studied. Although external pelvic dimensions at least do not seem to increase in concert with secular trends in height [51–53], external dimensions do not necessarily reflect the obstetrically relevant dimensions of the birth canal. Theoretically, more frequent Caesarean sections may be reducing the selective pressure against cephalopelvic disproportion where women can access this procedure, but simultaneously further increase the obstructed labour rates [54], and thus the need for Caesareans in subsequent generations.

Contemporary mothers in low- and middle-income countries probably face the greatest burden of obstructed labour as they are more likely to be shorter due to their own poorer growth environment, and simultaneously at increasing obesity risk with westernisation [55]. Low birthweight and short adult stature are associated with increased risk of obesity and type 2 diabetes, especially in an obesogenic environment [40, 56, 57], potentially further increasing the likelihood of obstructed labour.

3.3 Implications for Policy and Practice

An evolutionary perspective suggests that obstructed labour does not result from the 'classic' OD, but is temporally and environmentally contingent. It has been particularly exacerbated by more recent lifestyle changes and environmental impacts on maternal and/or offspring phenotype [12]. This has important implications for healthcare policy and practice in that the OD is not an unavoidable feature of our species, but one that can be managed through nutrition and healthcare. While Caesarean sections offer an immediate solution, other strategies to reduce obstructed labour rates are preferable in the light of limited healthcare access in poorer communities, the financial cost to health services, associated health risks for mother and child [58–62] (although see [63]), and the potential consequent increase obstructed labour rates in subsequent generations [54]. Therefore, focussing on diet and lifestyle factors to reduce obstructed labour rates is highly preferable, especially considering the health benefits, such changes bring to the wider community and across the life course.

3.3.1 Promotion of Healthy Weight and Lifestyle

In low- and middle-income countries, ensuring adequate maternal and child nutrition and health, especially for girls, will help to reduce the incidence of short maternal height and associated risks [9, 64], as well as bringing more widespread health improvements. Conversely, the associations between maternal obesity, excess pregnancy weight gain, diabetes, foetal macrosomia, obstructed labour and other birth complications give further urgency to the need to promote healthy lifestyles, especially among reproductive age women [47, 65–67] Given debates concerning the definition of appropriate pregnancy weight gain [65, 68, 69], more research is needed to enable medical practitioners to offer evidence-based guidance. The promotion of healthy maternal weight gain will also have wider-reaching benefits for offspring health across the lifespan [67, 70, 71].

3.3.2 Maternal Stature and Reducing Obstructed Labour

The higher plasticity of maternal body mass index compared with stature and pelvic morphology, and the link between high maternal body mass index and neonatal size, suggests that obstructed labour rates may take multiple generations to decline [12]. Secular trends in stature in high-income countries occurred incrementally over multiple decades [72]; thus, the alleviation of obstructed labour by increasing maternal stature in low- and middle-income countries is likely to take several generations. However, faster increases in obesity rates may perpetuate and even

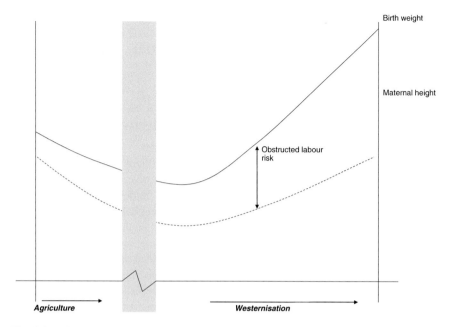

Fig. 3.2 Schematic representation of the relationship between maternal height, birthweight and obstructed delivery rates in low- and middle-income countries. After adopting agriculture, stature declined faster than birthweight and head circumference, and obstructed labour increased. With westernisation, maternal body mass index and diabetes increase rapidly, driving greater birthweight, while increased maternal stature takes multiple generations

exacerbate obstructed labour rates (Fig. 3.2). There is also evidence for a 'vicious cycle' of maternal and offspring obesity [73], potentially perpetuating and exacerbating obstructed labour across generations. Interventions during pregnancy to limit excess weight gain and regulate blood glucose can reduce foetal macrosomia and associated complications among overweight/obese mothers [69, 74, 75].

3.3.3 Childbirth Posture

In many traditional societies, childbirth positions were often upright, i.e. squatting, kneeling or sitting [4, 45, 76, 77], while the western supine or semi-recumbent position probably originated in the nineteenth- to twentieth-century medicalisation of childbirth [78, 79]. Upright birthing enlarges the bony birth canal by up to 28 % [80, 81] and may benefit both mother (e.g. reductions in pain, duration of the 2nd stage of labour, episiotomy, forceps assistance and perineal tearing [82–84]) and newborn (reduced incidence of abnormal heart rate, intensive care admissions [85]). However, some studies report increased maternal blood loss [85], and several report no difference in other neonatal outcomes with upright birth posture (e.g. Apgar scores [82, 83, 86]). A recent Cochrane review [85] indicated that the current evidence is low quality; more research is needed as an upright posture has potential to offer a simple means of reducing obstructed labour [12].

3.4 Future Directions: Identifying Mothers at Risk to Target Birth Intervention More Effectively

While maternal height, body mass index, pregnancy weight gain, and pelvic and other measurements show significant associations with obstructed labour [32, 87, 88], they have limited predictive power and current UK guidelines warn against the use of maternal measurements to predict Caesarean for cephalopelvic disproportion [89]. Combining foetal and maternal characteristics may be more accurate [32, 90], but predictive power remains relatively low. Promising areas for investigation include identification of large foetal head circumference by ultrasound in late pregnancy [91] and disproportionately large foetal mass relative to head circumference in macrosomic infants [4, 92], and predicting cephalopelvic disproportion risk from the ratio of parental head circumference to height [33, 93].

Glossary

Cephalopelvic disproportion	The infant's head is too large to pass through the bony birth canal of the mother's pelvis, necessitating Caesarean delivery
Obstructed labour	Where labour fails to progress as a result of mechanical problems, e.g., cephalopelvic disproportion, shoulder dystocia or malpresentation of the foetus, which prevent its passage through the birth canal [2]
Phenotype	The physical characteristics of an individual, population or species. They represent the interplay of genetic and environmental influences on the body
Plasticity	The degree to which a given biological characteristic can be modified in response to environmental factors. Characteristics that are more responsive to the environment, and so less strongly genetically determined, are described as plastic
Shoulder dystocia	During labour, the shoulder is trapped behind the maternal pelvis (typically the pubic symphysis, or the sacrum) following passage of the head

References

1. Washburn SL (1960) Tools and human evolution. Sci Am 203:63–75
2. Neilson J, Lavender T, Quenby S, Wray S (2003) Obstructed labour: reducing maternal death and disability during pregnancy. Br Med Bull 67(1):191–204. doi:10.1093/bmb/ldg018
3. Leutenegger W (1974) Functional aspects of pelvic morphology in simian primates. J Hum Evol 3(3):207–222. doi:10.1016/0047-2484(74)90179-1
4. Rosenberg K, Trevathan W (2002) Birth, obstetrics and human evolution. BJOG: Int J Obstet Gynaecol 109(11):1199–1206. doi:10.1046/j.1471-0528.2002.00010.x
5. Krogman WM (1951) The scars of human evolution. Sci Am 184:54–57
6. Lovejoy CO, Heiple KG, Burstein AH (1973) The gait of Australopithecus. Am J Phys Anthropol 38(3):757–779. doi:10.1002/ajpa.1330380315
7. Wittman AB, Wall LL (2007) The evolutionary origins of obstructed labor: bipedalism, encephalization, and the human obstetric dilemma. Obstet Gynecol Surv 62 (11):739–748. doi:10.1097/01.ogx.0000286584.04310.5c, pii: 0006254-200711000-00023
8. Weiner S, Monge J, Mann A (2008) Bipedalism and parturition: an evolutionary imperative for cesarean delivery? Clin Perinatol 35(3):469–478. doi:10.1016/j.clp.2008.06.003
9. Liston WA (2003) Rising caesarean section rates: can evolution and ecology explain some of the difficulties of modern childbirth? J R Soc Med 96(11):559–561
10. Trevathan WR (1996) The evolution of bipedalism and assisted birth. Med Anthropol Q 10(2):287–290. doi:10.1525/maq.1996.10.2.02a00100
11. Trevathan WR (1988) Fetal emergence patterns in evolutionary perspective. Am Anthropol 90(3):674–681. doi:10.2307/678231
12. Wells JCK, DeSilva JM, Stock JT (2012) The obstetric dilemma: an ancient game of Russian roulette, or a variable dilemma sensitive to ecology? Am J Phys Anthropol 149(S55):40–71. doi:10.1002/ajpa.22160
13. Shipman P (2013) Why is childbirth so painful? Am Sci 101(6):426
14. Dunsworth HM, Warrener AG, Deacon T, Ellison PT, Pontzer H (2012) Metabolic hypothesis for human altriciality. Proc Natl Acad Sci 109(38):15212–15216. doi:10.1073/pnas.1205282109
15. Rak Y (1991) Lucy's pelvic anatomy: its role in bipedal gait. J Hum Evol 20(4):283–290. doi:10.1016/0047-2484(91)90011-J
16. Wall-Scheffler CM (2012) Energetics, locomotion, and female reproduction: implications for human evolution. Ann Rev Anthropol 41(1):71–85. doi:10.1146/annurev-anthro-092611-145739
17. Kurki HK (2011) Pelvic dimorphism in relation to body size and body size dimorphism in humans. J Hum Evol 61(6):631–643. doi:10.1016/j.jhevol.2011.07.006
18. Kurki HK (2007) Protection of obstetric dimensions in a small-bodied human sample. Am J Phys Anthropol 133(4):1152–1165. doi:10.1002/ajpa.20636
19. Ruff CB (1991) Climate and body shape in hominid evolution. J Hum Evol 21(2):81–105
20. Weaver TD (2009) The meaning of Neandertal skeletal morphology. Proc Natl Acad Sci 106(38):16028–16033. doi:10.1073/pnas.0903864106
21. Betti L, von Cramon-Taubadel N, Manica A, Lycett SJ (2013) Global geometric morphometric analyses of the human pelvis reveal substantial neutral population history effects, even across sexes. PLoS One 8(2):e55909
22. Brown EA, Ruvolo M, Sabeti PC (2013) Many ways to die, one way to arrive: how selection acts through pregnancy. Trends Genet 29(10):585–592. doi:10.1016/j.tig.2013.03.001
23. Schultz AH (1949) Sex differences in the pelves of primates. Am J Phys Anthropol 7(3):401–424. doi:10.1002/ajpa.1330070307
24. Crelin ES (1969) Interpubic ligament: elasticity in pregnant free-tailed bat. Science 164(3875):81–82. doi:10.1126/science.164.3875.81
25. Epstein HT (1973) Possible metabolic constraints on human brain weight at birth. Am J Phys Anthropol 39(1):135–136. doi:10.1002/ajpa.1330390114

26. Grabowski M (2013) Hominin obstetrics and the evolution of constraints. Evol Biol 40(1):57–75. doi:10.1007/s11692-012-9174-7
27. Kurki HK (2013) Skeletal variability in the pelvis and limb skeleton of humans: does stabilizing selection limit female pelvic variation? Am J Human Biol 25:795. doi:10.1002/ajhb.22455
28. Tague RG (1995) Variation in pelvic size between males and females in nonhuman anthropoids. Am J Phys Anthropol 97(3):213–233. doi:10.1002/ajpa.1330970302
29. Tague RG (1989) Variation in pelvic size between males and females. Am J Phys Anthropol 80(1):59–71. doi:10.1002/ajpa.1330800108
30. Khan KS, Wojdyla D, Say L, Gülmezoglu AM, Van Look PFA (2006) WHO analysis of causes of maternal death: a systematic review. The Lancet 367(9516):1066–1074. doi:10.1016/S0140-6736(06)68397-9
31. Kjærgaard H, Dykes AK, Ottesen B, Olsen J (2010) Risk indicators for dystocia in low-risk nulliparous women: a study on lifestyle and anthropometrical factors. J Obstet Gynaecol 30(1):25–29. doi:10.3109/01443610903276417
32. Benjamin SJ, Daniel AB, Kamath A, Ramkumar V (2012) Anthropometric measurements as predictors of cephalopelvic disproportion. Acta Obstet Gynecol Scand 91(1):122–127. doi:10.1111/j.1600-0412.2011.01267.x
33. Connolly G, McKenna P (2001) Maternal height and external pelvimetry to predict cephalo-pelvic disproportion in nulliparous African women. Br J Obstet Gynaecol 108(3):338. doi:10.1016/S0306-5456(00)00081-4
34. Dunger DB, Petry CJ, Ong KK (2007) Genetics of size at birth. Diabetes Care 30(Suppl 2):S150–S155. doi:10.2337/dc07-s208
35. Wells JCK, Sharp G, Steer PJ, Leon DA (2013) Paternal and maternal influences on differences in birth weight between Europeans and Indians born in the UK. PLoS One 8(5):e61116
36. Veena SR, Kumaran K, Swarnagowri MN, Jayakumar MN, Leary SD, Stein CE, Cox VA, Fall CHD (2004) Intergenerational effects on size at birth in South India. Paediatr Perinat Epidemiol 18(5):361–370. doi:10.1111/j.1365-3016.2004.00579.x
37. Catalano PM, McIntyre HD, Cruickshank JK, McCance DR, Dyer AR, Metzger BE, Lowe LP, Trimble ER, Coustan DR, Hadden DR, Persson B, Hod M, Oats JJN, for the HAPO Study Cooperative Research Group (2012) The hyperglycemia and adverse pregnancy outcome study: associations of GDM and obesity with pregnancy outcomes. Diabetes Care 35(4):780–786
38. Kramer MS (1987) Determinants of low birth weight: methodological assessment and meta-analysis. Bull World Health Organ 65(5):663–737
39. Victora CG, Adair L, Fall C, Hallal PC, Martorell R, Richter L, Sachdev HS (2008) Maternal and child undernutrition: consequences for adult health and human capital. Lancet 371(9609):340–357
40. Barker D (ed) (1993) Fetal and infant origins of adult disease. British Medical Journal, London
41. Barbiro-Michaely E, Tolmasov M, Rinkevich-Shop S, Sonn J, Mayevsky A (2007) Can the "brain-sparing effect" be detected in a small-animal model? Med Sci Monit 13(10):BR211-219. Pii:502335
42. Roy RP (2003) A Darwinian view of obstructed labor. Obstet Gynecol 101(2):397–401. doi:10.1016/S0029-7844(02)02367-0
43. Mummert A, Esche E, Robinson J, Armelagos GJ (2011) Stature and robusticity during the agricultural transition: evidence from the bioarchaeological record. Econ Human Biol 9(3):284–301. doi:10.1016/j.ehb.2011.03.004
44. Cohen MN, Armelagos GJ (eds) (1984) Paleopathology at the origins of agriculture. Academic Press, London
45. Arriaza B, Allison M, Gerszten E (1988) Maternal mortality in pre-Columbian Indians of Arica, Chile. Am J Phys Anthropol 77(1):35–41. doi:10.1002/ajpa.1330770107

46. Kurki HK (2011) Compromised skeletal growth? Small body size and clinical contraction thresholds for the female pelvic canal. Int J Paleopathol 1(3–4):138–149. doi:10.1016/j.ijpp. 2011.10.004
47. Yu Z, Han S, Zhu J, Sun X, Ji C, Guo X (2013) Pre-pregnancy body mass index in relation to infant birth weight and offspring overweight/obesity: a systematic review and meta-analysis. PLoS One 8(4):e61627
48. Tsvieli O, Sergienko R, Sheiner E (2012) Risk factors and perinatal outcome of pregnancies complicated with cephalopelvic disproportion: a population-based study. Arch Gynecol Obstet 285(4):931–936. doi:10.1007/s00404-011-2086-4
49. Haerskjold A, Hegaard HK, Kjaergaard H (2012) Emergency caesarean section in low risk nulliparous women. J Obstet Gynaecol 32(6):543–547. doi:10.3109/01443615.2012.689027
50. Wu C-H, Chen C-F, Chien C-C (2013) Prediction of dystocia-related cesarean section risk in uncomplicated Taiwanese nulliparas at term. Arch Gynecol Obstet 288(5):1027–1033. doi:10. 1007/s00404-013-2864-2
51. Himes JH (1979) Secular changes in body proportions and composition. Monogr Soc Res Child Dev 44(3/4):28–58
52. Bowles GT (1932) New types of old Americans at Harvard. Harvard University Press, Cambridge
53. Ruff CB (1994) Morphological adaptation to climate in modern and fossil hominids. Am J Phys Anthropol 37(S19):65–107
54. Walsh JA (2008) Evolution and the cesarean section rate. Am Biol Teach 70(7):401–404. doi:10.1662/0002-7685(2008)70[401:etcsr]2.0.co;2
55. WHO (2005) Pocket book of hospital care for children: guidelines for the management of common illnesses with limited resources. World Health Organization, Geneva
56. Hales CN, Barker DJ (1992) Type 2 (non-insulin-dependent) diabetes mellitus: the thrifty phenotype hypothesis. Diabetologia 35(7):595–601
57. Wells JCK (2009) Thrift: a guide to thrifty genes, thrifty phenotypes and thrifty norms. Int J Obes 33(12):1331–1338
58. Souza J, Gulmezoglu A, Lumbiganon P, Laopaiboon M, Carroli G, Fawole B, Ruyan P, Maternal tWGSo, Group PHR (2010) Caesarean section without medical indications is associated with an increased risk of adverse short-term maternal outcomes: the 2004–2008 WHO Global Survey on Maternal and Perinatal Health. BMC Medicine 8(1):71
59. Villar J, Valladares E, Wojdyla D, Zavaleta N, Carroli G, Velazco A, Shah A, Campodónico L, Bataglia V, Faundes A, Langer A, Narváez A, Donner A, Romero M, Reynoso S, Simônia de Pádua K, Giordano D, Kublickas M, Acosta A (2006) Caesarean delivery rates and pregnancy outcomes: the 2005 WHO global survey on maternal and perinatal health in Latin America. Lancet 367(9525):1819–1829. doi:10.1016/S0140-6736(06)68704-7
60. Bodner K, Wierrani F, Grünberger W, Bodner-Adler B (2011) Influence of the mode of delivery on maternal and neonatal outcomes: a comparison between elective cesarean section and planned vaginal delivery in a low-risk obstetric population. Arch Gynecol Obstet 283(6):1193–1198. doi:10.1007/s00404-010-1525-y
61. Li H, Liu J, Blustein J (2013) Cesarean delivery on maternal request. Jama 310(9):977–978
62. B-s Wang, L-f Zhou, Coulter D, Liang H, Zhong Y, Y-n Guo, L-p Zhu, X-l Gao, Yuan W, E-s Gao (2010) Effects of caesarean section on maternal health in low risk nulliparous women: a prospective matched cohort study in Shanghai, China. BMC Pregnancy Childbirth 10(1):78
63. Lavender T, Hofmeyr GJ, Neilson James P, Kingdon C, Gyte Gillian ML (2012) Caesarean section for non-medical reasons at term. Cochrane Database Syst Rev (3). doi:10.1002/14651858.CD004660.pub3
64. Konje JC, Ladipo OA (2000) Nutrition and obstructed labor. Am J Clin Nutr 72(1):291S–297S
65. Siega-Riz AM, Gray GL (2013) Gestational weight gain recommendations in the context of the obesity epidemic. Nutr Rev 71:S26–S30. doi:10.1111/nure.12074
66. National Research Council (2009) Weight gain during pregnancy: reexamining the guidelines. The National Academies Press

67. Jungheim ES, Moley KH (2010) Current knowledge of obesity's effects in the pre- and periconceptional periods and avenues for future research. Am J Obstet Gynecol 203(6):525–530. doi:10.1016/j.ajog.2010.06.043

68. Rasmussen KM, Abrams B, Bodnar LM, Butte NF, Catalano PM, Maria Siega-Riz A (2010) Recommendations for weight gain during pregnancy in the context of the obesity epidemic. Obstet Gynecol 116(5):1191–1195

69. Artal R, Lockwood CJ, Brown HL (2010) Weight gain recommendations in pregnancy and the obesity epidemic. Obstet Gynecol 115(1):152–155

70. Ojha S, Saroha V, Symonds ME, Budge H (2013) Excess nutrient supply in early life and its later metabolic consequences. Clin Exp Pharmacol Physiol 40(11):817–823. doi:10.1111/1440-1681.12061

71. Kopp UMS, Dahl-Jorgensen K, Stigum H, Frost Andersen L, Nass O, Nystad W (2012) The associations between maternal pre-pregnancy body mass index or gestational weight change during pregnancy and body mass index of the child at 3 years of age. Int J Obes 36(10):1325–1331

72. Cole TJ (2003) The secular trend in human physical growth: a biological view. Econ Hum Biol 1(2):161–168

73. Cnattingius S, Villamor E, Lagerros YT, Wikstrom AK, Granath F (2012) High birth weight and obesity—a vicious circle across generations. Int J Obes 36(10):1320–1324

74. Kim C (2010) Gestational diabetes: risks, management, and treatment options. Int J Womens Health 2:339–351

75. Thangaratinam S, Rogozińska E, Jolly K, Glinkowski S, Roseboom T, Tomlinson JW, Kunz R, Mol BW, Coomarasamy A, Khan KS (2012) Effects of interventions in pregnancy on maternal weight and obstetric outcomes: meta-analysis of randomised evidence. BMJ 344. doi:10.1136/bmj.e2088

76. Russell JGB (1969) Moulding of the pelvic outlet. BJOG: Int J Obstet Gynaecol 76(9):817–820. doi:10.1111/j.1471-0528.1969.tb06185.x

77. Dunn P (1976) Obstetric delivery today: for better or for worse? Lancet 307(7963):790–793. doi:10.1016/S0140-6736(76)91624-X

78. Walrath D (2003) Rethinking pelvic typologies and the human birth mechanism. Curr Anthropol 44(1):5–31. doi:10.1086/344489

79. Dundes L (1987) The evolution of maternal birthing position. Am J Public Health 77(5):636–641. doi:10.2105/ajph.77.5.636

80. Michel SCA, Rake A, Treiber K, Seifert B, Chaoui R, Huch R, Marincek B, Kubik-Huch RA (2002) MR obstetric pelvimetry: effect of birthing position on pelvic bony dimensions. Am J Roentgenol 179(4):1063–1067. doi:10.2214/ajr.179.4.1791063

81. Russell JGB (1982) The rationale of primitive delivery positions. BJOG: Int J Obstet Gynaecol 89(9):712–715. doi:10.1111/j.1471-0528.1982.tb05096.x

82. Thies-Lagergren L, Kvist LJ, Sandin-Bojö A-K, Christensson K, Hildingsson I (2013) Labour augmentation and fetal outcomes in relation to birth positions: a secondary analysis of an RCT evaluating birth seat births. Midwifery 29(4):344–350. doi:10.1016/j.midw.2011.12.014

83. Gardosi J, Sylvester S, B-Lynch C (1989) Alternative positions in the second stage of labour: a randomized controlled trial. BJOG: Int J Obstet Gynaecol 96 (11):1290–1296. doi:10.1111/j.1471-0528.1989.tb03226.x

84. de Jong PR, Johanson RB, Baxen P, Adrians VD, van der Westhuisen S, Jones PW (1997) Randomised trial comparing the upright and supine positions for the second stage of labour. BJOG: Int J Obstet Gynaecol 104 (5):567–571. doi:10.1111/j.1471-0528.1997.tb11534.x

85. Gupta JK, Hofmeyr GJ, Shehmar M (2012) Position in the second stage of labour for women without epidural anaesthesia. Cochrane Database Syst Rev (5). doi:10.1002/14651858.CD002006.pub3

86. Bodner-Adler B, Bodner K, Kimberger O, Lozanov P, Husslein P, Mayerhofer K (2003) Women's position during labour: influence on maternal and neonatal outcome. Wien Klin Wochenschr 115(19–20):720–723. doi:10.1007/bf03040889

87. Melo B (2010) Intrapartum interventions for preventing shoulder dystocia. The WHO reproductive health library. World Health Organisation, Geneva
88. Mansor A, Arumugam K, Omar SZ (2010) Macrosomia is the only reliable predictor of shoulder dystocia in babies weighing 3.5 kg or more. Euro J Obstet Gynecol Reprod Biol 149(1):44–46. doi:10.1016/j.ejogrb.2009.12.003
89. National Collaborating Centre for Women's and Children's Health (2011) Caesarean section, 2nd edition edn. Royal College of Obstetricians and Gynaecologists, London
90. Mazouni C, Porcu G, Cohen-Solal E, Heckenroth H, Guidicelli B, Bonnier P, Gamerre M (2006) Maternal and anthropomorphic risk factors for shoulder dystocia. Acta Obstet Gynecol Scand 85(5):567–570. doi:10.1080/00016340600605044
91. Mujugira A, Osoti A, Deya R, Hawes S, Phipps A (2013) Fetal head circumference, operative delivery, and fetal outcomes: a multi-ethnic population-based cohort study. BMC Pregnancy Childbirth 13(1):106
92. Larson A, Mandelbaum D (2013) Association of head circumference and shoulder dystocia in macrosomic neonates. Matern Child Health J 17(3):501–504. doi:10.1007/s10995-012-1013-z
93. Connolly G, Naidoo C, Conroy RM, Byrne P, Mckenna P (2003) A new predictor of cephalopelvic disproportion? J Obstet Gynaecol 23(1):27–29. doi:10.1080/0144361021000043173

Chapter 4
Bottle Feeding: The Impact on Post-partum Depression, Birth Spacing and Autism

Gordon G. Gallup, Jr., Ph.D., Kristina N. Spaulding, Ph.D. and Fatima Aboul-Seoud, B.A.

Lay Summary Bottle feeding puts mothers out of phase with an integral feature of mammalian evolutionary history. In this chapter, we document some of the negative heath-related consequences that bottle feeding has not only for the baby but for the mother as well. Inspired by evolutionary theory, we make numerous suggestions for steps that could be taken to minimize some of these medical and psychological issues that have occurred as a result of bottle feeding as an alternative to breastfeeding.

4.1 Introduction

The existence of a growing number of discrepancies or mismatches between evolved human adaptations and the conditions many of us now confront as a consequence of recent technological changes is a focus point for those who aim to address the medical issues using "evolutionary thinking". As a result of the agricultural, industrial and technological revolution, we have increasingly emancipated ourselves from some of the conditions that gave rise to our existence.

G.G. Gallup Jr. (✉) · K.N. Spaulding
Department of Psychology, University at Albany, State University of New York, Albany, USA
e-mail: gallup@albany.edu

K.N. Spaulding
e-mail: kspaulding@albany.edu

F. Aboul-Seoud
Department of Psychology, University of Pennsylvania, Philadelphia, USA
e-mail: fa534176@gmail.com

© Springer International Publishing Switzerland 2016
A. Alvergne et al. (eds.), *Evolutionary Thinking in Medicine*,
Advances in the Evolutionary Analysis of Human Behaviour,
DOI 10.1007/978-3-319-29716-3_4

Body fat is a classic case in point. While seen as a risk factor for many medical problems, endogenous stores of body fat used to be adaptive. As illustrated by the phrase "feast or famine", variation in food supplies was undoubtedly a common recurring condition during most of human evolutionary history (see also Chap. 8). When food was available, it would have been adaptive to overindulge and consume excess calories that could be stored in the form of body fat and later utilized to enable humans to bridge periods when food was scarce. However, nowadays, as a result of modern agriculture, we have unwittingly created what amounts to a continuous feast in many parts of the world where food is widely available in almost limitless quantities, which puts us out of phase with the conditions that used to prevail during our evolutionary history. As a consequence of the evolved propensity to overeat in response to the abundance of food, this leads to obesity, diabetes, and cardiovascular problems. The same can be reasoned for the photoperiod. In spite of seasonal variation in the amount of daylight, through artificial illumination, we have created conditions nowadays in which many people spent 16 or more hours every day in the light. Not only has this unwittingly created what amounts to a continuous summer, but it has ushered in ways where more and more people have the option of working and being active under conditions that put them out of phase with the photoperiod, and there is growing evidence that this carries a number of health risks as well [1].

In this chapter, we focus on bottle feeding as an alternative to breastfeeding and attempt to identify and resolve some of the medical and psychological problems posed by the evolutionarily novel introduction of bottle feeding as a means of nourishing infants [2]. For most of human evolutionary history babies that were not breastfed would have perished and the decision not to breastfeed would have been tantamount to committing infanticide [3]. The technology needed to make bottle feeding an option is very recent and is a good example of an evolutionary mismatch. Not only is the formula that has been developed for bottle feeding a poor substitute for breast milk that may compromise the health of the infant [4], but mothers who bottle-feed their children also unwittingly put themselves at risk. Studies have shown that women who bottle-feed are at an elevated likelihood of becoming overweight [5] and developing breast cancer [6].

The decision to bottle-feed can be made for any number of reasons, including a concern about the effect breastfeeding might have on the mother's figure, embarrassment over breastfeeding in public, time and social constraints due to employment, or because of a physical inability to breastfeed or failure to produce adequate breast milk. These are all legitimate reasons upon which a mother is entitled to base her decision, and we stress that this decision is ultimately hers to make. Although this chapter highlights the potential problems that can arise with bottle feeding, there are always trade-offs with any decision. Our goal is to outline the possible adverse outcomes and suggest better alternatives to the ones that are widely used today so that women can make more informed decisions for themselves and their babies.

4.2   Research Findings

4.2.1   Bottle Feeding and Post-partum Depression

The absence of breastfeeding and the cessation of breastfeeding both trigger hormonal changes that function to terminate the production of breast milk (see Ref. [3] for details). During most of human history, the absence or sudden cessation of breastfeeding would often have been occasioned by the death of the child and as a consequence, it has been argued that at the level of the mother's basic biology, bottle feeding may simulate or approximate some of the conditions associated with the loss of a child [3]. The death of a child is a powerful stimulus for depression, especially among women [7], and growing evidence implicates bottle feeding as a significant risk factor for post-partum depression [8]. In a recent study, we discovered that in contrast to mothers who breastfed their babies, mothers who bottle fed were more likely to develop depressive symptoms during the post-partum period even after controlling for age, education, income, and the woman's relationship with the child's father [3]. We also found that mothers who bottle-fed reported wanting to hold their babies more. This apparent bottle feeding enhancement of infant holding may be reminiscent of common reports among many non-human primates where in response to the death of an infant, mothers often engage in excessive holding, carrying and clinging to the corpses of dead babies for prolonged periods of time. Consistent with our analysis, weaning also simulates child loss and weaning has been show to trigger maternal depressive episodes in many women [3].

4.2.2   Antidepressants and Breastfeeding

Women who develop post-partum depression and are treated with antidepressants are sometimes advised to discontinue breastfeeding because of a concern that the medication may get into the mother's milk and adversely affect the infant. The FDA does not recommend breastfeeding for women who are taking antidepressants, because of reported cases of adverse effects on infants [9]. However, most antidepressants have been demonstrated to be safe, only transmitting 1 % of the mother's dose per kilogram to the baby, which is neither toxic nor has it been shown to cause any developmental delays [10]. With few exceptions such as fluoxetine and nefazodone, most antidepressants have not been found to cause any negative effects to infants of mothers who were taking them while breastfeeding, and as a result, continuation of breastfeeding while taking antidepressants is now often encouraged [9, 11]. Thus, from the perspective we are taking, the decision to recommend that depressed mothers stop breastfeeding is ill-informed because it could exacerbate rather than resolving the problem by simulating the death/loss of the infant at time when the mother is already suffering from depression. The same may be true of other medical procedures. Antiquated hospital practices such as keeping newborns in nurseries that involve repeated separation of the mother from her infant may also unwittingly simulate child loss and contribute to the development of post-partum depression.

4.2.3 Bottle Feeding Breast Milk

With the availability of modern breast pumps for expressing breast milk, working mothers and others who cannot always be present to breastfeed now have the option of providing babies with the nutritional advantages of breast milk by using bottles.

The proximate mechanisms by which breastfeeding facilitates attachment to infants has been studied extensively. Oxytocin, a hormone that promotes bonding, is released by the mother as well as the infant when breastfeeding occurs [12, 13]. Mothers report an increase in positive affect and feelings of love for their baby after breastfeeding, which is thought to be influenced by the release of oxytocin [12], whereas mothers may experience a negative change in mood after bottle feeding [14]. While nourishment of the infant is an important function of breastfeeding, there are many signals that the mother receives, such as skin-to-skin contact and the scent of her baby, which, when paired with the act of breastfeeding, reinforce and promote bonding and caring for the child.

While breastfeeding can act as a buffer against post-partum depression, many women, especially those who work, do not have the option to breastfeed on demand but instead can opt to express milk and have a caretaker bottle-feed their breast milk to the baby while they are away. To minimize the adverse psychological effects that bottle feeding may have on mothers who express milk for their babies, and reduce cues that might otherwise approximate or simulate child loss, we recommend instituting the following procedures during times when mothers express breast milk. Using evolutionary history as a guide, it is possible to do this in ways that approximate the contexts in which breastfeeding typically occurs to achieve maximum benefits. When expressing milk ordinarily, the child is absent, so the mother is not receiving the breastfeeding signals she would otherwise. She does not experience skin-to-skin contact with her baby, nor does she smell or hear her child, nor is her nipple stimulated directly by her child. Oxytocin peaks when suckling starts just before the release of milk, but this does not always happen with other methods of expressing milk [15]. Some mothers have difficulty expressing milk because oxytocin stimulates the let-down reflex which makes milk available in the sinuses behind the nipples, and without the baby present, oxytocin is not as readily released. Consistent with an evolutionary approach, it has been recommended that mothers look at a picture of the baby, hear the baby cry or smell the baby (using an article of clothing), in order to stimulate the reflex and help the mother express milk [16]. From an evolutionary perspective, we would suggest that the closer the match between breastfeeding and the context in which milk is expressed, the more effective it will be. When expressing milk, we would predict that it would be even more effective to watch a video of the baby that was taped on an earlier occasion of breastfeeding. To maximize the effect, it would be important to position the camera over the mother's shoulder to approximate what the mother sees while she engages in breastfeeding. It might even be better if the videotape also included sounds that the baby typically makes while it is actually breastfeeding.

4.2.4 Bottle Feeding, Birth Spacing and Autism

Bottle feeding not only creates or approximates some of the conditions that used to be associated with the death of a child, but it also unwittingly undermines the existence of some important evolved birth spacing mechanisms [17]. Breastfeeding functions to promote birth spacing by creating hormonal changes that lead to lactational amenorrhoea (the absence of menstruation) and anovulation. In other words, among women who breastfed, the resumption of regular menstrual cycles and ovulation are typically delayed, and these changes function to reduce the likelihood of early/premature re-impregnation during the post-partum period. During human evolutionary history, it was common for breastfeeding to last 3–4 years following the birth of a child [18], and without bottles, early weaning was not an option.

Pregnancy and breastfeeding are nutritionally very expensive and taxing for the mother, and as a consequence, re-impregnation during the early post-partum period could adversely affect the availability of milk for the existing child and at the same time diminish the nutritional status of the prenatal environment for the developing foetus. Consistent with this analysis, recent evidence implicates birth spacing as a risk factor for autism. Research shows that children who are conceived within a year or two of their previous sibling are significantly more likely to develop autism [5]. Indeed, children who are conceived within a year of their next oldest sibling are three times more likely to develop autism. We have argued that these peculiar birth-order effects on the risk of autism may be a consequence of deficiencies in the prenatal environment that occur as a result of closely spaced pregnancies that are unwittingly due to bottle feeding and/or early weaning [17]. Thus, a mother's decision to bottle-feed her child may make re-impregnation more likely during the early post-partum period and thereby increases the risk of autism in her next child. One way doctors could reduce this risk would be to recommend birth control for mothers who are bottle-feeding as a method to promote planned birth spacing. Research showing a high degree of concordance in the incidence of autism not only among identical twins, but also among fraternal twins [19] is highly consistent with this bottle feeding/birth spacing hypothesis as it pertains to the prenatal intrauterine environment.

4.2.5   Other Bottle Feeding and Birth Spacing Problems

As further evidence that the prenatal/gestational environment is compromised by closely spaced pregnancies, children born within two years or less of one another are at increased risk of preterm birth, low birthweight and being small for gestational age [20, 21]. These results were based on a meta-analysis of studies conducted in the USA, Latin America, Asia, Europe and Australia. In developing countries, abbreviated periods of birth spacing between successive siblings are also associated with increased mortality for children less than 5 years of age [21], and these effects show a dose-dependent relationship, with shorter inter-birth intervals being associated with stronger negative effects. This study controlled for a number of possible confounding factors including maternal age at birth, survival of the preceding child, education, index of wealth and source of drinking water. Rutstein [21] argued that waiting 36 months to conceive again could reduce under-five-year-old deaths by as much as 25 %.

Also consistent with the evolutionary mismatch hypothesis, prolonged periods separating the birth of successive siblings have also been associated with negative outcomes. Inter-birth intervals exceeding five years are associated with an increase in the risk of pre-eclampsia [20] even after controlling for maternal age. The risk of preterm birth, low birthweight and small gestational age also increase after 59 months [20], and the same is true for the risk of mortality prior to age 5. The results of these studies suggest that optimal birth spacing for humans may be 3–4 years. These results map surprisingly well on to the typical amount of time breastfeeding lasts under preindustrial conditions that resemble those thought to be typical of human evolution. The connection between long intervals and poor outcomes is not surprising, since once lactation stops most women would have resumed normal cycling and become pregnant again shortly thereafter. Therefore, throughout most of human history, birth intervals of 5 years or more would have been the exception rather than the rule.

4.2.6   Bottle Feeding and Other Birth Spacing Anomalies

It is worth mentioning that birth spacing has a number of other behavioural side effects. Closely spaced pregnancies that occur as a consequence of bottle feeding elevate the risk of sibling rivalry [22]. Closely spaced births likewise appear to compromise the cognitive capacity of firstborn children by perhaps diminishing/diluting the intellectual environment of the family [23], and the same may hold true for cognitive impairment among subsequent children as a result of depleting long-chain polyunsaturated fatty acids in maternal gluteofemoral fat stores, which have been implicated as being important in healthy pre- and neonatal brain growth and development [24]. Birth spacing also has other behavioural effects. Intervals of less than 24 months have been associated with the increases in neglectful parenting, behavioural problems and a decrease in cognitive functioning in 1st grade of the younger child [25].

4.2.7 Obstacles to Breastfeeding in Medical Practice

Despite numerous benefits for both mothers and infants, rates of breastfeeding remain relatively low. Globally, less than 40 % of infants are exclusively breastfed for the first six months [26]. These numbers are further broken down into WHO regions as follows: African (35 %), the Americas (30 %), South-east Asia (47 %), European (25 %), Eastern Mediterranean (35 %) and Western Pacific (no data). Based on income, 47 % of low income, 38 % of lower middle income and 18 % of high income are exclusively breastfed for the first 6 months (no data for "upper middle income"). Evolutionary thinking can serve as a useful guide for increasing breastfeeding success. For most of human history, all babies would have been born vaginally, experienced almost immediate contact with their mothers and initiated breastfeeding shortly after birth. Caesarean sections put both mothers and infants out of phase with some of the critical features associated with the birthing process.

A large body of research has identified a negative correlation between Caesarean delivery (CD) and positive breastfeeding outcomes. In a study of American mothers, CD was associated with a decreased likelihood of breastfeeding initiation [27]. These results have been replicated in a number of other studies conducted elsewhere, including Perez-Escamilla et al. [28] (Mexico), Gubler et al. [29] (Switzerland) and Radwan [30] (United Arab Emirates). In some cases, CD has also been associated with a decreased duration of breastfeeding [29]. Even in CD mothers who initiate breastfeeding, there are differences compared to vaginal delivery (VD) mothers. In one study, the mean rate of breast-milk transfer among breastfeeding mothers was lower in CD mothers [31]. In addition, 25 % of CD infants had not suckled within the first four hours after delivery, whereas the number was only 3 % for VD mothers. CD mothers also received lower LATCH scores (rating quality of breastfeeding) than VD mothers [32].

The negative effects of CD on breastfeeding are found in both planned and emergency C-sections [31, 33, 34]. There are several possible explanations for these results, including delayed contact with the mother and negative side effects of drugs used during CD. CD not only appears to delay lactation [5, 35], but CD mothers may also show decreased levels of oxytocin and prolactin post-partum [36]; both of these hormones play an important role in lactation and maternal bonding.

Some hospital practices may encourage breastfeeding in VD and CD mothers alike. Rooming in, skin-to-skin contact and early initiation of breastfeeding are all associated with breastfeeding initiation and success [29, 30, 37]. Physical contact is particularly important. Babies that experience their first physical maternal contact within five minutes of birth and who room in with their mothers for 24 h show increased rates of nursing [29], increased likelihood of breastfeeding [30] and decreased use of bottle at discharge from the hospital [29]. Obviously, for babies that do not room in with the mother on a 24-h basis, the opportunity for physical contact is reduced. One study of women in Sweden found that only 37 % of the infants separated from their mothers for 1–6 days during the first week of life breastfed exclusively after three months, compared to 72 % for the non-separated group [38].

Research clearly shows that vaginal delivery and early contact with the mother are important for breastfeeding success. In 1991, WHO and United Nations Childrens' Fund (UNICEF) launched the Baby Friendly Hospital Initiative (BFHI) (see also Chap. 5) and recommended steps hospitals should take to increase the likelihood and duration of breastfeeding. These include helping mothers to breastfeed within 30 min of birth, rooming in 24 h a day and encouraging feeding on demand [39]. All three of these practices are consistent with what would have been typical during most of human evolutionary history. According to an Australian study, in one hospital implementing BFHI, 9 of 15 CD infants initiated breastfeeding within the first hour [33], whereas in three hospitals that did not participate in the BFHI, none of the 32 CD infants began breastfeeding within 1 h. In addition, the mean time until the first breastfeed was significantly shorter at the BFHI hospital, and the BFHI hospital also had a higher percentage of babies' breastfeeding after 1 h. In both cases, these differences remained regardless of the mode of delivery. Moreover, mothers from hospitals implementing BFHI practices in Hong Kong were also less likely to initiate premature weaning [40]. For most mothers (CD and VD), exclusive breastfeeding in the hospital is protective against early termination of breastfeeding after release [40]. However, even in BFHI hospitals, CD remains the most important barrier to breastfeeding. In an Italian hospital where optimization of skin-to-skin contact, early initiation of breastfeeding and rooming in are all standard practice, rates of breastfeeding at discharge in elective CD mothers (74.4 %) were still lower than those in VD mothers (87.8 %, [34]).

4.2.8 Obesity and Breastfeeding Success

Another factor that impacts breastfeeding success is the mother's body weight. Mothers who are overweight or obese are less likely to initiate and continue breastfeeding [41, 42]. Women who are overweight and obese are also more likely to be of lower socio-economic status and/or depressed, and these are features that have been associated with reduced incidences of breastfeeding. However, the effect of body weight persists even when such factors are taken into account [41]. Possible reasons for this include difficulty holding the infant in a way that would be conducive to breastfeeding [43], decreased prolactin response to suckling [44] and delayed lactation [5], as well as social factors such as greater embarrassment about breastfeeding in public and changes in breast and nipple configuration. In addition, obese women are more likely to have Caesarean sections [45], which, as detailed above, make breastfeeding less likely.

4.3   Implications for Policy and Practice

Breast feeding is good for mothers as well as babies. The risk of post-partum depression among mothers who bottle-feed is much higher than for those who breastfeed. From an evolutionary perspective, this may be because the failure to breastfeed simulates/approximates an instance of child loss. Other hospital practices that unwittingly simulate child loss may also contribute to post-partum depression; including periods of separating the infant from the mother, as well a discouraging breastfeeding among mothers being treated with antidepressant medication. In addition, mothers that do not breastfeed are at increased risk of becoming over-weight and developing breast cancer. Bottle feeding and early weaning also puts evolved birth spacing mechanisms (such as lactational amenorrhoea and lactational anovulation) on hold, and evidence shows that babies conceived within a year or two of their next oldest sibling are at a much higher risk of a number of medical and psychological problems, including autism. Caesarean sections likewise put mothers out of phase with the evolved features of the birthing process, and growing evidence shows that Caesarean sections are a major impediment to successful breastfeeding.

Glossary

Anovulation	Hormonal changes that inhibit the release of eggs
Lactational amenorrhoea	Hormonal changes that occur during breastfeeding and that suppress the release of eggs
Gluteofemoral fat	Fat stores of long-chain polyunsaturated fatty acids that are important for healthy brain growth and development

References

1. Vyas MV et al (2012) Shift work and vascular events: systematic review and meta-analysis. Br Med J 345:e4800
2. Gallup GG Jr, Reynolds CJ, Bak PA, Aboul-Seoud F (2014) Evolutionary medicine: the impact of evolutionary theory on research, prevention, and practice. The future of evolutionary studies in higher education
3. Gallup GG Jr, Nathan Pipitone R, Carrone KJ, Leadholm KL (2010) Bottle feeding simulates child loss: postpartum depression and evolutionary medicine. Med Hypotheses 74:174–176. doi:10.1016/j.mehy.2009.07.016
4. Heinig MJ, Dewey KG (1996) Health advantages of breast feeding for infants: a critical review. Nutr Res Rev 9:89–110
5. Chapman DJ, Perez-Escamilla R (1999) Identification of risk factors for delayed onset of lactation. J Am Diet Assoc 99:450–454. doi:10.1016/s0002-8223(99)00109-1

6. Stuebe AM, Willett WC, Xue F, Michels KB (2009) Lactation and incidence of premenopausal breast cancer: a longitudinal study. Arch Intern Med 169:1364–1371

7. Suarez SD, Gallup GG Jr (1985) Depression as a response to reproductive failure. J Soc Biol Struct 8:279–287

8. Dennis C-L, McQueen K (2009) The relationship between infant-feeding outcomes and postpartum depression: a qualitative systematic review. Pediatrics 123:E736–E751. doi:10.1542/peds.2008-1629

9. Gjerdingen D (2003) The effectiveness of various postpartum depression treatments and the impact of antidepressant drugs on nursing infants. J Am Board Fam Pract 16:372–382

10. Yoshida K, Smith B, Craggs M, Kumar RC (1997) Investigation of pharmacokinetics and of possible adverse effects in infants exposed to tricyclic antidepressants in breast-milk. J Affect Disord 43:225–237. doi:10.1016/s0165-0327(97)01433-x

11. Weissman AM et al (2004) Pooled analysis of antidepressant levels in lactating mothers, breast milk, and nursing infants. Am J Psychiatry 161:1066–1078. doi:10.1176/appi.ajp.161.6.1066

12. Klaus M (1998) Mother and infant: early emotional ties. Pediatrics 102:1244–1246

13. McNeilly AS, Robinson ICAF, Houston MJ, Howie PW (1983) Release of oxytocin and prolactin in response to suckling. Br Med J 286:257–259

14. Mezzacappa ES, Katkin ES (2002) Breast-feeding is associated with reduced perceived stress and negative mood in mothers. Health Psychol 21:187–193. doi:10.1037//0278-6133.21.2.187

15. Zinaman MJ, Hughes V, Queenan JT, Labbok MH, Albertson B (1992) Acute prolactin and oxytocin in responses and milk yield to infant suckling and artificial methods of expression in lactating women. Pediatrics 89:437–440

16. Spicer K (2001) What every nurse needs to know about breast pumping: instructing and supporting mothers of premature infants in the NICU. Neonatal Netw NN 20:35–41

17. Gallup GG Jr, Hobbs DR (2011) Evolutionary medicine: bottle feeding, birth spacing, and autism. Med Hypotheses 77:345–346. doi:10.1016/j.mehy.2011.05.010

18. Eaton SB, Shostak M, Konner M (1988) Paleolithic prescription. Harper and Row, New York

19. Ronald A, Happe F, Price TS, Baron-Cohen S, Plomin R (2006) Phenotypic and genetic overlap between autistic traits at the extremes of the general population. J Am Acad Child Adolesc Psychiatry 45:1206–1214

20. Conde-Agudelo A, Rosas-Bermudez A, Kafury-Goeta AC (2006) Birth spacing and risk of adverse perinatal outcomes—a meta-analysis. Jama J Am Med Assoc 295:1809–1823. doi:10.1001/jama.295.15.1809

21. Rutstein SO (2008) Further evidence of the effects of preceding birth intervals on neonatal infant and under-five-years mortality and nutritional status in developing countries: evidence from the Demographic and health surveys

22. Minnett AM, Vandell DL, Santrock JW (1983) The effects of sibling status on sibling interaction influence of birth order age spacing sex of child and sex of sibling. Child Dev 54:1064–1072. doi:10.1111/j.1467-8624.1983.tb00527.x

23. Zajonc RB, Markus GB (1975) Birth order and intellectual development. Psychol Rev 82:74–88. doi:10.1037/h0076229

24. Lassek WD, Gaulin SJC (2008) Waist-hip ratio and cognitive ability: is gluteofemoral fat a privileged store of neurodevelopmental resources? Evol Hum Behav 29:26–34. doi:10.1016/j.evolhumbehav.2007.07.005

25. Crowne SS, Gonsalves K, Burrell L, McFarlane E, Duggan A (2012) Relationship between birth spacing, child maltreatment, and child behavior and development outcomes among at-risk families. Matern Child Health J 16:1413–1420. doi:10.1007/s10995-011-0909-3

26. World Health Organization (2013) World health statistics 2013

27. Samuels SE, Margen S, Schoen EJ (1985) Incidence and duration of breast-feeding in a health maintenance organization population. Am J Clin Nutr 42:504–510

28. Perez-Escamilla R, Maulen-Radovan I, Dewey KG (1996) The association between cesarean delivery and breast-feeding outcomes among Mexican women. Am J Public Health 86:832–836. doi:10.2105/ajph.86.6.832

29. Gubler T, Krahenmann F, Roos M, Zimmermann R, Ochsenbein-Koelble N (2013) Determinants of successful breastfeeding initiation in healthy term singletons: a Swiss university hospital observational study. J Perinat Med 41:331–339. doi:10.1515/jpm-2012-0102

30. Radwan H (2013) Patterns and determinants of breastfeeding and complementary feeding practices of Emirati Mothers in the United Arab Emirates. Bmc Publ Health 13. doi:10.1186/1471-2458-13-171

31. Evans KC, Evans RG, Royal R, Esterman AJ, James SL (2003) Effect of caesarean section on breast milk transfer to the normal term newborn over the first week of life. Arch Dis Child 88:F380–F382. doi:10.1136/fn.88.5.F380

32. Cakmak H, Kuguoglu S (2007) Comparison of the breastfeeding patterns of mothers who delivered their babies per vagina and via cesarean section: An observational study using the LATCH breastfeeding charting system. Int J Nurs Stud 44:1128–1137. doi:10.1016/j.ijnurstu.2006.04.018

33. Rowe-Murray HJ, Fisher JRW (2002) Baby friendly hospital practices: cesarean section is a persistent barrier to early initiation of breastfeeding. Birth-Issues Perinat Care 29:124–131. doi:10.1046/j.1523-536X.2002.00172.x

34. Zanardo V et al (2010) Elective cesarean delivery: does it have a negative effect on breastfeeding? Birth-Issues Perinat Care 37:275–279. doi:10.1111/j.1523-536X.2010.00421.x

35. Vestermark V, Hogdall CK, Birch M, Plenov G, Toftager-Larsen K (1991) Influence of the mode of delivery on initiation of breast-feeding. Eur J Obstet Gynecol Reprod Biol 38:33–38

36. Nissen E et al (1996) Different patterns of oxytocin, prolactin but not cortisol release during breastfeeding in women delivered by Caesarean section or by the vaginal route. Early Human Dev 45:103–118. doi:10.1016/0378-3782(96)01725-2

37. Moore ER, Anderson GC (2007) Randomized controlled trial of very early mother-infant skin-to-skin contact and breastfeeding status. J Midwifery Women's Health 52:116–125. doi:10.1016/j.jmwh.2006.12.002

38. Elander G, Lindberg T (1984) Short mother infant separation during 1st week of life influences the duration of breast feeding. Acta Paediatr Scand 73:237–240. doi:10.1111/j.1651-2227.1984.tb09935.x

39. WHO & UNICEF (2009) Baby-friendly hospital initiative: revised, updated and expanded for integrated care

40. Tarrant M et al (2011) Impact of baby-friendly hospital practices on breastfeeding in Hong Kong. Birth-Issues Perinat Care 38:238–245. doi:10.1111/j.1523-536X.2011.00483.x

41. Amir LH, Donath S (2007) A systematic review of maternal obesity and breastfeeding intention, initiation and duration. BMC Pregnancy Childbirth 7:9

42. Wojcicki JM (2011) Maternal prepregnancy body mass index and initiation and duration of breastfeeding: a review of the literature. J Womens Health 20:341–347. doi:10.1089/jwh.2010.2248

43. Donath SM, Amir LH (2008) Maternal obesity and initiation and duration of breastfeeding: data from the longitudinal study of Australian children. Matern Child Nutr 4:163–170. doi:10.1111/j.1740-8709.2008.00134.x

44. Rasmussen KM, Kjolhede CL (2004) Prepregnant overweight and obesity diminish the prolactin response to suckling in the first week postpartum. Pediatrics 113:E465–E471. doi:10.1542/peds.113.5.e465

45. Sebire NJ et al (2001) Maternal obesity and pregnancy outcome: a study of 287 213 pregnancies in London. Int J Obes 25:1175–1182. doi:10.1038/sj.ijo.0801670

Part II
Paediatrics

Chapter 5
Sudden Infant Death Syndrome

Charlotte K. Russell, Ph.D., Lane E. Volpe, Ph.D.
and Prof. Helen L. Ball, Ph.D.

Lay summary Human infants have needs which are unique among primates. These are primarily a consequence of their comparatively poor neurological and muscular development at birth (compared to other primate infants who are able to cling and maintain proximity with their mother). This results in infants who would be very vulnerable in the absence of a caregiver and who rely on their mother to provide close contact for frequent feeds, safety and physiological regulation. Looking at human infants in this light allows us to think critically about the way parents in Western cultures care for their babies, and the possible consequences of these infant care practices for infant health, including the risk of sudden infant death syndrome (SIDS). In most traditional (non-Western) human societies, mothers keep their infants in close contact both during the day and the night. Infants in these societies wake frequently during periods of sleep, and breastfeed on demand. In contrast, in Western societies, infant care practices emphasise sleeping alone for extended periods of time, and early cessation of breastfeeding.

As both lone sleeping and use of infant formula are associated with increased incidence of SIDS, in human infants who have evolved to expect continuous physical contact with a caregiver, some researchers have suggested that SIDS (a syndrome characteristic of Western societies) may be a consequence of a mismatch between Western infant care practices, and the unique vulnerabilities of human infants. Where infants sleep has long been a focus of both parent-educators and campaigners trying to reduce the rate of SIDS.

C.K. Russell (✉) · L.E. Volpe · H.L. Ball
Parent-Infant Sleep Lab, Department of Anthropology, Durham University, Durham, UK
e-mail: c.k.russell@durham.ac.uk

H.L. Ball
e-mail: h.l.ball@durham.ac.uk

L.E. Volpe
The Implementation Group, Colorado, USA
e-mail: Lane@TheImplementationGroup.com

© Springer International Publishing Switzerland 2016
A. Alvergne et al. (eds.), *Evolutionary Thinking in Medicine*,
Advances in the Evolutionary Analysis of Human Behaviour,
DOI 10.1007/978-3-319-29716-3_5

Because some infants die of SIDS while sleeping in an adult bed, this sleep location has, for the past 20 years, been the focus of many anti-bed-sharing campaigns. However, because infant sleep location is also intimately related to infant feeding method (breast vs. formula), individual parenting ethos and cultural pressures, such campaigns have come to be regarded as being both ineffectual and unethical, and alternative approaches developed.

Alternate approaches focus on evidence-based education for parents, and culturally relevant interventions which facilitate close contact for infants and caregivers while providing safe sleeping spaces for infants who may be more vulnerable to SIDS. Such interventions, including the UNICEF UK Baby-Friendly Initiative, Infant Sleep Safety Tool (ISST) and Wahakura infant sleep basket, address infants' evolved needs while also acknowledging the trade-offs that parents consider in making decisions about infant care.

5.1 Introduction

5.1.1 Evolved Infant Biology and Infant Care

An evolutionary perspective on SIDS and night-time infant care considers the incongruity between an infant's evolved biological and behavioural needs, and culturally mediated twenty-first-century infant care practices. Understanding of human infants' evolved biological needs can be gained via the comparative perspective of human traits with those of other mammals and primates with whom we share a common ancestor.

As placental mammals, humans produce relatively well-developed live-born young who require maternal post-natal care and lactation. Developmental state at birth and gestational length vary among mammalian species, and infants can generally be categorised within two types. 'Altricial' species produce infants that are comparatively immature at birth; neuromuscular control is poor, infants are often blind and hairless and are 'cached', or sequestered in nests. They are fed infrequently with milk that is high in fat. Primates, along with many other mammals, produce 'precocial' infants—meaning they are well-developed at birth, able to see, hear and maintain proximity with their mothers via independent locomotion, or clinging. Mothers of precocial infants produce milk that provides energy, but little fat, that must be consumed frequently. Humans conform only partially to the typical 'precocial' primate pattern, with infants feeding frequently on milk that is relatively low in protein and fat, but high in sugar [1].

Unlike non-human primates, human infants also display secondarily altricial characteristics as a consequence of the limits placed either by the bipedally adapted pelvis on foetal brain growth (see also Chap. 3 on the obstetric dilemma) or by the constraints of maternal basal metabolic rate (BMR) to sustain a foetus for longer than 9 months [2, 3]. Regardless of the limiting factor, the net result is that human infants are born in a state of neurological immaturity with particularly poor neuromuscular control at birth, creating an inability to independently locomote or cling, and relatively poor homeostatic control [4, 5]. The high sugar content of human milk supplies the energy needed for fast brain growth in infancy; however, human infants' lack of neuromuscular control means that mothers are responsible for providing close physical contact for safety, frequent feeds and physiological regulation.

Cross-cultural studies demonstrate that in most traditional human societies, infants are maintained in constant physical contact with a caregiver—usually their mother—both day and night, experience frequent arousals during periods of sleep and suckle on demand, throughout the first year of life [6]. Care is therefore congruent with infants' evolved needs, providing close contact and responding to frequent feeds. In contrast, social and cultural changes occurring in industrial and post-industrial societies have resulted in infant care practices that encourage solitary and prolonged sleep bouts from an early post-natal age, and which are now considered characteristic of Western cultures [7]. Feeding artificial infant formula (see also Chap. 4 on bottle feeding)—composed largely of another quite different species' milk—and encouraging, or even training, infants to sleep without parental presence both affect normal patterns of early sleep development [8, 9]. Decisions about infant sleep location are both influenced by, and impact on, infant feeding practices—notably breastfeeding initiation and duration [10]. A vast and lucrative market promotes sleep training programmes based on behavioural modification, while numerous infant care products (e.g. dummies, swaddling wraps, rocking, swinging or bouncing cradles, white noise apps and soft toys which vibrate to mimic mothers' heartbeat) exist with the principal aim of allowing infants to be 'put down', self-sooth, sleep longer or wake less.

5.1.2   Sudden Infant Death Syndrome (SIDS)

SIDS was defined in 1965 under code 795 of the international classification of diseases (ICD-8; now ICD-10, code R95). SIDS is not a 'cause' of death; it is a category of exclusion used to designate the death of an infant that, following a review of clinical history, post-mortem examination and investigation of the death scene, remains unexplained [11]. Although rare, SIDS is the primary designation of death for infants between one month and one year of age, affecting approximately 1 in 3000 babies in the UK and 1 in 2000 in the USA annually. Deaths typically occur during night-time or daytime sleep, and prevalence peaks at 2–3 months.

SIDS is grouped with other sudden explainable infant deaths under the category SUDI—sudden unexpected death in infancy. There are clinical similarities between SIDS and explained SUDI: both groups of infants have poorer overall health, along with a history of apparent life-threatening events (ALTE) [12]. Differentiation of SIDS and explained SUDI can be problematic due to the lack of pathological markers distinguishing SIDS from soft suffocation. The evidence used to categorise such deaths, therefore, is often circumstantial. Coroners sometimes use the designation unexplained/unascertained, rather than SIDS or explained SUDI in response to contextual elements of the death scene, such as sleep-sharing at the time of death [13]. Local and national variation in designation can skew figures and distort comparisons between populations, particularly where population-level differences in the prevalence of certain risk factors—for example bed-sharing—exist [14].

Primarily, SIDS deaths are a phenomenon affecting Western post-industrial societies where prolonged and solitary infant sleeping has been promoted as a goal to be achieved early in infancy, and where parental behaviours (such as smoking or alcohol consumption) are incompatible with infant care. Alongside cultural variation in infant care practices, the incidence of SIDS varies dramatically, both on a global scale [15, 16], and within geographically local populations [17]. Typically, deaths occur during prolonged lone sleep bouts, or while sharing a sleep surface with an adult under dangerous circumstances. Studies in the UK found that 75 % of daytime SIDS occurred while infants were sleeping in a room alone [18]. Fifty-four per cent of SIDS infants died while sleep-sharing with an adult—however, only 6 % were sleep-sharing in the absence of cumulative risk factors including alcohol, illegal drugs, smoking and sofa-sharing [19]. Additionally, formula-feeding increases the risk of SIDS. A recent meta-analysis of 18 studies found that the risk of SIDS was lower for breastfed infants. This reduction was dose-responsive and may be explained in terms of decreased arousability from sleep, or immunological deficits associated with the absence of breastfeeding [20].

5.1.3   Cross-Cultural Perspectives

In the UK, studies of South Asian immigrants demonstrate an extremely low incidence of SIDS, with a death rate four times lower than their UK-born neighbours, despite residence in socio-economically disadvantaged areas typically associated with high SIDS rates. Studies comparing immigrant and Euro-born British families reveal substantially different infant care practices [17]. South Asian families employ 'proximal care' strategies, in which physical contact is maintained day and night between infants and one or more caregivers. Breastfeeding is typical, and frequent, and infant developmental trajectories are allowed to progress without interference [21]. Infant care therefore conforms to the pattern predicted by an evolutionary perspective based on comparative and cross-cultural patterns. In contrast, Euro-origin British families exhibit more 'distal care' strategies, encouraging self-soothing and sleeping alone from an early age, and earlier (within the first month) cessation of breastfeeding.

That an evolutionary mismatch exists between Western infant care practices and the unique vulnerabilities of human infants [22] may contribute to an explanation for the high rates of SIDS in Western cultures. McKenna and colleagues have argued that close mother–infant contact compensates for infants' developmental immaturity, and that the absence of such contact during extended bouts of sleep—a consequence of infant care practices that aim to promote early infant independence—may therefore be failing to support infants during early critical developmental periods [23, 24]. This hypothesis builds upon a body of work which identifies multiple deleterious consequences of early physical separation of infants from mother or other carer, in both humans and non-humans, in terms of infants' physiological regulation and development [4, 5, 25–30].

5.2 Research Findings

5.2.1 SIDS Epidemiology

Since the 1960s, researchers have been searching for factors that explain or are associated with such deaths. Much of this work has employed the case–control study design (see Box 5.1). While case–control studies cannot provide proof of causation, repeated findings of an association between exposure to a 'factor' and incidence of SIDS suggest links between characteristics of infants or their care and an increased or decreased risk, and allow public health advisers to make recommendations for ways in which risk might be reduced and arguably how deaths might be prevented. Generally, such hypotheses regarding risk reduction interventions would be tested via randomised control trials, but with SIDS cases are too infrequent for this to be a realistic proposition.

Box 5.1: What Is a Case–Control Study?
Case–control studies are used widely in epidemiology, as their retrospective design enables researchers to study factors associated with diseases which, because of their rarity, would be difficult to study prospectively. In the case of SIDS, researchers compare two groups: the 'cases'—infants who have died and whose deaths have been designated SIDS, and the 'controls'—infants who did not die, but are individually matched on multiple variables, which may include physical and socio-economic characteristics, to 'case' infants. Using data obtained via family interviews, and examination of the death scene (or a predetermined sleep scene for controls), researchers conduct a retrospective analysis of factors to which infants were, or were not, exposed. Analysis of the incidence of exposure to one or more risk factors in both 'case' and 'control' groups results in an 'odds ratio' (OR) which, together with associated confidence intervals, describe the direction, magnitude and statistical significance of the effect of exposure to risk factor(s) on incidence of SIDS within the study population.

Case–control studies have identified a range of factors—both intrinsic and extrinsic to the infant—that are associated with SIDS deaths. These include lower socio-economic status, male infant, premature or low birthweight infant, pre- or post-natal smoke exposure and absence of breastfeeding [20, 31]. Additionally, numerous aspects of the sleep environment that produce some form of physiological challenge on an infant have also been associated with SIDS: prone- or side sleep position, overwrapping [32, 33], overheating, soft bedding or sleep surfaces [34, 35], co-sleeping in some circumstances [19, 36], not using a dummy [35, 37, 38], and infant sleeping in a separate room from their parents [18, 36].

Research into the key mechanisms underlying SIDS has identified a number of potentially causal factors that may increase individual infant vulnerability. Primarily, theories have focused on deficits in autonomic control, genetic factors and infection. As noted above, many extrinsic risk factors have been found that provide a physiological challenge to the infant, or may be associated with a failure to arouse normally from sleep, and form the focus of much current SIDS reduction guidance under the assumption that these are modifiable factors. The commonly accepted triple-risk hypothesis proposed by Filiano and Kinney [39] describes the confluence of a vulnerable infant, at a critical stage of development, exposed to an external stressor (Fig. 5.1).

Fig. 5.1 The triple-risk model for SIDS

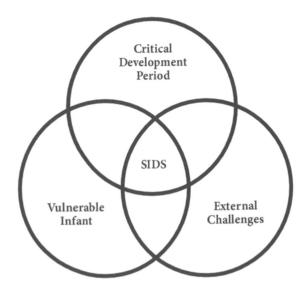

5.2.2   *The Question of Sleep Location*

The 'Back-to-Sleep' campaigns of previous decades were successful in modifying particular sleep behaviours (such as prone infant sleep position) and reducing SIDS rates in many countries across the Western world [40]. SIDS rates have now plateaued, and further reductions are proving difficult. The proportion of deaths occurring in a bed-sharing context and the absolute number of deaths occurring while an infant is sleep-sharing with an adult on a sofa have increased in recent years [19], prompting public health messages to focus on these issues as targets for further reduction. Such messages have so far had little impact on SIDS rates, however, suggesting that further change requires greater knowledge from parents and healthcare providers than can be achieved via a 'one-size-fits-all' infant care message [14]. It is now acknowledged in a variety of contexts that 'one-size-fits-all' messaging is ineffective and inappropriate, particularly for high-risk groups such as minority or low-income families [41].

Several reasons for the lack of further progress have been discussed. First, large-scale public health messages are inherently limited in their ability to effect widespread behaviour change, particularly when the targeted parenting behaviours relate to complex, multifaceted and culturally ingrained infant care practices. One-size-fits-all messages are not adequate to achieve the increased knowledge and commitment from parents and their healthcare providers, which are required for meaningful reductions in sleep-related mortality [14, 41]. Information is also directed to parents and providers without addressing their capacity to implement the recommended practices and is generally offered without contextual information about to whom the recommendations apply, under what circumstances and for how long [14]. Although studies have shown that most parents cannot eliminate all known risks to their infants for each sleep [42] and that there is a cost to parents in adhering to safe sleep advice [43], they are rarely assisted in prioritising risk factors or in developing strategies to ameliorate those costs.

Secondly, unlike infant sleep position, infant sleep location is embedded within a context of cultural and personal values and motivations. Where an infant sleeps is related to a deeply rooted belief system about the relationship of an infant and her caregiver, and about the very nature of infancy and parenting [44–48]. These beliefs may conflict with public health recommendations and policies, leading parents to reject paediatric advice in whole or in part.

Finally, infant sleep location is intricately related to other aspects of infant care, such as feeding method. Therefore, efforts to increase breastfeeding, which is also a protective factor for SIDS [20], represent a de facto intervention related to sleep location due to the nature and frequency of night-time feeding [49], yet the interplay of different safe sleep recommendations are rarely recognised or addressed in public health campaigns and parents are offered little guidance when various aspects of well-being and risk reduction agendas conflict [14]. Infant sleep location cannot be easily disarticulated from the larger behavioural context in which it exists.

5.3   Implications for Policy and Practice

5.3.1   Strategies to Reduce SIDS Based on an Evolutionary Perspective

Researchers who approach infant sleep behaviour and physiology using 'evolutionary thinking' consider the intersection between evolved infant biology and culturally determined patterns of night-time infant care described above. Using this, evolutionary framework [50] allows researchers to identify both proximate (immediate or mechanistic) and ultimate (evolutionary or adaptive) causes for individual parenting strategies [51, 52] including choice of infant sleep location. Existing studies have demonstrated the utility of evolutionary perspectives and methods for understanding the ways in which parents construct their infants' sleep environments [49, 53, 54]. These studies have documented how the complicated prospect of providing care to infants during the night causes parents to adopt strategies and behaviours that they had not planned [6], and to approach infant sleep location in a way that balances the constraints of infant physiology and development with parental goals and desires within a particular social and behavioural context [55, 56].

As part of the evolutionary approach to infant sleep environments, researchers and clinicians recognise that parent–offspring conflicts exist, producing a tension between the biological demands (maximising the reproductive success) of the infant and of the parent [52, 57, 58]. Individuals are repeatedly negotiating trade-offs between the benefit to infants derived from particular forms of care, and the costs to parents incurred by engaging in these forms of care [59, 60]. Parenting strategies are sensitive to individual circumstances [61] and calibrated to personal contexts and conditions, but are also adjusted over the lifetime of a single individual [62]. When applied to SIDS reduction and infant sleep location, this approach acknowledges the inherent trade-offs in providing night-time care to infants, and the costs to parents in following contemporary paediatric recommendations that lead them to pursue alternate strategies. Policy and practice interventions that take into account the biological needs of babies, and parent–infant trade-offs, can better understand the barriers to implementation of traditional SIDS reduction approaches and offer new innovations.

Although safe sleep guidelines and other public health messages suggest that risky infant sleep environments are potentially lethal to infants whenever they occur, in reality parents quickly learn that infants may be placed in a variety of sleep locations with no adverse outcomes. Therefore, parents may be willing to tolerate the risk of placing infants in a particular sleep location in exchange for other benefits; in so doing they weigh the costs and benefits of potentially risky sleep locations in ways that are complex and poorly understood [43]. Clinicians may use this awareness of trade-offs to engage in conversations with parents, such as by exploring the trade-offs that parents consider in selecting particular infant sleep locations or by examining the impact to the parents of attempting to implement sleep location recommendations. If clinicians want us to select certain infant sleep locations over others, they must assist us in identifying the costs of those sleep arrangements that dissuade us from following guidance (e.g. the implications for maternal sleep disruption) and help us implement strategies to decrease these costs [43].

5.3.2 Innovative Strategies for Practice

Recent evidence-to-practice initiatives for public health education have focused on presenting an evolutionary understanding of infant sleep to parents and health practitioners, the development of interactive tools and materials that encourage discussion of parent and infant needs and trade-offs, and the provision of simple and culturally appropriate infant sleep spaces that offer alternate solutions in trade-off situations. To illustrate these approaches, we give an overview of three initiatives: (a) UNICEF UK Baby-Friendly Initiative's approach to supporting night-time infant care; (b) the Infant Safe Sleep Tool, a risk-assessment and communication tool for parents and health professionals; and (c) a New Zealand initiative for the provision of alternate infant sleep spaces.

UNICEF UK Baby-Friendly Initiative In the UK, public health policy initiatives to promote breastfeeding initiation and continuation reflect the relevance of an evolutionary perspective in emphasising the importance of prolonged physical contact between babies and their carers to both facilitate normal infant development and optimise maternal milk production, and the avoidance of night-time separation of mothers and infants [63, 64]. All UK breastfeeding support organisations now endorse and signpost both parents and health professionals to the Infant Sleep Information Source Website which presents evolutionary perspectives on infant sleep development, sleep needs and managing sleep safety (www.isisonline.org.uk). The existence of the latter, which is signposted to by numerous NHS Trusts and Local Councils, is helping providers and parents access consistent and research-based information on the underlying evolutionary biology of their infant's sleep needs, development and safety.

The Infant Sleep Safety Tool The Infant Sleep Safety Tool (ISST) was developed, trialled and evaluated by Durham University Parent-Infant Sleep Lab in partnership with Blackpool and North Lancashire NHS Trusts in a service delivery project that aimed to discuss infant sleep location trade-offs with parents. The tool comprises a colourful illustrated booklet 'Where might my baby sleep?' and training materials for healthcare providers and peer supporters. The booklet presents information about the risks and benefits of infant sleep locations, along with a double-page checklist highlighting the factors which combine with bed-sharing to increase infant risk of SIDS and accidental SUDI, informs parents, enables them to identify their individual risk profiles and facilitates communication between healthcare providers and parents on topics that can prove difficult to initiate or explore. Evaluation found that compared to mothers who had received routine care, those receiving the ISST had improved knowledge of several aspects of SIDS risk and infant sleep safety, particularly relating to the roles of lone sleep, smoking, bed-sharing in hazardous circumstances and breastfeeding in affecting SIDS risk. Staff implementing the tool reported increased confidence in addressing issues relating to infant sleep safety with parents, and in discussing why parents implement hazardous night-time care strategies [65].

New Zealand Safe Sleep Spaces: Wahakura The disproportionately high SIDS rate among New Zealand's Maori population was found to be associated with parent–infant bed-sharing in the context of a high prevalence of maternal smoking in pregnancy and post-natally [66]. Due to the high degree of cultural value attached to bed-sharing in Maori culture, it was not considered a 'modifiable infant care practice' [14]. Tipene-Leach [67, 68] devised a safe sleep intervention involving infant sleep baskets woven from flax (via a traditional Maori technique) that he named Wahakura. These woven reed sleeping baskets are designed to support a modified form of bed-sharing—being placed on the parental bed—keeping the baby in close proximity while providing a safe sleep space. The provision of Wahakura produced by Maori weavers, and especially the process of teaching women to weave their own, provides an opportunity for discussion of safe infant sleep, and provision of information about SIDS risk reduction strategies [66]. The Wahakura programme is now in the process of evaluation, but since its inception the Maori SIDS rate has fallen substantially—while preserving parent–infant sleep proximity.

5.4   Conclusion

Humans are by nature adaptable, and critics of an evolutionary approach may point to humans' inherent capacity for both behavioural and biological plasticity, both of which have facilitated the cultural development of infant care practices including feeding of the milk of another species, and 'caching' of infants in specially designed cribs, cots and carriers. However, young infants have a limited ability to adapt, and the relative newness of cultural changes means that no evolutionary adaptations have occurred. By recognising that infant sleep locations are selected and modified as part of a larger behavioural repertoire, evolutionary perspectives on infant sleep location acknowledge that (a) infants biological needs are influenced by human evolutionary history; (b) human infants need close contact with a caregiver both day and night during early development; (c) when infant needs and parental (social and economic) needs conflict, parents will modify infant care practices in ways that trade off their own needs and those of their infants; (d) these trade-offs might increase or decrease infant exposure to SIDS risks, and sleep locations can be safer or unsafe according to the context in which they are implemented. By explicitly understanding these links, we can initiate more effective discussions with parents and develop new and innovative approaches to safe infant sleep that supersede one-size-fits-all recommendations that many parents are unable or unwilling to implement.

Glossary

Apparent life-threatening events (ALTE)	Incidents where babies cease breathing and become lifeless but are able to be resuscitated
Arousability	The ability to arouse from sleep easily. Poor or low arousability means an individual does not awaken upon application of normal stimuli
Back-to-Sleep campaign	A national population-based campaign of information and advertising to encourage parents to put infants in a supine position for sleep
Distal care strategies	Infant care behaviours that encourage separation of carer and baby: sleeping baby in room alone, use of bouncers and swings to sooth the baby, use of buggies to transport the baby
Proximal care strategies	Infant care behaviours that keep the baby close to a carer, e.g. carrying sling-use, sleeping within arm's reach
Risk factors	Characteristics found to be statistically associated with infant death outcomes in SIDS case–control studies
Sudden infant death syndrome (SIDS)	The death of an infant that following a review of clinical history, post-mortem examination and investigation of the death scene remains unexplained
Sudden unexpected death in Infancy (SUDI)	Sudden unexpected death in infancy includes SIDS and explained deaths occurring unexpectedly from illness accident, or deliberately
Triple-risk hypothesis	A model for explaining how SIDS occurs at the confluence of intrinsic, extrinsic and time-limited risk characteristics of an infant and his/her environment
UNICEF Baby-Friendly Initiative	A worldwide health promotion programme to encourage maternity practices that facilitate and support breastfeeding
Wahakura	A portable woven flax basket for use as an alternate strategy for bed-sharing in order to keep babies close but safe, when parents are smokers, was developed in NZ

References

1. Jelliffe DB, Jelliffe EFP (1978) Human milk in the modern world. Oxford University Press, London
2. Rosenberg KR, Trevathan WR (1995) Bipedalism and human birth: the obstetrical dilemma revisited. Evol Anthropol 4: 161–168
3. Dunsworth HM, Warrener AG, Deacon T, Ellison PT, Pontzer H (2012) Metabolic hypothesis for human altriciality. PNAS 109(38):15212–15216
4. Small MF (1999) Our babies ourselves—how biology and culture shape the way we parent. Doubleday Dell Publishing Group Inc, New York
5. Hrdy SB (1999) Mother nature: a history of mothers, infants, and natural selection. Ballantine, New York
6. Ball H (2007) Bed sharing practices of initially breastfed infants in the first six months of life. Infant Child Dev 16:387–401
7. McKenna JJ, Ball HL, Gettler L (2007) Mother-infant cosleeping, breastfeeding and sudden infant death syndrome: what biological anthropology has discovered about normal infant sleep and pediatric sleep medicine. Yearb Phys Anthropol 50:133–161
8. Cavkll B (1981) Gastric emptying in infants fed human milk or infant formula. Acta Paediatr 70:639–641
9. Mindell JA, Kuhn B, Lewin DS, Meltzer LJ, Sadeh A (2006) Behavioral treatment of bedtime problems and night wakings in infants and young children. Sleep 29(10):1263–1276
10. Russell CK, Robinson L, Ball HL (2013) Infant sleep development: location, feeding and expectations in the postnatal period. Open Sleep J 6 (Suppl 1: M9):68–76
11. Willinger M, James LS, Catz C (1991) Defining the sudden infant death syndrome (SIDS): deliberations of an expert panel convened by the national institute of child health and human development. Pediatric Pathol 11:677–684
12. Ward-Platt MP (2000) A clinical comparison of SIDS and explained sudden infant deaths: how healthy and how normal? Arch Dis Child 82(2):98e106
13. O'Hara M, Harruff R, Smialek JE, Fowler DR (2000) Sleep location and suffocation: how good is the evidence? (letter). Pediatrics 105(4 Pt 1):915–917
14. Ball HL, Volpe L (2013) Sudden infant death syndrome (SIDS) risk reduction and infant sleep location: moving the discussion forward. Soc Sci Med 79:84–91
15. Nelson EA, Taylor BJ (2001) International child care practices study: infant sleeping environment. Early Hum Dev 62(1):43–55
16. Hauck FR, Tanabe KO (2008) International trends in sudden infant death syndrome: stabilization of rates requires further action. Pediatrics 122(3):660–666
17. Ball HL, Moya E, Fairley L, Westman J, Oddie S, Wright J (2012) Bed and sofa-sharing practices in a UK bi-ethnic population. Pediatrics 129
18. Blair P, Ward-Platt M, Smith IJ, Fleming PJ, group. CSr (2006) Sudden infant death syndrome and the time of death: factors associated with night-time and day-time deaths. Int J Epidemiol 35 (6):1563–1569
19. Blair PS, Sidebotham P, Evason-Coombe C, Edmonds M, Heckstall-Smith EMA, Fleming P (2009) Hazardous cosleeping environments and risk factors amenable to change: case-control study of SIDS in south west England. Br Med J Online 339:1–11
20. Hauck F, Thompson J, Tanabe K, Moon R, Venetian M (2011) Breastfeeding and reduced risk of sudden infant death syndrome: a meta-analysis. Pediatrics 128:103–110
21. Crane D (2014) BradICS: bradford infant care study: a qualitative study of infant care practices and unexpected infant death in an urban multi-cultural UK population. Durham University
22. Konner MJ, Super CM (1987) Sudden infant death syndrome: an anthroplogical hypothesis. In: Super CM (ed) The role of culture in developmental disorder. Academic Press, San Diego, pp 95–108

23. McKenna JJ (1986) An anthropological perspective on the sudden infant death syndrome (SIDS): the role of parental breathing cues and speech breathing adaptations. Med Anthropol 10:9–53
24. McKenna JJ, McDade T (2005) Why babies should never sleep alone: a review of the co-sleeping controversy in relation to SIDS, bedsharing, and breastfeeding. Paediatr Respir Rev 6:134–152
25. Moore E, Anderson G, Bergman N (2007) Early skin-to-skin contact for mothers and their healthy newborn infants (review). Book: Cochrane Collab:1–63
26. Ferber SG, I.R. M (2004) The effect of skin-to-skin contact (Kangaroo Care) shortly after birth on the neurobehavioural responses of the term newborn: a randomized, controlled trial. Pediatrics 113(4):858–865
27. Bergman N, Linley L, Fawcus S (2004) Randomised controlled trial of skin-to-skin contact from birth versus conventional incubator for physiological stabilization in 1200–2199-gram newborns. Acta Paediatr 93:779–785
28. Ohgi S, Faked M, Moriuchi H, Akiyama T, Nugget J, Brazelton T, et al (2001) The effects of kangaroo care on neonatal neurobehavioral organization, infant development and temperament in healthy low-birth- weight infants through one year. J Perinatol 22 (374 –379)
29. Morgan BE, Horn AR, Bergman NJ (2011) Should neonates sleep alone? Biol Psychiatry:1–9
30. Middlemiss W, Granger DA, Goldberg WA, Nathans L (2012) Asynchrony of mother-infant hypothalamic-pituitary-adrenal axis activity following extinction of infant crying responses induced during the transition to sleep. Early Hum Dev 88:227–232
31. Leach CEA, Blair PS, Fleming PJ, Smith IJ, Ward-Platt M, Berry PJ, BCH, FRCP, Golding J, Group CSR (1999) Epidemiology of SIDS and explained sudden infant deaths. Pediatrics 104: e43–53
32. Fleming P, Sawczenko A (1996) Thermal stress, sleeping position, and the sudden infant death syndrome. Sleep 19(10):267–270
33. Fleming P, Blair P, Berry PJ, Tripp J (2003) How do environmental conditions and circumstances contribute to sudden unexpected death in infancy? A comparison of normal infant sleeping conditions and "death scene" investigations after Sudden Unexpected Death in Infancy (SUDI)—Research Protocol
34. Flick L, White K, D, Vemulapalli C, Stulac BB, Kemp JS (2001) Sleep position and the use of soft bedding during bed sharing among African American infants at increased risk for sudden infant death syndrome. J Pediatr 138 (3):338–343
35. Moon RY, Horne RS, Hauck FR (2007) Sudden infant death syndrome. Lancet 370:1578–1587
36. Blair P, Fleming P, Smith I, Platt M, Young J, Nadin P et al (1999) Babies sleeping with parents: case-control study of factors influencing the risk of the sudden infant death syndrome. BMJ 319:1457–1461
37. Vennemann MM, Bajanowski T, Brinkmann B, Jorch G, Sauerland C, Mitchell EA, Study G (2009) Sleep environment risk factors for sudden infant death syndrome: the German sudden infant death syndrome study. Pediatrics 123(4):1162–1170
38. Vennemann MM, Bajanowski T, Brinkmann B, Jorch G, Yücesan K, Sauerland C, Mitchell EA, Group GS (2009) Does breastfeeding reduce the risk of sudden infant death syndrome? Pediatrics 123 (3):e406–e410
39. Filiano J, Kinney H (1994) A perspective on neuropathologic findings in victims of the sudden infant death syndrome: the triple-risk model. Neonatology 65:194–197
40. Gilbert R, Salanti G, Harden M, See S (2005) Infant sleeping position and the sudden infant death syndrome: systematic review of observational studies and historical review of recommendations from 1940 to 2002. Int J Epidemiol 34:874–887
41. Fetherston C, Leach J (2012) Analysis of the ethical issues in the breastfeeding and bedsharing debate. Breastfeeding Rev 20(3):7–17
42. Ball HL, Moya E, Fairley L, Westman J, Oddie S, Wright J (2012) Infant care practices related to sudden infant death syndrome in South Asian and White British families in the UK. Paediatr Perinat Epidemiol 26(1):3–12

43. Volpe L, Ball H, McKenna J (2013) Nighttime parenting strategies and sleep-related risks to infants. Soc Sci Med 79:84–91
44. Abbott S (1992) Holding on and pushing away: comparative perspectives on an Eastern Kentucky child rearing practice. Ethos 20:33–65
45. Abel S, Park J, Tipene-Leach D, Finau S, Lennan M (2001) Infant care practices in New Zealand: a cross-cultural qualitative study. Soc Sci Med 53:1135–1148
46. Crawford C (1994) Parenting practices in the Basque country: implications of infant and childhood sleeping location for personality development. Ethos 22(1):42–82
47. Eades SJ, Read AW (1999) Infant care practices in a metropolitan aboriginal population bibbulung gnarneep team. J Paediatr Child Health 35(6):541–544
48. Gantley M, Davies D, Murcott A (1993) Sudden infant death syndrome: links with infant care practices. Br Med J 306:16–20
49. Ball HL (2002) Reasons to bed-share: why parents sleep with their infants. J Reprod Infant Psychol 20(4):207–222
50. Trevathan WR, Smith EO, McKenna JJ (2008) Introduction and overview of evolutionary medicine. In: Trevathan WR, Smith EO, McKenna JJ (eds) Evolutionary medicine and health: new perspectives. Oxford University Press, New York
51. Tinbergen N (1963) On aims and methods of ethology. Zeitschrift für tierpsychologie 20:410e433
52. Trivers R (1972) Parental investment and sexual selection. In: Campbell B (ed) Sexual selection and the descent of man. Aldine, Chicago
53. Ball H, Hooker E, Kelly P (1999) Where will the baby sleep? Attitudes and practices of new and experienced parents regarding cosleeping with their newborn infants. Am Anthropol 101(1):143–151
54. McKenna J, Mosko S, Richard C (1999) Breast-feeding and mother-infant cosleeping in relation to SIDS prevention. In: Trevathan WR, Smith EO, McKenna JJ (eds) Evolutionary medicine. Oxford University Press, New York
55. Anders TF, Taylor TR (1994) Babies and their sleep environment. Child Environ 11(2):123–134
56. Volpe L (2010) Using life-history theory to evaluate the nighttime parenting strategies of first-time adolescent and adult mothers. Durham University
57. Haig D (2014) Troubled sleep: night waking, breastfeeding and parent–offspring conflict. Evol Med Public Health 2014(1):32–39
58. Trivers R (2002) Natural selection and social theory: selected papers of Robert Trivers. Oxford University Press, New York
59. Altmann J (1987) Life span aspects of reproduction and parental care in anthropoid primates. In: Lancaster JB, Altmann J, Rossi AS, Sherrod LR (eds) Parenting across the life span: biosocial dimensions. Aldine de Gruyter, New York
60. Tully KP, Ball HL (2011) Trade-offs underlying maternal breastfeeding decisions: a conceptual model. Matern Child Nutr:1–8
61. Gross MR (2005) The evolution of parental care. Q Rev Biol 80(1):37–45
62. Daly M, Wilson M (1981) Abuse and neglect of children in evolutionary perspective. In: Alexander RD, Tinkle DW (eds) Natural selection and social behaviour: recent research and new theory. Chiron Press, New York
63. Entwistle F (2013) The evidence and rationale for the UNICEF UK Baby Friendly Initiative standards. UNICEF UK
64. Inch S, Blair P (2011) The health professional's guide to 'caring for your baby at night'
65. Russell C, Whitmore M, Burrows D, Ball H (2015) Where might my baby sleep? Design and evaluation of a novel discussion tool for parent education. Int J Birth Parenting Educ 2(2):12–16
66. Abel S, Tipene-Leach D (2013) SUDI prevention: a review of Maori safe sleep innovations for infants. New Zealand Med J 126(1379):86–94
67. Tipene-Leach D (2007) The wahakura: the safe bed-sharing project. http://www.birthcare.co.nz/assets/wahakura-project-(safe-bed-sharing).pdf
68. Tipene-Leach D, Abel S (2010) The wahakura and the safe sleeping environment. J Primary Health Care 2(1):81

Chapter 6
The Developmental Origins of Health and Disease: Adaptation Reconsidered

Ian J. Rickard, Ph.D.

Lay Summary The conditions that a foetus experiences in the womb leave their mark on that individual as they grow, and stay with it for its entire life. Evolutionary biologists study similar kinds of effects in animals. They think that the understanding of what the developing body does when it experiences something harmful, such as a reduction in nutrients, can be helped with evolutionary theory. Some argue that a foetus treats such harmful experiences as a forecast of what it will experience later in life, and that it prepares itself for that later life experience by changing its biology. However, there are other ways in which evolutionary theory can be used to understand these effects. For example, if a body cannot grow as big as it would ideally, it may need to change its biology to make the best out of the situation. Also, a foetus may need to adjust its growth when its mother is under stress and not able to provide many nutrients. Deciding which of these different possibilities are and are not true is important, because they mean different things for how we understand patterns of health and illness, and how we go about improving population health.

6.1 Introduction

The field of the Developmental Origins of Health and Disease (DOHaD) examines the consequences of early life conditions for health and disease risk [1]. DOHaD has its origins in some early work showing striking geographical correlations between mortality rates in infancy and—decades later—mortality due to coronary

I.J. Rickard (✉)
Department of Anthropology, Durham University, Durham, UK
e-mail: ian.rickard@durham.ac.uk

© Springer International Publishing Switzerland 2016
A. Alvergne et al. (eds.), *Evolutionary Thinking in Medicine*,
Advances in the Evolutionary Analysis of Human Behaviour,
DOI 10.1007/978-3-319-29716-3_6

Fig. 6.1 Standardised
mortality ratios for coronary
heart disease in England and
Wales during 1968–79 and
neonatal mortality during
1921–25 (*diamonds*—London
boroughs; *triangles*—county
boroughs; *circles*—urban
districts; *squares*—rural
districts). Reproduced from
Ref. [1]

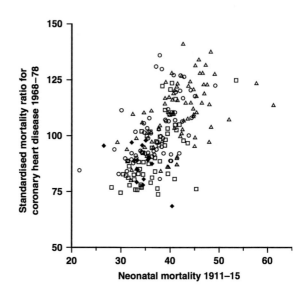

heart disease (CHD) [2] (see Fig. 6.1). It was subsequently hypothesised that the
same environmental conditions that contributed to high infant mortality in a cohort
of individuals also caused high rates of CHD in the survivors in their adult life, and
specifically that such conditions exerted themselves via intrauterine growth patterns
[1]. A prediction generated from this hypothesis was that individuals with lower
birthweight are at a greater risk of developing CHD as adults, due to a common role
of intrauterine growth restriction in promoting both. This prediction has been
supported empirically [3] and extended to other outcomes, including Type II diabetes [4, 5]. Causality of DOHaD effects is typically inferred from studies of animal
models [6] and that of cohorts exposed prenatally to famines [7], such as the Dutch
hunger winter [8], which experience higher incidences of schizophrenia, diabetes
and obesity in adulthood.

Several authors have suggested that applying the theory of evolution by natural
selection may improve our understanding of the above patterns [9–13]. This is
because natural selection can lead to the evolution of 'developmental plasticity':
selection favours genes that cause the developing bodies they are in to actively
respond to their environment in such a way that will improve their chances of
surviving and leaving behind a large number of descendants (having high evolutionary fitness). We should therefore expect the existence of developmental
responses to environmental conditions that cause the body to do better than it would
in the absence of such a response. For instance, 'brain sparing' is an apparently
adaptive process by which a foetus suffering from placental insufficiency undergoes
physiological and anatomical changes that appear to prioritise oxygen supply to the
developing central nervous system [14].

The predictive adaptive response (PAR) hypothesis argues that poor growth
conditions can be useful as a forecast of later life conditions, and that consequently

humans and other animals have evolved to adaptively change their phenotype in accordance with this information [9]. In particular, it is argued that poor intrauterine growth can be taken to indicate a low-nutrition environment, to which the foetus adapts by altering its insulin–glucose metabolism in such a way that would facilitate survival under nutritionally poor adulthood conditions. However, in the rapidly changing environments of industrialised societies over the course of the twentieth and twenty-first centuries, such forecasts have frequently been incorrect, leading to type II diabetes and cardiovascular diseases as well as other pathologies in later life [9].

Both the DOHaD and the PAR hypotheses hold implications for understanding patterns of health and disease, and which avenues for intervention might be usefully explored. If the first nine months of life, or those immediately preceding it, are of critical importance in determining later health, then it may be opportune to intervene during this period [15–18]. If the PAR hypothesis is correct, then we may infer that once the relevant mechanisms have been identified, it might be possible to change the predictive information that the foetus receives and thus improve its lifelong chances (e.g. through preventing altered insulin–glucose metabolism in individuals with intrauterine growth restriction). However, this requires critical evaluation.

In this chapter, I consider how an understanding of developmental plasticity (beginning with the PAR hypothesis) may be employed so as to understand relationships between early life environment and later life health outcomes. In 'Research Findings', I first evaluate evidence from a range of adjacent biological and biomedical fields that may be brought to bear in order to answer the question of whether developmentally plastic processes may be disrupted to therapeutic benefit, and highlight how much of this evidence points the way to accounts of developmental plasticity that differ significantly from the account given by the PAR hypothesis. In 'Implications for Policy and Practice', I point out how these different accounts are correspondingly associated with different implications for interpretations of patterns of health and disease. Finally, in 'Future Directions', I discuss some avenues of investigation that this alternative perspective suggests would be worthy of attention.

6.2 Research Findings

Beginning from the premise that adaptive developmental plasticity, here I critically evaluate what is arguably the dominant hypothesis that is used to account for DOHaD effects, the PAR hypothesis, as well as consider alternative adaptive perspectives.

6.2.1 Did Environmental Matching Select for Developmental Plasticity in Humans?

The PAR hypothesis gives rise to the clear prediction that the fitness of an individual will generally be higher when the environment it experiences as an adult matches that which it experienced through the maternal environment during gestation. As a rule, developmental adversity such as poor nutrition tends to permanent damage, or constrain, a body's ability to function optimally in later life [19–24], but it is nonetheless possible that adversity could also simultaneously act as a forecast of the external environment as well as being a constraint [25], allowing individuals to prepare for what is to come. A recent meta-analysis of the fitness benefits of environmental matching across a range of animal and plant species concluded that evidence for it is weak; in other words, individuals experiencing poor early conditions were not better able to later cope with such environments in later life [26]. Furthermore, the fact that a long time period in between development and adulthood increases the potential for the environment to change in the interim reduces the likelihood of beneficial matches occurring. It is therefore of particular importance that such evidence appears thus far absent for animals that are long-lived such as humans [27–29].

A further reason why PARs may be argued to be even more unlikely to be found in humans becomes apparent when we consider the reproductive ecology of humans in a comparative context. We should expect that the effects of the environment on developing offspring would differ dramatically between species according to the extent to which maternal reserves are called upon during reproduction. Rodents, from which comes most of the experimental evidence for DOHaD effects, are 'income breeders', meaning that foetal growth is largely dependent on resources that are consumed during the period of gestation [30]. On the other hand, human foetal growth is heavily reliant on accumulated reserves (capital). As a result, while rodent foetuses are inevitably greatly affected by variation in maternal nutrient consumption during gestation, the human foetus is relatively buffered from such short-term variation [31].

In fact, on close examination, the evidence for changes in maternal nutrition during pregnancy affecting offspring health outcomes does not appear to be particularly strong. Studies of cohorts that were in gestation during famine conditions certainly demonstrate 'proof of concept', i.e. that external nutritional conditions probably do have some effect on later life health outcomes [7]. However, their relevance to understanding relationships between birthweight and adult disease may be too easily overstated, as such extreme conditions are rarely likely to be analogous to the same external influences that give rise to the normal range of birth weights in non-famine conditions. Experimentally, while dietary restriction of pregnant animals can have dramatic effects on offspring phenotype [6], long-term follow-up of the offspring of Gambian, Bangladeshi and British South Asian women given protein-energy or micronutrient supplementation during pregnancy suggests few effects on offspring health outcomes measured in childhood [32–35]. There are thus good theoretical and empirical reasons why we might consider whether maternal factors set in motion long before pregnancy could be more relevant to long-term health than those acting during pregnancy itself. This would make it seem unreasonable to expect that environmental matching via maternal nutrition during pregnancy has been a significant force shaping developmental plasticity in the evolutionary history of humans and human ancestors.

6.2.2   Reconsidering the 'Environment' that Selects for Developmental Plasticity

How then, in the absence of selection by matching/mismatching, can the responses of developing individuals to adverse environments be understood? The answer may lie in re-evaluating the premises of the question. Our understanding of developmental plasticity can be limited when we consider the process of selection as something that emerges from a dichotomy between the individual organism and the 'environment' in which it lives (e.g. the physical landscape, the elements of heat and water, the prospects of either eating or being eaten). We may consider the process more inclusively by recognising that selection acts on genes, through their trait products that emergently form cells, tissues, systems, and individuals, and that the 'environment' is simply the context in which that selection takes place. Thus, life is organised hierarchically, and at each level of this hierarchy, the ability of each unit to do its job is dependent on characteristics of other units. One of the best-studied animal models in this context is the mouse mandible, in which bone, muscle and teeth are integrated together into a 'functional complex' [36]. The different components work together to influence the organism's fitness, with a change in the character of one component ideally requiring a change in the character of the others, in order to maintain function. At this level, the phenomenon is known as 'phenotypic integration' [37].

Considering this changes how we look at developmental plasticity. In this light, we may explain many changes within the body as being necessary to maintain overall function when one or more other traits are influenced by external factors. For instance, in DOHaD animal model research, maternal dietary restriction leads to not only altered glucose–insulin metabolism, but also: reduced pancreatic beta cell function [38], reduced hepatic gluconeogenesis [39], altered intestinal enzyme activity [40], a tendency to lay down visceral adipose tissue [41] and increased dietary preference for fatty food [42]. It is possible that such traits facilitate matching between early and late life environment when the environment is nutrition-poor, in accordance with the PAR hypothesis. Alternatively, it is also possible that such traits are in fact optimal in any later environment for an individual whose fitness potential is constrained by their poor start in life. Because the negative effects of poor nutrition in early life are inevitable, some features of the eventual adult body (e.g. skeletal size) can be predicted accurately (see Box: Internal Prediction). Thus, whereas the PAR hypothesis proposes that the aforementioned traits develop in order to facilitate the adaptation of the individual to its environment, the alternative is that such traits develop in order to facilitate adaptation to itself.

Equivalents to phenotypic integration may be observed at other levels of the organisational hierarchy. The foetus experiences its world not directly, but entirely through the prism of the maternal environment. In addition, acting as a buffer between the foetus and the wider environment, the maternal environment also constitutes a powerful selective force, leading to co-adaptation between foetal growth and maternal provisioning [31]. This co-adaptation is mediated by the placenta, which matches foetal growth with maternal nutrient supply through actively regulating its growth, nutrient transport to the foetus and maternal physiology itself [43]. This dynamic materno-foetal complex must be functionally integrated and responsive to the constraints of the maternal resource base. Maternal endocrine factors act as signals of those constraints [44] and may therefore facilitate adaptation to (integration with) them. Phenotypic integration across diverse and remote tissues and systems is also likely to be controlled by endocrine factors at the within-individual level [45].

6.3   Implications for Policy and Practice

The point of introducing a new theoretical framework to account for natural processes is not merely an intellectual exercise in conjecture. If evolutionary theory is to have any significance for understanding and predicting patterns of health and well-being, then different evolutionary hypotheses should carry different implications for how societies should think about these issues. In this penultimate section, I outline three implications of the preceding arguments that suggest themselves as being of particular importance.

6.3.1   Maternal Diet During Pregnancy Itself May Have Relatively Little Impact on Offspring Health

The PAR hypothesis proposes that environmental matching between the prenatal environment and the postnatal environment determines health, the corollary being that intervention during or before pregnancy may provide a useful avenue for intervention. However, as argued here, the maternal environment is not restricted to what a woman's body does and experiences from the time of conception onwards, nor even by her adult health and behaviour in general, but by the sum of the environmental experiences that she has had up until that point, including her own experience in the womb [31]. Thus, it is likely that any environmental conditions that are responsible for the relationship between birthweight and health in non-famine conditions largely exert themselves through the mother's body long before conception. As we have seen, there is little evidence that even in developing countries, changes to maternal diet during pregnancy have beneficial effects on offspring health outcomes. Clearly, public health interventions to encourage the consumption of balanced diets in new mothers should be pursued where effective, since the consequences of these can potentially extend far beyond pregnancy. However, rather than 'blaming the mothers' [46] we should expect that a more effective way of improving the lifelong health of newborns would be to ensure the health and well-being of their mothers while they themselves are still growing.

6.3.2   *Endocrine Signal Disruption May Not Be a Useful Avenue for Intervention*

Other than altering maternal diet during or before pregnancy, another mode of intervention that the PAR hypothesis suggests is alteration of endocrine signalling in order to prevent development of particular traits harmful to health later in life. For example, signalling via leptin, produced by fat cells, is a prospect much discussed in the context of interventions to improve offspring health, due to its important role in programming the metabolic profile of individuals with poor intrauterine growth [18, 47]. However, if natural selection has shaped developmental plasticity in order that an individual's phenotype may be well integrated, then their overall function may be compromised if a therapeutic intervention then disrupts this integration. Recent work on leptin programming in rats favours the rejection of the hypothesis that postnatal administration of leptin can be used to reverse the metabolic consequences of intrauterine growth restriction [48]. Instead, the effects of postnatal leptin may depend on a highly complex interaction between leptin, the degree of growth restriction and the sex of the individual. This complexity might be intelligible if rather than providing information about the nutritional state of the environment, as has been proposed [47], leptin levels were considered to be providing internal information about the animal's phenotype, e.g. level of nutritional reserves. Therapeutic administration of leptin could lead to a disintegrated phenotype, with unpredictable and potentially adverse consequences as a result.

6.3.3   *DOHaD Effects May Be Properly Situated Within the Wider Sphere of the Social Determinants of Health*

That there are relationships between the early life environment and later life health outcomes is very clear, and despite pleiotropic effects of some alleles on both intrauterine growth and later health outcomes [49], this must be in part ultimately due to the wider environment. Given the arguments above about problems with nutritional intervention and endocrine single disruption, we may ask how we might usefully intervene along this pathway. This leads to the question of what are the major determinants of women's health that could lead to both early life outcomes such as low birthweight and high infant mortality risk (Fig. 6.1) and adverse later life health outcomes, and the hypothesis that intergenerational social determinants of health are what are primarily responsible. Arguing against this, meta-analyses of the relationships between birthweight and Type II diabetes and coronary heart disease have found that they did not change after adjustment for socioeconomic status [4, 5]. However, as the authors of these studies discuss, there are numerous limitations inherent in attempts to adjust for socioeconomic status [50, 51] and this crudity will be exacerbated if the process by which the social determinants of health begin to come into play occur decades before an individual is even born. An analogy may be found in the existence of both lower birthweight and increased adverse life health outcomes in African Americans compared with European Americans, which is not accounted for after adjustment for socioeconomic status [52, 53]. That such differences are not observed between non-American Africans and European Americans suggests that social structural factors engender long-run aspects of the experience of being African American that are not captured by conventional socioeconomic status measures [54]. This raises the prospect that, beyond the social construct of race, there are also deeply ingrained multigenerational social determinants of health that may affect socially stratified societies broadly, but be elusive currently.

6.4   Future Directions

The PAR hypothesis may be argued to have advanced the cause of population health by articulating to non-evolutionary scientists the value of the predictive framework that evolutionary theory provides. However, there are theoretical and empirical objections to this specific adaptive explanation for human developmental plasticity. I have argued for the value of exploring alternative explanations, and the particular value of reconsidering what kind of adaptation might be driving such plasticity. As highlighted in the previous section, the kind of adaptation that has driven the evolution of developmental plastic processes holds implications for interpretation of patterns of health and disease, as well as which kinds of interventions might ultimately be effective in improving health and well-being. I conclude by suggesting some future research directions that can be used to help answer these questions.

6.4.1   Experimental and Population Studies of Phenotypic Integration

As selection for developmental plasticity acts upon combinations of traits, the limitations of studying small numbers of traits at a time is very clear. There would thus seem to be potential for DOHaD researchers to interact with those studying phenotypic integration. Placing the PAR hypothesis within the framework of this field leads to the very general, but strong, prediction that environmental, developmental and functional patterns of trait integration will correspond closely to one another [37]. This can be studied experimentally—a recent study manipulated the consistency of food received by mice and found that those raised on a hard (versus soft) diet had adaptively altered mandible shape and biomechanical profile, implicating an integrated pattern of developmental plasticity across a number of different, functionally related tissues [55]. Such work shows how taking a multi-trait and even multi-system approach can enrich understanding of the function of trait plasticity, although caution needs to be exercised in extrapolating too readily between animal models and humans, for reasons outlined in this chapter. Although experimental control cannot be employed easily in human studies, large multi-trait databases (e.g. such as those held by the Danish National Patient Registry [56]) may be analysed to infer the adaptive value of different combinations of traits. One explanation for trait combinations that appear more frequently than would be expected by chance is that such combinations are beneficial for the overall function of the organism [57]. The relationships between traits combination and function/health can also be explored directly.

6.4.2   Glucocorticoids as Intergenerational Transducers of Social Determinants of Health

A variety of laboratory treatments of pregnant animals (caloric restriction, protein restriction, iron restriction, fat-feeding, stressors) give rise to similar phenotypic consequences in adult offspring, suggesting a role for common causal pathways [18]. Glucocorticoid (in humans, cortisol) signalling has been implicated in this respect, thus marking it as having a special role in foetal programming, mediating effects not only of nutrient insufficiency, but also other adverse environmental conditions [58, 59]. Glucocorticoids act partially via the placenta through changes in amino acid transport [43] although excess maternal glucocorticoids can bypass the placental barrier [59]. Thus, activation of the maternal stress response system (e.g. through psychosocial stress) can directly affect foetal growth. One study of pregnant German women found that maternal cortisol levels accounted for as much as 19.8 % of the variance in birthweight, after controlling for confounders [60].

From an adaptive perspective, this fits within the wider picture of the functional role of glucocorticoids. In general, glucocorticoid activity serves the adaptive function of temporarily allocating resources away from non-essential activities towards those that are more important in the short term. To the placenta and foetus, high levels of glucocorticoids may act as an internal signal of the short-term level of available circulating resources. Downregulation of foetal growth rate in response to high levels of glucocorticoids can therefore be understood as part of an adaptive strategy. However, if mothers are chronically stressed, then the consequences for the foetus of doing so may be pathological. The experiences of chronic psychosocial stress and ill health are intimately entwined with one another [61]. Individuals with low birthweight have altered cortisol profiles as adults [59], raising the possibility that glucocorticoid transfer from mother to foetus could constitute a major mechanism by which the social determinants of health are transmitted more than two generations. This possibility that the maternal stress response is a major mediator of DOHaD effects hints at opportunities for intervention to improve lifelong health and well-being through social, rather than nutritional or pharmacological intervention.

6.5 Conclusion

Recourse to evolutionary accounts for DOHaD effects may be due to the insufficiency of explanatory models that rely primarily on mechanism and pathology. In
this chapter, I have discussed what is arguably the dominant evolutionary explanation employed in this respect, detailed some problems with it, and provided some
guidelines for expanding the way DOHaD researchers think about adaptation.
I have also argued that these considerations have very real implications for how we
think about and understand health and well-being. Thus, when considering the
evolutionary story behind DOHaD effects, researchers, policy makers and practitioners should always keep an open mind about the nature of adaptation.

> **Box 6.1 Internal Prediction**
>
> Predicting the future is a tricky business. It may be argued that the more time
> that elapses between development and adulthood, the more likely it is that the
> environment will change, and therefore, the more likely that any prediction
> made about the environment will be invalid by the time an organism reaches
> adulthood [10–13]. Such individuals will be 'mismatched' to their environ-
> ment, and run a greater risk of losing out in the evolutionary game of survival
> and reproduction. However, if we relax our expectations about what consti-
> tutes 'the environment', specifically so that it includes internal factors, then
> the prospect of accurate prediction becomes a lot more realistic [62]. As we
> might expect, it is reliably the case that adversity in early life compromises
> the ability of an organism to function well in later life [19–24], e.g. through
> lower nutrition during development leading to smaller adult size. Therefore,
> early adversity has predictable consequences for the individual's internal
> environment. These internal characteristics partially constitute the selective
> pressures to which developing individuals should adapt. We should therefore
> expect that natural selection will act on (will have acted on) developmentally
> plastic processes that facilitate this. Unlike when it requires a match between
> the individual organism and the external environment, and when develop-
> mental plasticity requires co-adaptation between traits, mismatches are pre-
> vented. In the parasitoid wasp *Aphaerta genevensis*, experimentally inducing
> small body size through food restriction limits reproductive potential and
> causes individuals to change their reproductive scheduling accordingly [63].
> In many species, individuals develop profoundly different physical charac-
> teristics depending on internal traits [64]. We may ask, does a trait that
> emerges a consequence of poor foetal growth facilitate the adaptation of an
> individual to its external environment, or does it instead adapted to its internal
> environment?

Acknowledgments I thank the editors and members of the Evolutionary Anthropology Research Group and the Anthropology of Health Research Group in the Department of Anthropology at Durham University.

References

1. Barker DJP (2007) The origins of the developmental origins theory. J Intern Med 261:412–417. doi:10.1111/j.1365-2796.2007.01809.x
2. Barker DJP, Osmond C (1986) Infant mortality, childhood nutrition, and ischaemic heart disease in England and Wales. Lancet 327:1077–1081. doi:10.1016/S0140-6736(86)91340-1
3. Huxley R, Owen CG, Whincup PH et al (2007) Is birth weight a risk factor for ischemic heart disease in later life? Am J Clin Nutr 85:1244–1250
4. Harder T, Rodekamp E, Schellong K et al (2007) Birth weight and subsequent risk of type 2 diabetes: a meta-analysis. Am J Epidemiol 165:849–857. doi:10.1093/aje/kwk071
5. Whincup PH, Kaye SJ, Owen CG et al (2008) Birth weight and risk of type 2 diabetes: a systematic review. JAMA-J Am Med Assoc 300:2886–2897. doi:10.1001/jama.2008.886
6. Bertram CE, Hanson MA (2001) Animal models and programming of the metabolic syndrome. Brit Med Bull 60:103–121. doi:10.1093/bmb/60.1.103
7. Lumey LH, Stein AD, Susser E (2011) Prenatal famine and adult health. Annu Rev Public Health 32:237–262. doi:10.1146/annurev-publhealth-031210-101230
8. Roseboom TJ, Painter RC, van Abeelen AFM et al (2011) Hungry in the womb: What are the consequences? Lessons from the Dutch famine. Maturitas 70:141–145. doi:10.1016/j.maturitas.2011.06.017
9. Bateson P, Gluckman PD, Hanson M (2014) The biology of developmental plasticity and the Predictive Adaptive Response hypothesis. J Physiol-London 592:2357–2368. doi:10.1113/jphysiol.2014.271460
10. Jones JH (2004) Fetal programming: adaptive life-history tactics or making the best of a bad start? Am J Hum Biol 17:22–33. doi:10.1002/ajhb.20099
11. Kuzawa CW, Quinn EA (2009) Developmental origins of adult function and health: evolutionary hypotheses. Annu Rev Anthropol 38:131–147. doi:10.1146/annurev-anthro-091908-164350
12. Rickard IJ, Lummaa V (2007) The predictive adaptive response and metabolic syndrome: challenges for the hypothesis. Trends Endocrinol Metab 18:94–99. doi:10.1016/j.tem.2007.02.004
13. Wells JCK (2003) The thrifty phenotype hypothesis: thrifty offspring or thrifty mother? J Theor Biol 221:143–161. doi:10.1006/jtbi.2003.3183
14. Baschat DAA (2004) Fetal responses to placental insufficiency: an update. BJOG-Int J Obstet Gy 111:1031–1041. doi:10.1111/j.1471-0528.2004.00273.x
15. Gluckman PD, Hanson MA (2008) Developmental and epigenetic pathways to obesity: an evolutionary-developmental perspective. Int J Obes 32(Suppl 7):S62–S71. doi:10.1038/ijo.2008.240
16. Gluckman PD, Hanson M, Zimmet P, Forrester T (2011) Losing the war against obesity: the need for a developmental perspective. Sci Transl Med 3:1–4. doi:10.1126/scitranslmed.3002554
17. Hanson MA, Gluckman PD, Ma RC et al (2012) Early life opportunities for prevention of diabetes in low and middle income countries. BMC Public Health 12:1025. doi:10.1017/S0029665112000055
18. Vickers MH (2011) Developmental programming of the metabolic syndrome—critical windows for intervention. World J Diabetes 2:137–148. doi:10.4239/wjd.v2.i9.137

19. Doblhammer G, van den Berg GJ, Lumey LH (2013) A re-analysis of the long-term effects on life expectancy of the Great Finnish Famine of 1866–68. Pop Stud-J Demog 67:309–322. doi:10.1080/00324728.2013.809140

20. Ekamper P, van Poppel F, Stein AD, Lumey LH (2014) Independent and additive association of prenatal famine exposure and intermediary life conditions with adult mortality between age 18–63 years. Soc Sci Med 232–239. doi:10.1016/j.socscimed.2013.10.027

21. Rickard IJ, Holopainen J, Helama S et al (2010) Food availability at birth limited reproductive success in historical humans. Ecology 91:3515–3525. doi:10.1890/10-0019.1

22. Song S (2010) Mortality consequences of the 1959–1961 Great Leap Forward famine in China: debilitation, selection, and mortality crossovers. Soc Sci Med 71:551–558. doi:10.1016/j.socscimed.2010.04.034

23. van Abeelen AF, Veenendaal MV, Painter RC et al (2011) Survival effects of prenatal famine exposure. Am J Clin Nutr 95:179–183. doi:10.3945/ajcn.111.022038

24. van den Berg GJ, Lindeboom M, Portrait F (2013) The Dutch Potato Famine 1846–1847: a study on the relationships between early-life exposure and later-life mortality. In: Lumey LH, Vaiserman A (eds) Early Life Nutrition and Adult Health Development. Nova Science Publishers, New York, pp 229–249

25. Monaghan P (2008) Early growth conditions, phenotypic development and environmental change. Philos Trans R Soc Lond B Biol Sci 363:1635–1645. doi:10.1016/j.biopsycho.2006.01.009

26. Uller T, Nakagawa S, English S (2013) Weak evidence for anticipatory parental effects in plants and animals. J Evol Biol 26:2161–2170. doi:10.1111/jeb.12212

27. Douhard M, Plard F, Gaillard JM et al (2014) Fitness consequences of environmental conditions at different life stages in a long-lived vertebrate. P Roy Soci B-Biol Sci 281:20140276. doi:10.1530/eje.1.02233

28. Hayward A, Rickard IJ, Lummaa V (2013) Influence of early-life nutrition on mortality and reproductive success during a subsequent famine in a pre-industrial population. P Natl Acad Sci USA 110:13886–13891. doi:10.1073/pnas.1301817110

29. Hayward A, Lummaa V (2013) Testing the evolutionary basis of the predictive adaptive response hypothesis in a preindustrial human population. Evolution 106–117. doi:10.1093/emph/eot007

30. Drent RH, Daan S (1980) The prudent parent: energetic adjustments in avian breeding. Ardea 68:225–252

31. Wells JCK (2010) Maternal capital and the metabolic ghetto: an evolutionary perspective on the transgenerational basis of health inequalities. Am J Hum Biol 22:1–17. doi:10.1002/ajhb.20994

32. Hawkesworth S, Prentice AM, Fulford AJ, Moore SE (2009) Maternal protein-energy supplementation does not affect adolescent blood pressure in The Gambia. Int J Epidemiol 38:119–127. doi:10.1093/ije/dyn156

33. Hawkesworth S, Walker CG, Sawo Y et al (2011) Nutritional supplementation during pregnancy and offspring cardiovascular disease risk in The Gambia. Am J Clin Nutr 94:1853S–1860S. doi:10.3945/ajcn.110.000877

34. Hawkesworth S, Wagatsuma Y, Kahn AI et al (2013) Combined food and micronutrient supplements during pregnancy have limited impact on child blood pressure and kidney function in rural Bangladesh. J Nutr 143:728–734. doi:10.3945/jn.112.168518

35. Macleod J, Tang L, Hobbs FDR et al (2013) Effects of nutritional supplementation during pregnancy on early adult disease risk: follow up of offspring of participants in a randomised controlled trial investigating effects of supplementation on infant birth weight. PLoS ONE 8: e83371. doi:10.1371/journal.pone.0083371.s001

36. Zelditch ML, Wood AR, Bonett RM, Swiderski DL (2008) Modularity of the rodent mandible: integrating bones, muscles, and teeth. Evol Dev 10:756–768. doi:10.1111/j.1525-142X.2008.00290.x

37. Klingenberg CP (2014) Studying morphological integration and modularity at multiple levels: concepts and analysis. Philos Trans R Soc Lond B Biol Sci 369:20130249. doi:10.1111/j. 1558-5646.2009.00857.x
38. Petrik J, Reusens B, Arany E et al (1999) A low protein diet alters the balance of islet cell replication and apoptosis in the fetal and neonatal rat and is associated with a reduced pancreatic expression of insulin-like growth factor-II. Endocrinology 140:4861–4873. doi:10. 1210/endo.140.10.7042
39. Burns SP, Desai M, Cohen RD et al (1997) Gluconeogenesis, glucose handling, and structural changes in livers of the adult offspring of rats partially deprived of protein during pregnancy and lactation. J Clin Invest 100:1768
40. Pinheiro DF, Pacheco PDG, Alvarenga PV et al (2013) Maternal protein restriction affects gene expression and enzyme activity of intestinal disaccharidases in adult rat offspring. Braz J Med Biol Res 46:287–292. doi:10.1590/1414-431X20122561
41. Guan H (2004) Adipose tissue gene expression profiling reveals distinct molecular pathways that define visceral adiposity in offspring of maternal protein-restricted rats. Am J Physiol-Endoc M 288:E663–E673. doi:10.1152/ajpendo.00461.2004
42. Bellinger L, Lilley C, Langley-Evans SC (2007) Prenatal exposure to a maternal low-protein diet programmes a preference for high-fat foods in the young adult rat. Brit J Nutr 92:513. doi:10.1079/BJN20041224
43. Díaz P, Powell TL, Jansson T (2014) The role of placental nutrient sensing in maternal-fetal resource allocation. Biol Reprod 91:82. doi:10.1095/biolreprod.114.121798
44. Fowden AL, Forhead AJ (2009) Hormones as epigenetic signals in developmental programming. Exp Physiol 94:607–625. doi:10.1113/expphysiol.2008.046359
45. Lancaster LT, McAdam AG, Sinervo B (2010) Maternal adjustment of egg size organizes alternative escape behaviour, promoting adaptive phenotypic integration. Evolution 64:1607–1621. doi:10.1111/j.1558-5646.2009.00941.x
46. Richardson SS, Daniels CR, Gillman MW et al (2014) Society: don't blame the mothers. Nature 512:131–132. doi:10.1038/512131a
47. Gluckman PD, Lillycrop KA, Vickers MH et al (2007) Metabolic plasticity during mammalian development is directionally dependent on early nutritional status. P Natl Acad Sci USA 104:12796–12800. doi:10.1073/pnas.0705667104
48. Ellis PJI, Morris TJ, Skinner BM et al (2014) Thrifty metabolic programming in rats is induced by both maternal undernutrition and postnatal leptin treatment, but masked in the presence of both: implications for models of developmental programming. BMC Genom 15:49. doi:10. 1186/1471-2164-15-49
49. Shields BM, Freathy RM, Hattersley AT (2010) Genetic influences on the association between fetal growth and susceptibility to type 2 diabetes. J Devel Orig Health Dis 1:96. doi:10.1017/S2040174410000127
50. Braveman PA, Cubbin C, Egerter S et al (2005) Socioeconomic status in health research—One size does not fit all. JAMA 294:2879–2888. doi:10.1001/jama.294.22.2879
51. Shavers VL (2007) Measurement of socioeconomic status in health disparities research. J Natl Med Assoc 99:1013–1023
52. Jasienska G (2009) Low birth weight of contemporary African Americans: an intergenerational effect of slavery? Am J Hum Biol 21:16–24. doi:10.1002/ajhb.20824
53. Geronimus AT (2013) Deep integration: Letting the epigenome out of the bottle without losing sight of the structural origins of population health. Am J Public Health 103:S53–S56. doi:10. 2105/AJPH.2013.301380
54. Kaufman JS, Cooper RS, McGee DL (1997) Socioeconomic status and health in blacks and whites: the problem of residual confounding and the resiliency of race. Epidemiology 8:621–628
55. Anderson PS, Renaud S, Rayfield EJ (2014) Adaptive plasticity in the mouse mandible. BMC Evol Biol 14:85. doi:10.1186/1471-2148-14-85

56. Hollegaard B, Byars SG, Lykke J, Boomsma JJ (2013) Parent-offspring conflict and the persistence of pregnancy-induced hypertension in modern humans. PLoS ONE 8:e56821. doi:10.1371/journal.pone.0056821
57. Schluter D, Nychka D (1994) Exploring fitness surfaces. Am Nat 143:597–616
58. Cottrell EC, Holmes MC, Livingstone DE et al (2012) Reconciling the nutritional and glucocorticoid hypotheses of fetal programming. Faseb J 26:1866–1874. doi:10.1096/fj.12-203489
59. Reynolds RM (2013) Glucocorticoid excess and the developmental origins of disease: two decades of testing the hypothesis—2012 Curt Richter Award Winner. Psychoneuroendocrino 38:1–11. doi:10.1016/j.psyneuen.2012.08.012
60. Bolten MI, Wurmser H, Buske-Kirschbaum A et al (2011) Cortisol levels in pregnancy as a psychobiological predictor for birth weight. Arch Womens Ment Health 14:33–41. doi:10.1007/s00737-010-0183-1
61. Thoits PA (2010) Stress and health: major findings and policy implications. J Health Soc Behav 51:S41–S53. doi:10.1177/0022146510383499
62. Rickard IJ, Frankenhuis WE, Nettle D (2014) Why are childhood family factors associated with timing of maturation? A role for internal prediction. Perspect Psychol Sci 9:3–15. doi:10.1177/1745691613513467
63. Thorne AD, Pexton JJ, Dytham C, Mayhew PJ (2006) Small body size in an insect shifts development, prior to adult eclosion, towards early reproduction. P Roy Soci B-Biol Sci 273:1099–1103. doi:10.1093/beheco/12.5.577
64. Gross MR (1996) Alternative reproductive strategies and tactics: diversity within sexes. Trends Ecol Evol 11:92–98

Chapter 7
Is Calculus Relevant to Survival? Managing the Evolutionary Novelty of Modern Education

P. Douglas Sellers II, Ph.D., Karin Machluf, Ph.D. and Prof. David F. Bjorklund, Ph.D.

Lay Summary Efficient and effective education is important not simply for producing the next generation of scientists, engineers, doctors, or skilled laborers, but for issues of child health, public health, and health education. Many approaches and theories contribute to education policies; however, an often overlooked dynamic is incorporating human evolutionary history into education practice, especially considering the historical novelty of formal education in human life. Evolution has physiologically, cognitively, and socially shaped human children and adults, making it an important consideration for any large-scale education endeavor. Through consideration of evolutionary and developmental psychology, we propose that education practices can become more child-centered by considering specific traits in children selected for by evolution.

7.1 Introduction

7.1.1 The Advent of Standardized Education

For most of human history, and for all of human prehistory, children learned the skills necessary to be competent adults in their society "in the context" of daily life.

P.D. Sellers II (✉) · K. Machluf
Department of Psychology, Pennsylvania State University,
State College, PA 18512, USA
e-mail: pds5183@psu.edu

K. Machluf
e-mail: kxm5600@psu.edu

D.F. Bjorklund
Department of Psychology, Florida Atlantic University, Boca Raton, FL 33431, USA
e-mail: dbjorklu@fau.edu

© Springer International Publishing Switzerland 2016
A. Alvergne et al. (eds.), *Evolutionary Thinking in Medicine*,
Advances in the Evolutionary Analysis of Human Behaviour,
DOI 10.1007/978-3-319-29716-3_7

As such, the best available proxy of early human life is modern hunter-gatherers. We must take care to acknowledge that current research cannot assume perfect correspondence between modern groups and those living across evolutionary time, but the opportunity for comparison is relevant and advantageous [1]. Hunter-gatherer children spend most of their days freely playing with other children in mixed-age groups, learning practical skills from their older peers, and occasionally learning important skills by watching and interacting with their parents or other adults. Modern hunter-gatherer adults rarely directly instruct children in any skill, and it is likely that this was also true for our ancestors [2–4].

With the advent of agriculture and a more sedentary lifestyle, children continued to learn necessary life skills by performing them, often in the company of older and more accomplished individuals, and they often practiced such skills in their play. The need to be literate (and later numerate) changed not only what children needed to learn but also how they learned it. Although writing dates back nearly 6000 years, for millennia only the elite (priests, members of the ruling class, some merchants) were literate, and such advanced knowledge was typically passed on to others via tutoring. Until the invention of the printing press in the 1400s, the written word could not be mass produced (and thus mass consumed). Cultural changes accompanying the ability to economically print books made reading an important adult skill, one that differentiated illiterate children from literate adults [5]. Subsequently, learning shifted to schooling, which is done "out of context," with children having to master tasks that have no immediate relevance for their daily lives.

In contrast to children from traditional societies, modern technological skills are usually learned to meet modern cultural needs, not to solve any pressing problem of survival, making this style of learning an evolutionary novelty. Modern children must also learn these skills and related information whether they are interested and motivated for such learning or not. Although some children will learn to read and calculate on their own via discovery learning, many will not, and direct instruction and tedious practice, to some degree, are inevitable.

Most developed nations see having an educated populace as the backbone of a successful society, and, around the world, nations vie to develop curricula that will produce intelligent and productive citizens. Although America currently leads the world in worker productivity and scientific accomplishments, the achievement gap between children from middle-class versus lower-income homes is substantial, and parents, educators, and government officials have focused their attention on improving American children's academic accomplishments while reducing the learning gap between the "haves" and "have nots." As a result, the USA, as well as other developed countries, has increased its emphasis of academic learning, often at the expense of other non-tested subjects, such as art, music, and physical education, and activities such as recess [6]. The important skills of mathematics and reading are increasingly stressed, as required by the 2002 Federal No Child Left Behind Act, which mandated assessment of progress in these core subjects for children in the grades 3–8.

Although there is substantial controversy on the success of No Child Left Behind on children's academic performance, some gains are found at some ages for some abilities, but not at other ages for other abilities [7–9]. One consequence of No

Child Left Behind is the use of standardized assessments as a yardstick for a school's (or classroom's) progress, often with monetary rewards and punishments linked to children's scores on the tests. However, even if we grant the gains in academic abilities, there is a potential downside. Academic subjects not explicitly assessed may not receive the same classroom attention as the "tested" subjects, with the majority of resources devoted to rigorous instruction on these tested subjects. Because of these performance demands, school has increasingly become a high-stress environment, for both children and teachers. According to developmental psychologist William Crain [10],

> Historically, children seem to have never liked school very much. It has always taken a toll on the natural curiosity and enthusiasm for learning with which children began life. But today, as the standards movement rolls on, the pressure on children is becoming quite oppressive

(p. 10) Crain further writes

> ... [W]e are, in effect, stunting their growth, and future research may show that the effects show up in increased depression, suicidal ideation, restlessness, and other symptoms of unfulfilled lives" (p. 6).

These concerns were echoed a decade later by Gray [11], among others, who posits that contemporary education practices have produced a generation of highly stressed, unenthusiastic students, who are ill-prepared for the challenges that a changing economy presents. The push by modern schools for early and invariant achievement may be mismatched with the often slow-paced cognitive and social development of human children. Gray [11], among others [12–15], suggests that modern educational practices can be made more effective by considering the environment in which learning had taken place over evolutionary history.

7.1.2 Evolutionary Educational Psychology

David Geary [13, 16, 17], a vocal advocate of an evolutionarily informed approach to education, proposed that humans evolved intuitive cognitive systems for managing their physical, biologic, and social worlds (folk physics, folk biology, and folk psychology) that develop over childhood (see Fig. 7.1). Geary referred to these as biologically primary abilities. They are universal and develop in a species-typical pattern given a species-typical environment, and children are intrinsically motivated to engage in them. Language is perhaps the prototypical example of a biologically primary ability; specific neurological architecture is responsible for its functioning and, with appropriate input, is a universal human trait. These are contrasted with biologically secondary abilities, which do not have an evolutionary history but rather are based on biologically primary abilities. Reading is a clear example of a biologically secondary ability; based on language abilities, it is a non-essential cultural extension of a primary domain. Biologically secondary abilities could just as easily be called "culturally primary abilities," as

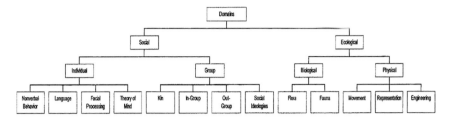

Fig. 7.1 Proposed domains of mind. From Ref. [51]. Reprinted with permission

they are invented by each culture to deal with ecological demands. When these ecological demands are similar among different cultures (e.g., the need for communication between group members), the produced ability is likely to be similar. However, when demands vary with culture or ecology, the resulting abilities are also likely to vary. Thus, unlike biologically primary abilities, biologically secondary abilities are not universal and children are not necessarily intrinsically motivated to achieve proficiency. As such, biologically secondary abilities are most often those that require formal teaching and education. For example, the use of language (speaking and listening; a primary ability) is not explicitly taught in school to the same degree as reading and writing (secondary abilities).

Geary proposed that humans' inventions of new technologies often result in gaps between intuitive folk knowledge (acquired via biologically primary abilities) and the skills needed to be successful in modern societies (acquired via biologically secondary abilities). Based on these premises, Geary presented six principles of evolutionary educational psychology, and these are displayed in Table 7.1. As you can see, Geary argues that the role of schools is to fill the gap between folk knowledge and needed technological skills. Moreover, because children are not inherently motivated to exercise biologically secondary skills, direct instruction is often needed. Yet, children's acquisition of a biologically secondary skill can be facilitated if the content is related to their biologically primary interests. For example, with respect to reading, Geary [16: 28] states:

> The motivation to read … is probably driven by the content of what is being read rather than by the process itself. In fact, the content of many stories and other secondary activities (e.g., video games, television) might reflect evolutionary-relevant themes that motivate engagement in these activities (e.g., social relationships, competition…).

With respect to reading, this may be especially important for boys, who, compared to girls, have lower levels of reading comprehension for low-interest stories [18]. As a result of less motivation for acquiring biologically secondary skills, explicit instruction (as opposed to "discovery learning") is typically necessary for many children to acquire these skills.

Therefore, an evolutionarily informed theory of education must translate three major themes into applicable pedagogy: human evolutionary history, modern technological necessities, and practices for best bridging the gaps between

Table 7.1 Principles of evolutionary educational psychology. Adapted from Ref. [13]

Principles of evolutionary educational psychology
1. Biologically secondary abilities associated with scientific, technological, and academic advances emerged from the biologically primary abilities associated with folk physics, folk biology, and folk psychology. As a society's knowledge increases, the gap between folk knowledge and the skills necessary to acquire the technological skills of society widens
2. Schools emerged in societies to fill the gap between folk knowledge and needed technological skills
3. The purpose of schools is to organize the activities of children, so they can acquire the biologically secondary abilities that close the gap between folk knowledge and the occupational and social demands of their society
4. Biologically secondary abilities are built from biologically primary abilities and components of general intelligence and evolved to deal with environmental variation and novelty
5. Children are inherently motivated to engage in activities that promote their folk knowledge, but this sometimes conflicts with the need to engage in activity that will promote secondary learning (e.g., reading), because children are not inherently motivated to engage in biologically secondary abilities
6. There is a need for direct instruction for children to learn most biologically secondary abilities

children's intuitive capabilities and those required by their ecology. The remainder of this chapter will evaluate the merits of such a perspective through specific examples.

7.2   Research Findings

7.2.1   *Educating* Homo Ludens

In his 1944 book, *Homo Ludens* (Playing Man), Johan Huizinga [19] argued that play serves an important function throughout the human life span, pointing out that humans are the only species who plays for the sake of playing into adulthood. In adulthood, we think of play as recreational and socially affiliative; however, childhood play is, in many respects, work. When children engage in symbolic play in early childhood, or dramatic role-playing in middle childhood, they are not simply entertaining themselves. The boundaries of new cognitions and social relationships are tested, mental and physical muscles are trained and flexed, and adult roles are imitated in low-risk environments. Such behaviors are vital for learning and testing species-wide abilities (e.g., maneuvering the social hierarchy) while also discovering culturally specific technologies and roles (e.g., how to use a computer). For instance, many children in traditional societies spend much of their time engaging in pretend versions of typical adult behavior, such as hunting or cooking [20]. Taking an evolutionary perspective on these behaviors allows us to make note of their functional elements as potential adaptations. Deferred adaptations are those behaviors in childhood that serve to prepare children for adulthood, whereas ontogenetic adaptations serve a beneficial purpose in childhood itself (for a more general discussion on the application of an evolutionary perspective to childhood [21].

Much of what our evolutionary forechildren learned would have been in the context of play. As we noted, formal education is evolutionarily novel in that it is a recent cultural invention that departs radically from the historically typical experiences of humans. Konner [2] has proposed that the modern hunter-gatherer childhood can be viewed as the cultural baseline, or model, for comparisons across evolutionary history. Some characteristics of hunter-gatherer childhoods are that children have substantial freedom to engage in activities within the community with little adult supervision, children of different ages interact together, and there is little direct teaching by adults [2, 3]. Extending this to education, Wilson and his colleagues [14] proposed that departures from ancestral environments can create unintended consequences and that efforts should be made to emulate traditional learning environments whenever possible. This includes learning in mixed-aged settings and making learning spontaneous, playful, and child-driven [11, 22].

This concept is not necessarily new, although it previously existed in a slightly different form. Vygotsky [23] proposed in his sociocultural theory of development that children acquire knowledge and develop skills necessary for survival and success within their social world by interacting with others, preferably others who are more knowledgeable and competent than they are. Central to this theory is the zone of proximal development, which refers to the difference between what a child can achieve on his or her own compared to what he or she can accomplish when scaffolded by someone who is more skilled in that domain. Children often perform new tasks in collaboration with others. But independent functioning is achieved when the scaffolding (or assistance) is slowly removed until children are able to perform the action on their own. According to this theory, most of the skills acquired in development occur most rapidly, effectively, and efficiently when children collaborate with others within their zone of proximal development, taking advantage of the assistance of more competent individuals.

Blank and White [24] suggest that although scaffolding would be ideal in an educational setting, it seems unlikely to occur in modern schools. For example, given the standardization of learning criteria, teachers must adhere to the curriculum schedule to ensure that all students receive the required information. Furthermore, for scaffolding to occur, a single teacher (the competent other in modern educational settings) must be able to extensively interact with no more than a few students at a time. Given the class sizes in most contemporary educational settings, this is not possible. The further segregation of mixed-age peers seems to eliminate another opportunity for scaffolding. Although such traditional pedagogy may seem unrealistic in today's competitive and high-stress educational systems, some schools have adopted such procedures with considerable success.

7.2.1.1   The Sudbury Valley School

An educational program that adopts a traditional approach to pedagogy is the Sudbury Valley School, located in Framingham, Massachusetts, for children between 4 years of age and high school age. The premise behind the school's educational philosophy is guided by the idea that each child is solely responsible for his or her own education. According to this perspective, if children are given a supportive opportunity-filled environment, they will, through self-directed play and exploration with peers across various ages, educate themselves or will be motivated to request the information from the staff at the school (see Gray [11] for a description of Sudbury Valley School and its philosophy). Peter Gray has described the Sudbury Valley School in detail and posits that the school's organization more successfully addresses the evolutionary novelty of modern education. Children in the Sudbury Valley School interact in mixed-age groups, engage in activities of their choosing, acquire most new information through play or games, and receive little or no unrequested direct adult instruction. These practices take into account the social and cognitive characteristics under which human children have learned for thousands of years, making the process of education consistent with the evolved neurocognitive architecture that allows such incredible amounts of learning in childhood. A central tenant of evolutionary developmental psychology is that the brain has not evolved in isolation from environmental input; rather, appropriate input is as vital as appropriate biology. Given the close relationship between the principles that govern Sudbury Valley's environment and the environment of human evolution, it is possible that an educational advantage could exist from this evolutionary model. Some readers may find this description ostensibly similar to Montessori education practices; however, Sudbury Valley School differs in many critical domains including true age mixing, pure democratic decision making, and child-driven curricula.

Formal classes at Sudbury Valley School are offered only in response to students' requests, but even then, there are no requirements for attending class. Books, materials, and knowledgeable staff members are available to aid in the learning of any subjects and skills, but students are always free to use or not use these resources. Additionally, there are no examinations or formal assessments. Most importantly, children are not assigned to grades or classes. They are encouraged to move through the school buildings as they wish. Children are often found playing, talking, and engaging in a range of self-directed activities. In this way, younger children interact with older children on a regular basis.

Allowing children to interact with peers of different ages seems to be essential for successful social learning. Evidence for this is seen through research with hunter-gatherer and other traditional, preindustrial societies that allow older children to guide younger children in their exploration of culturally relevant activities [25]. In fact, these studies find that self-directed age mixing has apparently been the primary vehicle of education throughout human history [11]. Although it is minimal, some research in non-traditional societies also suggests qualitative differences between children's interactions with adults and with their peers. This research is mainly viewed through the lens of play. Given the minimal contact of mixed ages in modern schools, it is difficult to examine these interactions within educational settings. Elias and Berk [26] found the amount of time 3-year-olds spend talking with peers, while pretending is positively associated with the size of their vocabularies at age 5. Furthermore, spontaneous sociodramatic play with peers improved children's abilities to remember, reproduce, and comprehend stories [27, 28]. These findings suggest that using peers as scaffolds assists in the development and efficacy of learning.

According to Greenberg and Sadoffsky [29], this combination of free age mixing and the democratic ethos of the Sudbury Valley School is the key to success of its educational approach. Allowing children to choose what and when they want to learn fosters intrinsic motivation, creating a positive, self-motivated learning environment. Furthermore, the benefits of children of various ages interacting with one another are reciprocal. For younger children, it allows them to be able to watch older children's behaviors, learning from them through observation. Helping with the acquisition of new skills by younger children allows older children the opportunity to practice difficult skills that they themselves are perhaps still struggling to master. Additionally, it promotes empathy, critical thinking, and pride [11].

7.2.1.2   Too Good to Be True?

Given the description of Sudbury Valley School, one is inclined to question how well these children fare when they graduate and enter the "real" world. Follow-up studies of children who attended Sudbury Valley School suggest that students who graduate from this school perform just as well as students from other educational institutions [20, 29]. In these surveys, roughly 75 % of the graduates went on to pursue a higher education degree. Some even attended Ivy League institutions, with great success. Regardless of whether they attended college, on average, graduates of Sudbury Valley School were highly successful in attaining employment in their chosen careers [22, 30].

Greenberg and colleagues [30] and Gray [22] consolidated benefits of this program into four distinct categories. The first category was self-direction and responsibility. Given the minimal structure at Sudbury Valley and the personal responsibility instilled in the students there, graduates were able to take responsibility for their higher education and direct themselves into filling the gaps necessary for their success in higher education or the workplace. A second category is their high motivation for their field or path of education. The graduates expressed great interest and intrinsic motivation in continuing their education or entering the workplace in the field they chose to pursue. The third benefit concerned their skill in the given area of work or education chosen by graduates. These students chose to study or work in the area that they showed interest in while in school. Through their self-directed play and exploration, they acquired great skill in their area of interest and almost always went on to study or work in the field that corresponded to that interest. Lastly, graduates seemed to lack a fear of authority reported by many traditional students. Sudbury Valley graduates reported having positive relationships with their professors or employers, communicated with them well, and were able to take directions without being defensive. Allowing children to direct their own learning, enlisting older children to help and teach along with able-bodied and sympathetic staff, and the incorporation of play and self-exploration are surprisingly effective education features. Even in the modern world with more standardization in the traditional classroom and longer school days and homework, children from Sudbury Valley School are able to compete and succeed in the social and professional world.

Evidence from education intervention programs also confirms the value of evolutionary principles. Wilson, Kauffman, and Purdy [31] created the Regents Academy program for academically at-risk 9th and 10th graders within their Binghamton, New York school system. The program's choice of guiding evolutionary principles focused largely on how to make cooperative small-group interactions as productive and efficient as possible, stemming specifically from the work of Ostrom [32, 33]. Using cumulative year-end grade point average as a comparison, the Regents Academy group significantly outperformed their matched sample who remained enrolled in the regular school. In fact, they were indistinguishable from the not-at-risk sample also enrolled in the regular school [31].

7.2.2 The Adaptationist Perspective

Adopting a large-scale naturalistic educational program, however, might be unrealistic in modern times. Therefore, others have argued that preschool education is particularly important and should reflect children's learning propensities, specifically, that beginning rigorous educational practices too early can be detrimental [34, 35]. This is consistent with the cognitive immaturity hypothesis [36, 37] which argues that infants' and young children's cognitive and perceptual abilities are well suited for their particular time in life and are not simply incomplete versions of the adult form. Along these lines is research with infants [38] showing that beginning a task too early in development can actually hinder subsequent learning. For example, Papousek [38] presented infants with an operant conditioning task (turn head in one direction to a bell, the other to a buzzer) beginning either at birth, 31 days or 44 days. He reported that infants who began training at birth required more trials (814) and more days of training (128) to reach criterion than the infants who started training at 31 or 44 days (278 and 224 trials, and 71 and 72 days for the 31- and 44-days-old infants, respectively). Papousek concluded that

> beginning too early with difficult learning tasks, at a time when the organism is not able to master them, results in prolongation of the learning process.

Although few people engage newborn infants in formal education, there are commercially available DVDs and educational software aimed at enriching infants' cognitive experiences. However, despite the testimonials on their Web sites, there is no evidence that these products enhance cognitive abilities, and recent research indicates that they either provide no cognitive advantage [39] or may actually be detrimental to children's cognitive development [40]. Zimmerman and colleagues [41] reported that the amount of time 8- to 16-month-old infants spent watching "baby" DVDs such as Baby Einstein® was negatively associated with receptive vocabulary: Each hour children watched baby DVDs/videos was associated with 6–8 fewer vocabulary words. Moreover, although infants are often attentive to and seem to enjoy these DVDs as well as television, it is not until 18 months that the content of the video, rather than the physical stimulus qualities of the display, will hold a child's attention [42]. Although the research is admittedly scant, the evidence is consistent with the position that stimulation in excess of the species norm early in development can have detrimental consequences [43].

More research is available assessing preschool educational programs, contrasting the effect of formal instruction (termed direct instructional programs) versus programs that take children's "natural" propensities for play and activity into consideration (termed developmentally appropriate programs). There are few consistent differences in terms of academic performance at the end of a year between children who attend these two types of programs, with some researchers finding small advantages for the direct instructional programs, some for the developmentally appropriate programs, and others no differences [34]. When researchers look at long-term effects, however, different patterns emerge, with more studies reporting greater cognitive gains for children who attended the developmentally appropriate programs [44, 45]. In addition to the cognitive differences, children who attend the developmentally appropriate programs tend to experience socioemotional benefits; they experience less stress, like school better, are more creative, and have less test anxiety than children attending direct instructional programs [34]. Although the differences are small, most clearly favor the developmentally appropriate programs. In other words, any academic benefits gained from a teacher-directed program had its costs in terms of motivation. According to the authors of one study [46] that contrasted developmentally appropriate and direct instructional programs,

> If it has no clear benefit to the child's development, and if it may hinder development, there may be no defensible reason to encourage the introduction of formal academic instruction and adult-focused learning during the preschool years

Although these findings and interpretations may on the surface appear to contradict Geary's evolutionarily informed observation that direct instruction is necessary for children to acquire many of the biologically secondary abilities so important in modern societies, they do not. Instead, consistent with Geary, the findings show that young children should receive instruction compatible with their intuitive learning biases, which are well suited to the niche of early childhood. The ideas of Geary and those who advocate a form of education that minimizes the mismatch between modern learning and traditional learning also demonstrate that evolutionary thinking as applied to education is not monolithic, but rather that different Darwinian-influenced hypotheses can be generated and tested with the goal of improving education for the most educable of animals.

7.3   Implications for Policy and Practice

The obvious critique most academics or educators are likely to have of the addition of evolutionary reasoning to education is that the principles underlying their implementation can be derived just as easily without invoking evolution. Indeed, it is appealing to consider sharpening Occam's Razor and simply settle upon a system of "socially informed education." Shifting evolution into the background may be preferred and necessary in many occasions that may be encountered during the construction of a system for education. Day-to-day operations are, in fact, likely to be run along the lower-order level of adherence to rules that simply promote success among group relations. This is largely due to the evolutionary novelty of modern education. For example, in-group preference enabled small bands of hunter-gatherers to help protect the survival interests of their close kin and thus raise inclusive fitness. But in a modern educational setting, in-group preference may serve to foster trust and closeness for helping each other solve complex mathematics problems. However, these social principles function as they do because the human brain has evolved to place import on social stimuli, social situations, and social cognitions. We are nothing if not social animals. Given the increasing specificity with which cognitive and behavioral neurosciences are uncovering the brain's dedication to socialness [47], it seems advisable to retain evolution as a guiding theory for a social-based education. The provision of distal causation to proximate behaviors allows for integration and coordination of seemingly disparate aims related to development as a whole. As mentioned previously, evidence from evolutionary developmental psychology shows how the hasty introduction of certain stimuli or learning situations can lead to ultimately deleterious effects on cognitive development. Moreover, a new model was recently put forth reconceptualizing adolescent risky behavior (and intervention) based on evolutionary principles [48]. If education is to successfully integrate with these (and other) ideas, then an evolutionary perspective is insightful.

However, the goal of a traditional Western-style education system is to provide access to standardized education services to the entire populace, a difficult proposition given the size and diversity of most developed countries. In spite of the obvious difficulties, the current system works for a majority of children, as 78.2 % of American 9th graders in 2006 graduated from high school in 2010 [49]. Additionally, 65.9 % of 2013 American high school graduates enrolled in college, evidence that experience in the public school system translates, for most children, into higher education [50], although there are substantial ethnic, racial, and income differences. This system is certainly not perfect, but would it be worth the time, effort, and dollars to incorporate evolutionary-based education practices into the public school system? Is there evidence to suggest that the new system would be demonstrably more effective for the majority of children? Or is this style of education better suited for private schools or specific intervention programs?

Definitive answers to these questions will require years more research and assessment. Large-scale feasibility and longitudinal testing would be required before any systemic changes can be reasonably proposed. However, we are confident that there is utility in an evolutionary perspective on education, even if it is not invoked in day-to-day decision making. The available evidence suggests that a system of evolutionary-informed education and intervention is just as successful (if not more so) as traditional education for professional outcomes while fostering greater positivity, enjoyment, and intrinsic motivation [11]. There seems to be little risk in incorporating these ideas into practice and few potential hurdles to implementation, with serious potential upside for academic performance and child emotional welfare. As society continues to search for the best manner in which to educate its members, it should, at the very least, give due process to incorporating evolutionary theory into the current functional system.

References

1. Foley R (1995) The adaptive legacy of human evolution: a search for the environment of evolutionary adaptedness. Evol Anthropol Issues News Rev 4:194–203. doi:10.1002/evan. 1360040603
2. Konner M (2010) The evolution of childhood: relationships emotions, mind. Belknap Press, Cabridge, MA
3. Lancy D (2014) The anthropology of childhood, 2nd edn. Cambridge University Press, Cambridge
4. Lancy D, Grove MA (2010) The role of adults in children's learning. In: Lancy DF, Bock J, Gaskins S (eds) The anthropology of learning in childhood. AltaMira Press, New York, pp 145–179
5. Postman N (1982) The disappearance of childhood. Vintage Books, New York. doi:10.1080/ 00094056.1985.10520201
6. Pellegrini AD (2005) Recess: its role in education and development. Erbaum, Mahwah, NJ
7. Center on Education Policy (2008) Has student achievement increased since 2002: state test score trends through 2006–2007. CEP, Washington, DC
8. Dee TS, Jacob B (2011) The impact of no child left behind on student achievement. J Policy Anal Manage 30:418–446. doi:10.1002/pam.20586
9. Fuller B, Wright J, Gesicki K, Kang E (2007) Gauging growth: how to judge no child left behind? Educ Res 36:268–278. doi:10.3102/0013189X07306556
10. Crain WC (2003) Reclaiming childhood: letting children be children in our achievement oriented society. Times Books, New York
11. Gray P (2013) Free to learn: why unleashing the instinct to play will make our childrenhappier, more self-reliant, and better students for life. Basic Books, New York
12. Bjorklund DF, Bering JM (2002) The evolved child: applying evolutionary developmental psychology to modern schooling. Learn Individ Differ 12:1–27. doi:10.1016/S1041-6080(02) 00047-X
13. Geary DC (2007) Educating the evolved mind: conceptual foundations for an evolutionary educational psychology. In: Carlson JS, Levin JR (eds) Educating the evolved mind: conceptual foundations for an evolutionary educational psychology. Information Age Publishing, Charlotte, NC
14. Wilson DS, Geher G, Waldo J (2009) EvoS: completing the evolutionary synthesis in higher education. J Evol Stud Consortium 1:3–10
15. Carlson JS, Levin JR (eds) (2007) Educating the evolved mind: Conceptual foundations for an evolutionary educational psychology. Information Age Publishing, Charlotte, NC

16. Geary DC (1995) Reflections of evolution and culture in children's cognition: implications for mathematical development and instruction. Am Psychol 50:24–37. doi:10.1037/0003-066X. 50.1.24

17. Geary DC (2005) The origin of mind: evolution of brain, cognition, and general intelligence. Am Psychol Assoc, Washington, DC. doi:10.1037/10871-000

18. Renninger K, Hidi SE, Krapp AE (1992) The role of interest in learning and development. Lawrence Erlbaum Associates Inc

19. Huizinga J (1944; reprinted 1970) Homo ludens: a study of the play-element in culture. Paladin, London

20. Sutton-Smith B, Roberts JM (1970) The cross-cultural and psychological study of games. In: Lüschen G (ed) The cross-cultural analysis of sport and games. Stipes, Champagne, IL, pp 100–108). doi:10.1177/101269027100600105

21. Bjorklund DF, Sellers PD II (2011) The evolved child: adapted to family life. In: Roberts SC (ed) Applied evolutionary psychology. Oxford University Press, Oxford, pp 55–77

22. Gray P (2011) The decline of play and the rise of psychopathology in children and adolescents. Am J Play 3:443–463

23. Vygotsky LS (1978) Interaction between learning and development. In: Cole M, Joh Steiner V, Scribner S, Souberman E (eds) Mind in society: the development of higher psychological processes. Harvard University Press, Cambridge, MA

24. Blank M, White S (1999) Activating the zone of proximal development in school: obstacles and solutions. In: Lloyd P, Fernyhough C (eds) Lev Vygotsky: critical assessments, vol 3. The zone of proximal development. Routledge, London

25. Rogoff B (1990) Apprenticeship in thinking: cognitive development in social context. Oxford University Press, New York

26. Elias CL, Berk LE (2002) Self-regulation in young children: Is there a role for sociodramatic play? Early Child Res Q 17:216–238. doi:10.1016/S0885-2006(02)00146-1

27. Smilansky S (1968) The effects of sociodramatic play on disadvantaged preschool children. Wiley, New York

28. Pellegrini AD, Galda L (1982) The effects of thematic-fantasy play training on the development of children's story comprehension. Am Educ Res J 19:443–452. doi:10.3102/00028312019003443

29. Greenberg D, Sadofsky M (1992) Legacy of trust: life after the Sudbury Valley School experience. The Sudbury Valley School, Sudbury, MA

30. Greenberg D, Sadofsky M, Lempka J (2005) The pursuit of happiness: the lives of Sudbury Valley alumni. The Sudbury Valley School, Sudbury, MA

31. Wilson DS, Kauffman RA Jr, Purdy MS (2011) A program for at-risk high-school students informed by evolutionary science. PLoS ONE 6:e27826. doi:10.1371/journal.pone.0027826

32. Ostrom E (1990) Governing the commons: the evolution of institutions for collective, action. Cambridge University Press, Cambridge, UK

33. Ostrom E (2005) Understanding institutional diversity. Princeton University Press, Princeton

34. Bjorklund DF (2007) The most educable of species. In: Carlson JS, Levin JR (eds) Psychological perspectives on contemporary educational issues. Information Age Publishing, Greenwich, CT, pp 119–129

35. Hirsh-Pasek K, Hyson MC, Rescorla L (1990) Academic environments in preschool: do they pressure or challenge young children. Early Educ Dev 1:401–423. doi:10.1207/s15566935eed0106_1

36. Bjorklund DF (1997) The role of immaturity in human development. Psychol Bull 122:153–169. doi:10.1037/0033-2909.122.2.153

37. Bjorklund DF, Green BL (1992) The adaptive nature of cognitive immaturity. Am Psychol 47:46–54. doi:10.1037/0003-066X.47.1.46

38. Papousek H (1977) The development of learning ability in infancy (Entwicklung de Lernfähigkeit im Säuglingsalter). In: Nissen G (ed) Intelligence, learning, and learning disabilities (Intelligenz, Lernen und Lernstörungen). Springer, Berlin

39. Richert R, Robb MB, Fender JG, Wartella E (2010) Word learning from baby videos. Arch Pediatr Adolesc Med 164 (No. 5). Published online, March 1. WWW.ARCHPEDIATRICS. COM. doi:10.1001/archpediatrics.2010.24

40. Courage ML, Murphy AN, Goulding S, Setliff AE (2010) When the television is on: the impact of infant directed video on 6-and 18-month-olds' attention during to play and on parent–infant interaction. Infant Behav Dev 33:176–188. doi:10.1016/j.infbeh.2009.12.012

41. Zimmerman FJ, Christakis DA, Meltzoff AN (2007) Television and DVD/video viewing in children younger than 2 years. Arch Pediatr Adolesc Med 161:473–479. doi:10.1001/archpedi. 161.5.473

42. Courage ML, Setliff AE (2010) When babies watch television: attention-getting, a tention-holding, and the implications for learning from video material. Dev Rev 30:220–238. doi:10.1016/j.dr.2010.03.003

43. Turkewitz G, Kenny PA (1982) Limitations on input as a basis for neural organization and perceptual development: a preliminary theoretical statement. Dev Psychobiol 15:357–368. doi:10.1002/dev.420150408

44. Burts DC, Hart CH, Charlesworth R, DeWolf DM, Ray J, Manuel K, Fleege PO (1993) Developmental appropriateness of kindergarten programs and academic outcomes in first grade. J Res Childhood Educ 8:23–31. doi:10.1080/02568549309594852

45. Marcon RA (1999) Differential impact of preschool models on development and early learning of inner-city children: a three-cohort study. Dev Psychol 35:358–375. doi:10.1037/0012-1649. 35.2.358

46. Hyson MC, Hirsh-Pasek K, Rescorla L (1990) The classroom practices inventory: an observation instrument based on NAEYC's guidelines for developmentally appropriate practices for 4-and 5-year-old children. Early Childhood Res Q 5:475–494. doi:10.1016/0885-2006(90)90015-S

47. Baron-Cohen S, Lombardo M, Tager-Flusberg H, Cohen D (eds) (2013) Und standing other minds: perspectives from developmental social neuroscience. Oxford University Press, Oxford

48. Ellis BJ, Del Giudice M, Dishion TJ, Figueredo AJ, Gray P, Griskevicius V, Wilson DS (2012) The evolutionary basis of risky adolescent behavior: implications for science, policy, and practice. Dev Psychol 48:598–623. doi:10.1037/a0026220

49. U.S. Department of Education (2014) Public high school four-year on-time graduation rates and event dropout rates: school years 2010–11 and 2011–2012. Retrieved May 1, 2014. http://nces.ed.gov

50. Bureau of Labor Statistics, United States Department of Labor (2014) College Enrollment and Work Activity of 2013 High School Graduates. Retrieved May 1, 2014, www.bls.gov

51. Geary DC (1998) Male, female: the evolution of human sex differences. American Psychological Association, Washington D. C., p 180)

Part III
Nutrition

Chapter 8
Binge Eating, Disinhibition and Obesity

Prof. Stanley Ulijaszek, Ph.D. and Eleanor Bryant, Ph.D.

Lay summary Obese people are more likely to eat at every opportunity (display disinhibition) and often binge on food. We argue in this chapter that binge eating and disinhibition are evolved mechanisms for dealing with one of the most fundamental of insecurities, that of food, especially in seasonal and unpredictable environments. It is only in recent decades, with improved food security in industrialized nations and the emergence of obesity at the population level, that they have become deleterious for health and have been medically pathologized. Binge eating and disinhibition are no longer responses to uncertainty in food availability as they would have been across evolutionary history. Rather, uncertainty and insecurity in everyday life in present-day society are likely to lead to disinhibition, binge eating and obesity, through the linked physiology of stress and appetite.

8.1 Introduction

Obesity is new in human evolutionary history. It has become possible at the population level since the increase in food security and has risen in most countries on most continents since the 1980s [1]. While prevalence rates of obesity have since levelled off in a small number of regions [2], they have risen steadily in most

S. Ulijaszek (✉)
Unit for Biocultural Variation and Obesity, School of Anthropology,
University of Oxford, Oxford, UK
e-mail: stanley.ulijaszek@anthro.ox.ac.uk

E. Bryant
Division of Psychology, Richmond Building, University of Bradford,
Bradford, UK

© Springer International Publishing Switzerland 2016
A. Alvergne et al. (eds.), *Evolutionary Thinking in Medicine*,
Advances in the Evolutionary Analysis of Human Behaviour,
DOI 10.1007/978-3-319-29716-3_8

industrialized countries. In the USA, the prevalence of obesity (body mass index greater than 30 kg/m^2) and extreme obesity (body mass index greater than 40 kg/m^2) has shown rapid increases from the 1980s to the 2000s. In 2003–2004, the national prevalence of obesity was 30 %, while that of extreme obesity was 5 % [3, 4], rising to 34 % (obesity) and 6 % (extreme obesity) in 2008 [5, 6]. In the UK, the prevalence rates of obesity and extreme obesity have been much lower than in the USA, but have followed similar trajectories [7, 8].

There are many pathways, still poorly understood, to the positive energy balance (due either to high dietary energy intake, low energy expenditure or both) that leads to obesity. Among the eating disorders associated with obesity (and extreme obesity especially), binge eating disorder (BED) is particularly important. This was first described by Stunkard [9] and linked to obesity by Spitzer et al. [10, 11]. Studies of obese individuals who sought treatment identified subgroups that experienced distress and dysfunction due to binge eating [12]. The diagnostic and statistical manual of mental disorders 4 (DSM4) [13] defines binge eating as a series of recurrent binge episodes in which each episode is defined as eating a larger amount of food than normal during a short period of time (usually within any two-hour period). Characterized by economists as "high time preference", the desire to binge-eat might reflect the desire for palatable food now, rather than have a slim body, or the health that can go with it, later [14]. Binge eating behaviour is not always associated with obesity and does not necessarily have to be practised at extreme levels to be of importance for the causation of obesity.

The vast majority of binge eating remains undiagnosed or outside of clinical parameters [15], for a range of reasons. These include the lack of valid and reliable instruments for its measurement, and the likelihood that many people who experience binge eating does not come for help, often because of embarrassment, or because they do not see their behaviour as problematic. Binge eating disorder is usually identified when people seek help, as for example when obese subjects put themselves forward for bariatric surgery. Regular overeating within meals and snacking between meals are behaviours that are strongly linked with both obesity and subsequent health issues, and have similar underlying mechanisms to BED. Such behaviours are well captured by the psychological measure of disinhibition, as assessed by the Three-Factor Eating Questionnaire [16]. An individual scoring highly on the disinhibition scale of this measure is characterized as having very opportunistic eating behaviours and expressing a readiness to eat [17], both being associated with BED.

Binge eating (whether associated with obesity or not) can be thought of as a response to insecurity with deep evolutionary roots. The greatest cause of insecurity among past populations was that related to food and its provisioning and of climatic seasonality and variation in which food availability could not be guaranteed across the year, or sometimes from day-to-day [18]. Across evolutionary time, environmental variability has been the norm, and modern humans are very capable of responding physiologically, behaviourally and culturally to such variability [19, 20].

Such variability would have led to variation in dietary possibilities, food intake and uncertainty in food availability. We argue that binge eating and disinhibition are mechanisms that have been selected for by natural selection as they enabled individuals dealing with one of the most fundamental of insecurities, that of food, especially in the seasonal and unpredictable environments during the evolution of early Homo, from about 2–4 million years ago [21]. High reactivity to external cues such as colour, taste and smell [21] and subsequent disinhibited eating under conditions of food scarcity and/or high levels of competition would have favoured survivorship for most people and populations until fairly recent times. Only with improved food security in industrialized nations, the decline in price of energy dense foods and the emergence of obesity at the population level that disinhibited eating and binge eating would have become deleterious in terms of health outcomes. The chapter begins with descriptions of binge eating disorder and disinhibited eating and then considers the evolutionary importance of rapidly consuming large quantities of food in seasonal environments. It concludes by examining the evolutionary significance of binge eating and disinhibition and their implications for medicine in an age of obesity.

8.2   Research Findings

8.2.1   Binge Eating Disorder and Disinhibition

According to the DSM4 [13], binge eating episodes are associated with 3 or more of the following presentations: (1) eating until feeling uncomfortably full, (2) eating large amounts of food when not physically hungry, (3) eating much more rapidly than normal, (4) eating alone because of embarrassment about how much is eaten in a single eating episode, (5) feeling disgusted, depressed or guilty after overeating or (6) marked distress or anxiety regarding eating in such a way. Aside from those who are clinically diagnosed as having binge eating disorder, there are far more people who often binge on food, but perhaps not regularly enough to warrant a clinical diagnosis [15]. In line with this, disinhibition describes similar eating behaviour patterns as BED, although it is not used to diagnose clinical symptomatology; it is related to eating disorder severity [22]. Disinhibition is an eating behaviour trait measured by the Three-Factor Eating Questionnaire [16] which has been validated among a number of populations in Europe, the USA and Australia. It measures enduring characteristics of an individual's behaviour towards food. Disinhibition is defined by a readiness to eat [17] and is associated with frequent overeating and/or binging [23], frequent snacking [24], a high liking of all food groups with a particular preference for high-fat sweet foods [22], a tendency to gain weight and regain weight quickly following weight loss [25] and a tendency to be engaged in sedentary behaviour [26].

Although there is no clear line dividing eating a large meal and a pathological binge, people usually binge on highly palatable energy-rich food, typically high in fats, sugars or both [15]. It is impossible to know whether prehistoric foragers had cuisine, but they are certain to have had food preferences, if only on the basis of energy density, sweetness or fattiness of foods [18]. Individuals that binge on sweet foods tend to do so more frequently than those binging on other types of foods, and there is the potential for abuse of carbohydrates among individuals who claim to have cravings for them [27]. Similar addictive-like states can be created for fat consumption in animal models [28, 29], which also engage the dopaminergic reward system [30]. This system has been invoked for both addictions and obesity [31], and binge eating and disinhibition may be the link. From an evolutionary perspective, it is reasonable to think of addiction to palatable energy dense foods as being an evolved predisposition, if binge eating, by maximizing energy intake, also maximized reproductive success.

8.2.2 The Evolutionary Importance of Binge Eating and Disinhibition in Seasonal Environments

If the environment is uncertain or unstable, being able to eat a lot of energy dense foods very quickly when they are available is expected to have favoured survivorship and, by extension, reproduction. Although many foraged foods may not deteriorate very quickly, competition for food in uncertain or unstable environments would drive the need for fast eating. There is a physiological ceiling on how much protein an individual can eat, because of the finite ability of the liver to up-regulate enzymes necessary for urea synthesis in the face of increasing dietary protein intake [32]. According to Cordain et al. [32], hunter-gatherers would have had several options to circumvent the dietary protein ceiling. They could have eaten proportionately more plant food energy; hunted larger animals because percentage body fat increases with increasing body size; hunted smaller animals in seasons of plentiful forage, when body fat is maximized; selectively eaten only the fattier portions of the carcass; and/or increased their intake of concentrated sources of carbohydrate such as honey. Human feeding adaptations are likely to have arisen in the context of resource seasonality in which diet choice for energy dense and palatable foods would have been selected for by way of foraging strategies that maximized energy intake [21]. This may have been facilitated by heightened responses to visual cues associated with foods of high hedonic value when hungry [33]. These visual cues may have included aspects of colour, shape and size that reflect the relative ripeness, palatability and gustatory satisfaction to be gained from fruits and vegetables, for example, and would have been learned from previous exposure to, and consumption of, such foods [33]. Although primates respond to visual cues of food palatability much as do humans [28], there is no evidence that they have heightened responses to them when hungry.

Evidence that BED is a physiological feeding adaptation to food uncertainty comes from a mix of psychological and physiological studies of human feeding. Food uncertainty may have benefitted those able to consume great amounts of food in one go, potentially resulting in natural selection for this ability. Such selection would have been in addition to the strategies for protecting energy reserves shared by all human beings by being more sedentary, by readily storing body fat, being resistant to weight loss and quickly regaining weight lost, and the ability to binge on large volumes of food by some individuals points to this combination of characteristics being one type of thrifty phenotype [17]. Physiological differences in the appetite systems of individuals with a high disinhibition score and among binge eaters could also point to adaptations that enable consumption of greater amounts of food. For example, obese individuals with high disinhibition have high serum ghrelin and insulin levels [34]. These hormones regulate and initiate hunger, respectively. Furthermore, the cholecystokinin response of the gut, which promotes satiety in response to having eaten, is blunted in people who score highly on the tests of disinhibition [35]. On the other hand, serum ghrelin levels have been found to be lower in individuals with BED [36], suggesting a down-regulation of hunger in response to binge eating behaviour, much as obese people experience physiological down-regulation of hunger. Binge eating can take place in the absence of a hunger stimulus from the gut, especially of highly palatable foods for pleasure [37]. Food consumption under such conditions has been termed hedonic hunger [38].

Differences between those with BED and high disinhibition and those without have also been found in neurophysiology. For example, increased activity in the insular cortex (an area involved in emotion, motor function and homeostasis; [39]) and the prefrontal cortex (an area associated with executive function such as emotion regulation and general processing) when full [40] has been observed in individuals with high disinhibition scores in response to ingestion of high-fat food. Furthermore, those with BED show a higher activation in the orbitofrontal cortex (an area involved in sensory integration, emotion and decision-making [41]) in response to images of food. Those neurological differences could lend at least a partial explanation to individual response to food and ability to overeat, although the direction of the relationships between behaviours and imaged patterns of brain activity is not resolved (brain image patterns may be either the cause or consequence of food-related behaviours). While the mechanisms that these findings represent are not clear, they indicate a very strong emotional involvement with food and a higher degree of integration of sensory information associated with seeing, eating and digesting food among those with BED and high disinhibition scores than among those who do not display these feeding traits. The evidence for food uncertainty among prehistoric humans is indirect, but compelling [18]. Dispersal of anatomically modern human populations out of Africa into Europe and Asia would have exposed them to new stresses, physical, biological and social, which would have varied over time and place. In all places, tropical and temperate, humans may have experienced a double burden of resource fluctuation through seasonality as well as shifts in their environment caused by abrupt climate change [18]. At points in the Pleistocene, especially during the short-lived warming events that occurred within glacial periods, climatic transitions that potentially affected food security and resource distribution may have taken place over timescales of decades rather than many generations [42]. With less time to adapt, it is easy to imagine the starvation- and cold-related mortality that resulted and the food uncertainty that would have been a daily reality, especially among those that colonized the higher latitudes in temperate regions where extreme temperature and precipitation seasonality would have been the norm. Binge eating when food was available would have been an advantageous trait under such circumstances. According to Wisman and Capehart [43: 24]

> putting on weight -banking calories- is a 'natural' response to insecurity, a genetic trigger selected during the course of evolution. When sustenance, life-and-limb, or social positions were threatened, favouring fat-rich morsels and banking energy in the form of fat reserves would have been adaptive.

The amount of weight banking among contemporary foraging and agricultural societies can be as high as 5.9 kg body weight, equivalent to over 40,000 kcals dietary energy [44]. This would meet daily energy needs under complete starvation for at least 16 days if there were no physiological adaptation to starvation. However, such adaptation, as down-regulation of energy metabolism and reduced physical activity, is well documented [45], and weight banking to this extent would more likely cover the energy cost of complete starvation for double this time, or just over a month. Total food shortages among societies practicing foraging or tradi- tional agriculture are rare, however, and putting on weight when food was plentiful would have provided a reliable fallback to subsequent food shortages in most years.

Eating, dietary manners and restraint are unlikely to have been favoured under such conditions. Common to many mammalian species, gorging and binging among early humans and their hominin ancestors when the circumstances allowed would have been the norm:

> humans are evolved to over indulge when surpluses are available. Because surpluses of this kind are rare in the natural environment, such a strategy does not commonly lead to obesity in traditional societies. However, in such conditions, it pays to be attracted to sweet, sugary foods and to carbohydrates, and to gorge on them whenever they happen to be available, since these provide the primary sources of free energy. In the 'natural state', this is not a problem, since the feast-times do not occur all that often. It is only in modern post-industrial economies, where we live in a state of permanent feast, that it becomes an issue [46: 56].

Seasonality continues to influence activity and behaviour, even though humans, particularly in the industrialized world, are buffered against many aspects of environmental seasonality. This is especially so with respect to the availability of cheap, high-energy-density foods based on refined carbohydrates and fats, upon which most of the nutrition transition has been based [47] and which dominates diet in the industrialized and post-industrial world.

8.3   Implications for Policy and Practice

In contemporary society, the structures of food and of human feeding are at least partly incompatible and, at the extreme, pathological. In considering binge eating and disinhibition, it is important to think about the structures that human societies, and individuals in them, have in place to regulate eating, be it enough, too little, or too much. It is often not clear what the distinction is between eating a large meal and binge eating [15], but it might be that the structure of eating, social and individual, is a defining principle often missed by biomedicine. There are many social conventions surrounding food and its consumption in all societies [48, 49], mediating the physiological drive to eat, whether for hunger or pleasure (or both), and mediating individual responses to the dopaminergic reward system. Eating while listening to the radio or watching television increases the amount of food consumed [50], and the presence of familiar others is particularly potent at increasing food intake [51]. Both allow diversion of attention from food consumption and reduced self-monitoring of intake, allowing eating beyond immediate dietary energy needs [52]. Engaging in other tasks, such as working at the computer, also reduces self-monitoring of food intake. Modern humans in industrialized and post-industrial societies consume high-energy-density diets, often in unstructured ways [21], often under conditions of distraction. Modern life and cheap palatable foods make disordered, disregulated and distracted eating easier to practice, and extreme bodily phenotypes signal disordered eating to others, even in societies where generosity and plenty are valued, and large bodies are traditionally seen as carrying prestige [53].

The processing of agricultural raw materials by the food industry has produced a huge range of food products with reduced micro-nutrient content and high energy density. It has also resulted in the production of foods with altered physical matrices and new combinations of ingredients that make them more pleasurable to eat [18]. Food technology has been used to both respond to existing markets for products and create new market niches. This has involved the production of cheaper existing ingredients, completely new ingredients, new food types with existing ingredients, existing food types with proportions of new ingredients and new food types with new ingredients. Because food products are placed in competitive marketplaces in most countries, they must maximize some combination of novelty, status, price and palatability that will ensure their economic success. The cheaper raw ingredients are overwhelmingly cereal-based commodities and fats, and it is unsurprising that the majority of food products on the market involve some palatable combination of these ingredients [18]. Where palatability leads, increased consumption follows. The drive to create new, highly palatable high-fat and/or high sugar food products to develop and maintain market share by transnational companies has had profound effects on human food consumption, including ambivalence about eating [54] and possibly disordered eating [55]. Obesity may be a result of incremental overeating across many years or of periodic or sporadic binging, dieting, restriction, loss of control in response to guilt and ambivalence about foods and food types and distorted self-perception of body weight.

Binge eating, BED, high disinhibition and obesity all approach or fall into the category of biomedical pathology. But from an evolutionary perspective, binge eating and disinhibition are evolved predispositions. Among contemporary industrial and post-industrial society, they are deleterious, largely because of the plentiful nature of, and lack of seasonality in, the foods that promote gorging, however categorized. The sheer abundance of high-energy-density foods in industrial and post-industrial societies means that it is almost inevitable that people experience binge eating to some extent and at some time or another. It is social convention that categorizes a food consumption binge as taking place when it is solitary and involves high-energy-density foods, for example, and not taking place when a substantial symbolic meal like Christmas or thanksgiving dinner is consumed with family and friends. Feasting across Christmas and thanksgiving has been shown to be associated with significant weight gain [56, 57], but there have been no serious moves anywhere to pathologize either of these meals. Binge eating and disinhibition are no longer a response to uncertainty in food availability, as they are likely to have been across evolutionary history. Rather, there may be other types of uncertainty and insecurity that lead to disinhibition, binge eating and obesity. Stress-related chronic stimulation of the hypothalamic–pituitary–adrenal axis and resulting excess glucocorticoid exposure is tightly intertwined with the endocrine regulation of appetite, as well as orchestrating appropriate physiological response to stress [58]. This is hardly surprising given the importance of the regulation of energy and food intake under stress for survival [58].

Stress for survival takes on a different meaning in the contemporary world, but the mechanism for dealing with it is the same. Overeating is often a personal response to chronic life stress, and market liberalism, especially in countries such as the USA, UK and Australia, creates environments of great individual insecurity that cuts across socio-economic position, and this is the source of stress that drives higher levels of obesity in such countries [14]. The structures that neoliberal societies put in place promote insecurity and inequality, while work-related insecurity, including low income, poor job mobility and absence of union protection, raises the likelihood of stress and ill health. Responses to stress, in turn, include overeating and preferences for high-energy-density foods, both of which are implicated in the causation of obesity. Not everyone becomes obese, however, and the mechanism whereby insecurity and obesity are linked through overeating may well be binge eating, which at its extreme has been classified as the pathology of BED. Although it is not usually the place of clinical practice to question the ways in which society is run, it can at least engage with some of the health consequences of living in a neoliberal world. Stresses at work and in everyday life are both higher-level factors that structure health and illness of patients and communities, and downstream consequences of living insecure lives.

8.4 Future Directions

An important future direction would involve researchers and clinicians identifying the extent of stress-related binge eating. General practice researchers in the UK have made a strong case for brief interventions in primary care settings for weight management [59]. Furthermore, the National Obesity Observatory [60] has compiled good evidence that brief interventions can lead to at least short-term changes in weight-related behaviour and body weight if they meet a number of criteria. These are that they focus on both diet and physical activity; are delivered by practitioners trained in motivational interviewing; incorporate behavioural techniques, especially self-monitoring; are tailored to individual circumstances; and encourage the individual or patient to seek support from other people. From an evolutionary perspective, this seems strange and difficult: the major weight management task in prehistory was to keep body weight as high as possible. Binge eating, then and now, was and is a response to stress. To ask people to manage their weight, especially if the emphasis is placed on self-monitoring and individual responsibility, is to also ask them to manage their stress. It might be more effective if clinical practitioners could take the lead in linking awareness of stress and binge eating through their brief interventions.

Another future direction would be to identify how evolved predispositions to overeat could be located within frameworks that encourage increased physical activity, rather than reduced food intake and the self-control that this requires.

Glossary

Binge eating	When a larger than normal amount of food is consumed in one sitting
Disinhibition	An eating behaviour trait measuring a readiness for eating or opportunistic eating: a "see food and eat it" response. This trait is assessed through the Three-Factor Eating Questionnaire [16]
Dopamine reward system	Systems in the brain (mesolimbic pathway and mesocortical pathway in the ventral tegmental area); when these areas are stimulated, reward is perceived
Glucocorticoid	A corticoid substance which increases gluconeogenesis, increasing blood glucose levels. The main glucocorticoid is cortisol, which is responsible for regulating metabolism of protein, lipids and carbohydrates

Hedonic hunger	When an individual experiences frequent thoughts, feelings and urges towards food, while they have no physiological need for food
Hypothalamic–pituitary–adrenal axis	Feedback interactions among the hypothalamus, pituitary gland and the adrenal glands that control reactions to stress and, among other things, regulate digestion, mood, emotions and energy balance
Nutrition transition	Shifts in dietary consumption towards higher-energy-density foods, reduced physical activity and energy expenditure that have coincided with economic, demographic and epidemiological changes across the world
Psychological ambivalence	Describes a situation where an individual holds both positive and negative attitudes towards a food (e.g. I love the taste of this food, but I hate it as it makes me fat)
Three-Factor Eating Questionnaire	A psychometric tool to assess the eating behaviour traits of disinhibition (tendency to overeat and eating opportunistically), restraint (restricting intake to control body weight) and hunger (the extent to which feeling of hunger elicit eating episodes)
Thrifty Phenotype	The thrifty phenotype hypothesis suggests that early-life metabolic adaptations help in the survival of the organism by selecting an appropriate trajectory of growth in response to environmental cues (See also Chap. 6)

References

1. Stevens GA, Singh GM, Lu Y, Danaei G, Lin JK, Finucane MM, Bahalim AN, McIntire RK, Gutierrez HR, Cowan M, Paciorek CJ, Farzadfar F, Riley L, Ezzati M (2012) National, regional, and global trends in adult overweight and obesity prevalences. Popul Health Metrics 10:22
2. Rokholm B, Baker JL, Sorensen TIA (2010) The levelling off of the obesity epidemic since the year 1999—a review of evidence and perspectives. Obes Rev 11:835–846
3. Ogden CL, Carroll MD, Curtin LR, McDowell MA, Tabak CJ, Flegal KM (2006) Prevalence of overweight and obesity in the United States, 1999–2004. J Am Med Assoc 295:1549–1555
4. Wang Y, Beydoun MA (2007) The obesity epidemic in the United States—gender, age, socioeconomic, racial/ethnic, and geographic characteristics: a systematic review and meta-regression analysis. Epidemiol Rev 29:6–28
5. Ogden CL, Carroll MD (2010) Prevalence of overweight, obesity, and extreme obesity among adults: United States, trends 1960–1962 through 2007–2008. Centers of Disease Control: NCHS Health and Stats

6. Friedman DS (2011) Obesity—United States, 1988–2008. Centers for Disease Control and Prevention Morbidity and Mortality Weekly Report Supplements 60:73–77
7. Rennie KL, Jebb SA (2005) Prevalence of obesity in Great Britain. Obes Rev 6:11–12
8. Zaninotto P, Head J, Stamatakis E, Wardle H, Mindell J (2009) Trends in obesity among adults in England from 1993 to 2004 by age and social class and projections of prevalence to 2012. J Epidemiol Community Health 63:140–146
9. Stunkard AJ (1959) Eating patterns and obesity. Psychol Bull 33:284–294
10. Spitzer RL, Devlin M, Walsh BT, Hasin D, Wing R, Marcus M, Stunkard A, Wadden T, Yanovski S, Agrass S, Mitchell J, Nonas C (1992) Binge eating disorder: a multisite field trial of the diagnostic criteria. Int J Eat Disord 11:191–203
11. Spitzer RL, Yanovski S, Wadden T, Wing R, Marcus MD, Stunkard A, Devlin M, Mitchell J, Hasin D, Horne RL (1993) Binge eating disorder: Its further validation in a multisite study. Int J Eat Disord 13:137–153
12. Yanovski SZ (2003) Binge eating disorder and obesity in 2003: could treating an eating disorder have a positive effect on the obesity epidemic? Int J Eating Dis 34(S1):S117–S120
13. American Psychiatric Association (1994) Diagnostic and statistical manual of mental disorders, 4th edn. American Psychiatric Association, Washington DC
14. Offer A, Pechey R, Ulijaszek SJ (eds) (2012) Insecurity, inequality and obesity. The Clarendon Press, Oxford
15. Avena NM, Rada P, Hoebel BG (2009) Sugar and fat bingeing have notable differences in addictive-like behavior. J Nutr 139:623–628
16. Stunkard AJ, Messick S (1985) The three-factor eating questionnaire to measure dietary restraint, disinhibition and hunger. J Psychosom Res 29:71–83
17. Bryant E, King N, Blundell J (2008) Disinhibition: its effects on appetite and weight regulation. Obes Rev 9:409–419
18. Ulijaszek SJ, Mann N, Elton S (2012) Evolving human nutrition: implications for public health. Cambridge University Press, Cambridge
19. Elton S (2008) Environments, adaptation, and evolutionary medicine: Should we be eating a stone age diet? In: Elton S, O'Higgins P (eds) Medicine and evolution: current applications, future prospects. CRC Press, Boca Raton, pp 9–33
20. Elton S (2008) The environmental context of human evolutionary history in Eurasia and Africa. J Anat 212:377–393
21. Ulijaszek SJ (2002) Human eating behaviour in an evolutionary ecological context. Proc Nutr Soc 61:517–526
22. Bryant E (2006) Understanding disinhibition and its influences on eating behaviour and appetite. PhD dissertation, University of Leeds
23. Mailloux G, Bergeron S, Meilleur D, D'Antono B, Dube I (2014) Examining the associations between overeating, disinhibition and hunger in a nonclinical sample of college women. Int J Behav Med 21:375–384
24. Chaput JP, Depres JP, Bouchard C, Tremblay A (2011) The association between short sleep duration and weight gain is dependent on disinhibited eating behaviour in adults. Sleep 34:1291–1297
25. Niemeier HM, Phelan S, Fava JL, Wing RR (2007) Internal disinhibition predicts weight regain following weight loss and weight maintenance. Obesity 15:2485–2494
26. Bryant EJ, Kiezebrink K, King NA, Blundell JE (2010) Interaction between disinhibition and restraint: implications for body weight and eating disturbance. Eat Weight Disord 15:43–51
27. Spring B, Schneider K, Smith M, Kendzor D, Appelhans B, Hedeker D, Pagoto S (2008) Abuse potential of carbohydrates for overweight carbohydrate cravers. Psychopharmacology 197:637–647
28. Le Magnen J (1990) A role for opiates in food reward and food addiction. In: Capaldi PT (ed) Taste, experience, and feeding. American Psychological Association, Washington DC, pp 241–252
29. Teegarden SL, Bale TL (2007) Decreases in dietary preference produce increased emotionality and risk for dietary relapse. Biol Psychiatry 61:1021–1029

30. Arias-Carrion O, Stamelou M, Murillo-Rodriguez E, Menendez-Gonzalez M, Poppel E (2010) Dopaminergic reward system: a short integrative review. Int Arch Med 3:24
31. Volkow ND, Wang GJ, Fowler JS, Telang F (2008) Overlapping neuronal circuits in addiction and obesity: evidence of systems pathology. Philos Trans R Soc Lond B Biol Sci 363:3191–3200
32. Cordain L, Brand-Miller J, Eaton SB, Mann N, Holt SHA, Speth JD (2000) Macronutrient estimations in hunter-gatherer diets. Am J Clin Nutr 72(6):1589–1592
33. Cornier M-A, von Kaenel SS, Bessesen DH, Tregellas JR (2007) Effects of overfeeding on the neuronal response to visual food cues. Am J Clin Nutr 86:965–971
34. Bryant EJ, King NA, Falken Y, Hellstrom PM, Holst JJ, Blundell JE, Naslund E (2013) Relationships among tonic and episodic aspects of motivation to eat, gut peptides and weight before and after surgery. Surg Obes Relat Dis 9:802–808
35. Burton-Freeman BM, Kiem NL (2008) Glycemic index, cholecyctokinin, satiety and disinhibition: is there an unappreciated paradox for overweight women? Int J Obes 32:1647–1654
36. Geliebter A, Ladell T, Logan M, Schweider T, Sharafi M, Hirsch J (2006) Responsivity to food stimuli in obese and lean binge eaters using functional MRI. Appetite 46:31–35
37. Lowe MR, Butryn ML (2007) Hedonic hunger: a new dimension of appetite? Physiol Behav 91:432–439
38. Witt AA, Lowe MR (2014) Hedonic hunger and binge eating among women with eating disorders. Int J Eat Disord 47:273–280
39. Delparigi A, Chen K, Salbe AD, Reiman EM, Tataranni A (2005) Sensory experience of food and obesity: a positron emission tomography study of the brain regions affected by tasting a liquid meal after a prolonged fast. Neuroimage 24:436–443
40. Lee Y, Chong MFF, Liu JCJ, Libedinsky C, Gooley JJ, Chen S, Wu T, Tan V, Zhou M, Meaney MJ, Lee YS, Chee MWL (2013) Dietary disinhibition modulates neural valuation of food in ed and fasted states. Am J Clin Nutr 97(5):919–925
41. Schienle A, Scafer A, Hermann A, Vaitl D (2009) Binge-eating disorder: reward sensitivity and brain activation to images of food. Biol Psychiatry 65:654–661
42. Dowdeswell JA, White JWC (1995) Greenland ice core records and rapid climate change. Philos Trans Phys Sci Eng 352:359–371
43. Wisman JD, Capeheart HW (2012) Creative destruction, economic and security, stress, and epidemic obesity. In: Offer A, Pechey R, Ulijaszek SJ (eds) Insecurity, inequality and obesity. Oxford University Press, Oxford, pp 5–53
44. Ulijaszek SJ, Strickland SS (1993) Nutritional anthropology: prospects and perspectives in human nutrition. Smith-Gordon and Company Ltd, London
45. Ulijaszek SJ (1996) Energetics, adaptation, and adaptability. Am J Human Biol 8:169–182
46. Dunbar RIM (2012) Obesity: an evolutionary perspective. In: Offer A, Pechey R, Ulijaszek SJ (eds) Insecurity, inequality and obesity. Oxford University Press, Oxford, pp 5–53
47. Popkin BM (2009) The world is fat. The fads, trends, policies, and products that are fattening the human race. Avery, New York
48. Coveney JD (2000) Food, morals and meaning. The pleasure and anxiety of eating. Routledge, Abingdon
49. Lupton D (1996) Food, the body and the self. Sage Publications, London
50. Bellisle F, Dalix AM, Slama G (2004) Non food-related environmental stimuli induce increased meal intake in healthy women: comparison of television viewing versus listening to a recorded story in laboratory settings. Appetite 43:175–180
51. de Castro JM (1994) Family and friends produce greater social facilitation of food intake than other companions. Physiol Behav 56:445–455
52. Hetherington MM, Anderson AS, Norton GNM, Newson L (2006) Situational effects on meal intake: a comparison of eating alone and eating with others. Physiol Behav 88:498–505
53. de Garine I, Pollock N (eds) (1995) Social aspects of obesity and fatness. Gordon and Breach, New York

54. Maio GR, Haddock GG, Jarman HL (2007) Social psychological factors in tackling obesity. Obes Rev 8(Supplement 1):123–125
55. Ifland JR, Preuss HG, Marcus MT, Rourke KM, Taylor WC, Burau K, Jacobs WS, Kadish WM, Manso G (2009) Refined food addiction: a classic substance use disorder. Med Hypotheses 72:518–526
56. Reid R, Hackett AF (1999) Changes in nutritional status in adults over Christmas 1998. J Hum Nutr Diet 12:513–516
57. Hull HR, Radley D, Dinger MK, Fields DA (2006) The effect of the thanksgiving holiday on weight gain. Nutr J 5:29
58. Adam TC, Epel ES (2007) Stress, eating and the reward system. Physiol Behav 91:449–458
59. Lewis A, Jolly K, Adab P, Daley A, Farley A, Jebb S, Lycett D, Clarke S, Christian A, Jin J, Thompson B, Aveyard P (2013) A brief intervention for weight management in primary care: study protocol for a randomized controlled trial. Trials 14:393
60. Cavill N, Hillsdon M, Anstiss T (2011) Brief interventions for weight management. National Obesity Observatory, Oxford

Chapter 9
Evolutionary Aspects of the Dietary Omega-6/Omega-3 Fatty Acid Ratio: Medical Implications

Artemis P. Simopoulos, M.D.

Lay summary Modern agriculture, in particular agribusiness and its emphasis on production, changed animal feeds from grass to seeds and favoured the production of vegetable oils rich in ω-6 fatty acids. This has led to a pro-inflammatory diet characterized by an increased consumption of both linoleic acid (ω-6) and arachidonic acid (ω-6) and a decreased consumption of alpha-linoleic acid and its metabolites (ω-3). In the last 100–150 years, the absolute and relative change of ω-6 and ω-3 fatty acids in the food supply of Western societies led the Western diet to reach a ratio of 20:1 ω-6/ω-3. This ratio is at odds with human evolutionary history, during which a balance (1:1) existed between ω-6 and ω-3 fatty acids as ω-3 fatty acids were found in all foods consumed: meat, eggs, fish, wild plants, nuts, and berries. While a balance between the ω-6 and ω-3 fatty acids is consistent with human evolution and normal human development, a number of studies suggest that an imbalance lead to an increase in the risk of cardiovascular and other chronic diseases, particularly in those genetically predisposed. Therefore, we recommend that food labels distinguish between ω-6 and ω-3 fatty acids. In addition, in order to balance the consumption of ω-6 and ω-3 fatty acids, we advocate a decrease in the intake of vegetable oils high in ω-6 (corn oil, sunflower oil, safflower oil, cottonseed oil, and soybean oil) and an increase in the intake of oils both high in ω-3 fatty acids (flaxseed, perilla, chia, rapeseed, and walnuts) and low in ω-6 fatty acids (olive oil, macadamia nut oil, and hazelnut oil). Finally, we suggest people should eat more fish and less meat.

A.P. Simopoulos (✉)
The Center for Genetics, Nutrition and Health,
4330 Klingle Street NW, Washington, DC 20016, USA
e-mail: cgnh@bellatlantic.net

© Springer International Publishing Switzerland 2016
A. Alvergne et al. (eds.), *Evolutionary Thinking in Medicine*,
Advances in the Evolutionary Analysis of Human Behaviour,
DOI 10.1007/978-3-319-29716-3_9

119

9.1 Introduction

There are two classes of essential fatty acids (EFA), omega-6 (ω-6) and omega-3
(ω-3). The distinction between ω-6 and ω-3 fatty acids is based on the location of
the first double bond, counting from the methyl end of the fatty acid molecule
(Fig. 9.1). ω-6 and ω-3 fatty acids are essential because humans, like all mammals,
cannot synthesize them and must obtain them from the diet. ω-6 fatty acids are
represented by linoleic acid (LA) (18:2w6) and ω-3 fatty acids by alpha-linolenic
acid (ALA) (18:3w3). LA is plentiful in nature and is found in the seeds of most
plants except for coconut, cocoa, and palm. ALA, on the other hand, is found in the
chloroplasts of green leafy vegetables, and in the seeds of flax, rape, chia, perilla,
and walnuts. Both essential fatty acids are metabolized to longer-chain fatty acids of
20 and 22 carbon atoms. LA is metabolized to arachidonic acid (AA) (20:4w6),
while ALA is metabolized to eicosapentaenoic acid (EPA) (20:5w3) and docosahexaenoic acid (DHA) (22:6w3). This is achieved by increasing the chain length
and the degree of unsaturation by adding extra double bonds to the carboxyl end of
the fatty acid molecule [1] (Fig. 9.2). AA is found predominantly in the phospholipids of grain-fed animals, dairy, and eggs. EPA and DHA are found in the oils
of fish, particularly fatty fish.

In mammals, including humans, the cerebral cortex, retina, testis, and sperm are
particularly rich in DHA. DHA is one of the most abundant components of the
brain's structural lipids. DHA, like EPA, can be derived only from direct ingestion
or by synthesis from dietary EPA or ALA: Humans and other mammals, except for
certain carnivores such as lions, can convert LA to AA and ALA to EPA and DHA,
although it is slow [2, 3]. There is competition between ω-6 and ω-3 fatty acids for
the desaturation enzymes. Both fatty acid desaturase 1 (D-5) and fatty acid

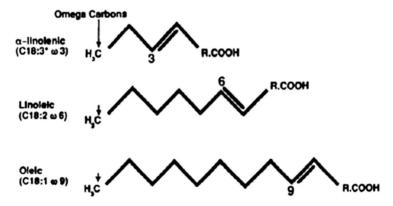

Fig. 9.1 Structural formulas for ω-3 (a-linoleic = ALA), ω-6 (linoleic = LA), and ω-9 (oleic) fatty
acids. The first number (*before the colon*) gives the number of carbon atoms in the molecule, and
the second gives the number of double bonds. ω-3, ω-6, and ω-9 indicate the position of the first
double bond in a given fatty acid molecule [1]

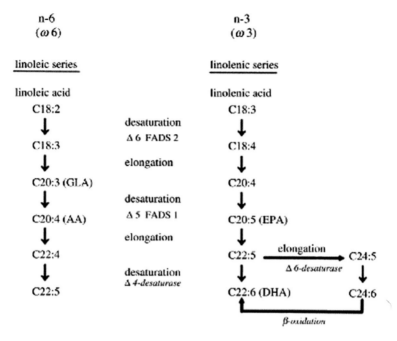

Fig. 9.2 Desaturation and elongation of ω-3 and ω-6 fatty acids by the enzymes fatty acid desaturases FADS2 (D6) and FADS1 (D5) [10]

desaturase 2 (D-6) prefer ωALA to ωLA [2, 4, 5]. However, a high LA intake, such as that characterizing Western diets, interferes with the desaturation and elongation of ALA [3–6]. Similarly, trans-fatty acids interfere with the desaturation and elongation of both LA and ALA.

In addition, fatty acid desaturase 2 is the limiting enzyme and there is some evidence that it decreases with age [2]. Premature infants [7], hypertensive individuals [8], and some diabetics [9] are limited in their ability to make EPA and DHA from ALA. These findings are important and need to be considered when making dietary recommendations.

Mammalian cells cannot convert ω-6 to ω-3 fatty acids because they lack the converting enzyme, ω-3 desaturase. Ω-6 and ω-3 fatty acids are not interconvertible, are metabolically and functionally distinct, and often have important opposing physiological effects; therefore, their balance in the diet is important. When humans ingest fish or fish oil, the EPA and DHA from the diet partially replace the ω-6 fatty acids, especially AA, in the membranes of probably all cells, but especially in the membranes of platelets, erythrocytes, neutrophils, monocytes, and liver cells (reviewed in Refs. [10, 11]). AA and EPA are two alternative parent compounds for eicosanoid production (Fig. 9.3). Because of the increased amounts of ω-6 ω in the Western diet, the eicosanoid metabolic products from AA, specifically prostaglandins, thromboxanes, leukotrienes, hydroxy fatty acids, and lipoxins, are

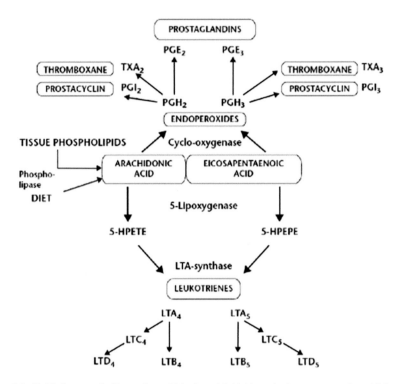

Fig. 9.3 Oxidative metabolism of arachidonic acid (AA) and eicosapentaenoic acid by the cyclooxygenase and 5-lipoxygenase pathways. 5-HPETE denotes the 5-hydroperoxyeicosatetranoic acid, and 5-HPEPE denotes the 5-hydroxyeicosapentaenoic acid [10]

formed in larger quantities than those derived from ω-3 fatty acids, specifically EPA (Fig. 9.3) [10]. The eicosanoids from AA are biologically active in very small quantities and, if they are formed in large amounts, they contribute to the formation of thrombus and atheromas; to allergic and inflammatory disorders, particularly in susceptible people; and to proliferation of cells [12] (Table 9.1). Thus, a diet rich in ω-6 fatty acids shifts the physiological state to one that is proinflammatory, pro-thrombotic, and proaggregatory, with increases in blood viscosity, vasospasm, and vasoconstriction.

9.1.1 The Importance of the ω-6/ω-3 Fatty Acid Ratio: And Evolutionary Aspects

A balance existed between ω-6 and ω-3 fatty acids during the long evolutionary history of the genus Homo [13]. During evolution, ω-3 fatty acids were found in all foods consumed: particularly meat, fish, wild plants, nuts, and berries [13–29]. Recent studies by Cordain et al. [30] on the composition of the meat of wild animals

Table 9.1 Effects of ingestion of EPA and DHA from fish or fish oil on eicosanoids

Decreased production of prostaglandin E_2 (PGE$_2$) metabolites
A decrease in thromboxane A_2, a potent platelet aggregator and vasoconstrictor
A decrease in leukotriene B_4 formation, an inducer of inflammation, and a powerful inducer of leukocyte chemotaxis and adherence
An increase in thromboxane A_3, a week platelet aggregator and weak vasoconstrictor
An increase in prostacyclin PGI$_3$ leading to an overall increase in total prostacyclin by increasing PGI$_3$ without a decrease in PGI$_2$, and both PGI$_2$ and PGI$_3$ are active vasodilators and inhibitors of platelet aggregation
An increase in leukotriene B_5, a weak inducer of inflammation and a weak chemotactic agent

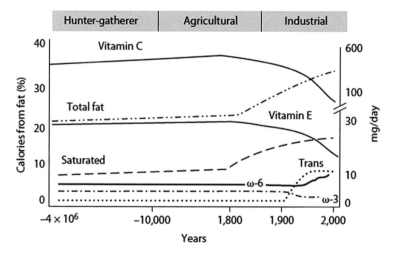

Fig. 9.4 Hypothetical scheme of fat, fatty acid (ω6, ω3, trans, and total) intake (as percent of calories from fat), and intake of vitamins E and C (mg/d). Data were extrapolated from cross-sectional analyses of contemporary hunter-gatherer populations and from longitudinal observations and their putative changes during the preceding 100 years [10]

confirm the original observations of Crawford [15] and Sinclair et al. [31]. However, rapid dietary changes over short periods of time as have occurred over the past 100–150 years is a totally new phenomenon in human evolution (Fig. 9.4). A balance between the ω-6 and ω-3 fatty acids is a physiological state that is less inflammatory in terms of gene expression [32], prostaglandin and leukotriene metabolism, and interleukin-1 (IL-1) production [10]. The current recommendation to substitute vegetable oils (ω-6) for saturated fats leads to increases in IL-1, prostaglandins, and thromboxanes of the 2 series from LA that are prothrombotic and leukotrienes of the 3 series that are proinflammatory. It is not consistent with human evolution and may lead to maladaptation in those genetically predisposed (Table 9.1; Figs. 9.2 and 9.3).

Prior to the agricultural revolution, humans ate a wide variety of wild plants, whereas today about 17 % of plant species provide 90 % of the world's food supply, with the greatest percentage contributed by cereal grains [33]. Agribusiness and modern agriculture contributed to the decrease in ω-3 fatty acids in animal carcasses. Wild animals and birds who feed on wild plants are very lean, with a carcass fat content of only 3.9 % [34], and contain about five times more polyunsaturated fatty acids (PUFAs) per gram than those found in domestic live-stock [15, 16]. Most importantly, 4 % of the fat of wild animals contains EPA. Domestic beef contains very small or undetectable amounts of ALA because cattle are fed grains rich in ω-6 fatty acids and poor in ω-3 fatty acids [17], whereas deer that forage on ferns and mosses contain more ω-3 fatty acids in their meat.

Although diets differed in the various geographic areas [35], a number of investigators including Crawford [15], Cordain et al. [30], Eaton et al. [36], and Kupiers et al. [37] have shown that during the Paleolithic period, the diets of humans included equal amounts of ω-6 and ω-3 fatty acids from both plants (LA + ALA) and from the fat of animals in the wild and fish (AA + EPA + DHA). Recent studies by Kuipers et al. [37] estimated macronutrient and fatty acid intakes from an East African Paleolithic diet in order to reconstruct multiple Paleolithic diets and thus estimated the ranges of nutrient intakes on which humans evolved. They found (range of medians in energy %) intakes of moderate-to-high protein (25–29), moderate-to-high fat (30–39), and moderate carbohydrates (39–40). Just as others have concluded previously, Kuipers et al. [37] stated "compared with Western diets, Paleolithic diets contained consistently high-protein and long-chain PUFA and lower LA." Guil-Guerrero et al. [38] determined the fat composition of frozen mammoths (from 41,000 to 34,000 years BP), Bisons from Early Holocene (8200–9300 years BP) and horses from Middle Holocene (4600–4400 years BP), often consumed by Paleolithic/Neolithic hunters-gatherers, and concluded. It "contained suitable amounts of ω-3 and ω-6 fatty acids, possibly in quantities sufficient to meet today's dietary requirements for good health." The elucidation of sources of ω-3 fatty acids available for the humans who lived in the Paleolithic and Neolithic is highly relevant to ascertain the availability of nutrients at that time and to draw conclusions about healthy dietary habits for present-day humans. As in previous studies, the amount of ALA was higher than LA in the fat of the frozen specimens [39, 40] (Tables 9.2 and 9.3).

Modern agriculture, by changing animal feeds as a result of its emphasis on production, has decreased the ω-3 fatty acid content in many foods: green leafy vegetables, animal meats, eggs, and even fish [18–21]. Foods from edible wild plants contain a good balance of ω-6 and ω-3 fatty acids. Purslane, a wild plant, in comparison with spinach, red leaf lettuce, buttercrunch lettuce, and mustard greens, has eight times more ALA than the cultivated plants [25]. Modern aquaculture produces fish that contain less ω-3 fatty acids than do fish grown naturally in the ocean, rivers, and lakes [20]. The fatty acid composition of egg yolk from free-ranging chicken has an ω-6:ω-3 ratio of 1.3, whereas the US Department of

Table 9.2 Estimated ω-3 and ω-6 fat acid intake in the Late Paleolithic period

Plants	
LA	4.28
ALA	11.40
Animals	
LA	4.56
ALA	1.21
Total	
LA	8.84
ALA	12.60
Animals	
AA (ω6)	1.81
EPA (ω3)	0.39
DTA (ω6)	0.12
DPA (ω3)	0.42
DHA (ω3)	0.27
Ratios of ω6/ω3	
LA/ALA	0.70
AA + DTA/EPA + DPA + DHA	1.79
Total ω6/ω3	0.79

Data from Eaton et al. [39]: assuming an energy intake of 35:65 of animal: plant sources

Table 9.3 Ω-6 Ω-3 ratios in different populations

Population	ω-6/ω-3
Paleolithic	0.79
Greece prior to 1960	1.00–2.00
Current Japan	4.00
Current India, rural	5–6.1
Current UK and northern Europe	15.00
Current USA	16.74
Current India, urban	38–50

Adapted from Ref. [10]

Agriculture (USDA) egg has a ratio of 19.9 [21]. By enriching the chicken feed with fishmeal or flaxseed, the ratio of ω-6:ω-3 decreased to 6.6 and 1.6, respectively.

9.1.2 Genetic Adaptation of Fatty Acid Metabolism

There are important genetic variables in fatty acid biosynthesis involving fatty acid desaturase 1 (FADS1) and fatty acid desaturase 2 (FADS2), which encode rate-limiting enzymes for fatty acid metabolism (Fig. 9.2). Ameur et al. [41]

performed genome-wide genotyping ($n = 5,652$ individuals) of the FADS region in five European population cohorts and analyzed available genomic data from human populations, archaic hominins, and more distant primates. Their results show that present-day humans have two common FADS haplotypes A and D that differ dramatically in their ability to generate LC-PUFAs. The most common haplotype, denoted haplotype D, was associated with high lipid levels ($P = 1 \times 10^{-65}$), whereas the less common haplotype (haplotype A) was associated with low levels ($P = 1 \times 10^{-52}$). The haplotype D associated with the enhanced ability to produce AA and DHA from their precursors LA and ALA, respectively, is specific to humans and has appeared after the split of the common ancestor of humans and Neanderthals. This haplotype shows evidence of a positive selection in African populations in which it is presently almost fixed and it is less frequent outside of Africa. Haplotype D provides a more efficient synthesis of LC-PUFAs, and in today's high LA ω-6 dietary intake from vegetable oils, it leads to increased synthesis of AA from LA. Thus, it represents a risk factor for coronary heart disease (CHD), cancer, obesity, diabetes, and the metabolic syndrome, thus adding further to health disparities in populations of African origin living in the West, in addition to lower socioeconomic status.

As estimated from the human genome diversity panel (HGDP-CEPH) [42], the geographic distributions of haplotypes A and D differ dramatically among continents. In African populations, HGDP populations haplotype A is essentially absent (has a frequency of 1 %), whereas in Europe, West, South and East Asia, and Oceania, it occurs at a frequency of 25–50 %. Among the 126 Native Americans included in HGDP, haplotype A occurs at a frequency of 97 %. Among individuals of African ancestry, 49 % carry mixed FADS haplotypes with a higher resemblance to haplotype D than to haplotype A, consistent with a decay of haplotype D by recombination in African populations.

It has been proposed that a shift in diet, characterized by access to food sources that are rich in essential LC-PUFAs, was initiated about 2 million years ago [43]. This change in the availability of LC-PUFAs might have been important for maintaining the proportionally large hominoid brain relative to body size. Humans use a very large portion of dietary fats, predominantly AA and DHA to feed the brain [43]. Consequently, human's ability to more efficiently synthesize LC-PUFAs from their precursors might have played an important role in their ability to survive in periods where AA and DHA rich diets were not available. Therefore, haplotype D is likely to have been advantageous to humans living in environments with limited access to these fatty acids, and this could explain the positive selection for haplotype D in African populations. However, as the present Western diet is high in LA due to the elevated intake of vegetable oils high in ω-6 fatty acids, the advantage of having faster biosynthesis of LC-PUFAs for carriers of haplotype D has turned into a disadvantage: High intake of LA leads to an increased production of AA and thereby increases the synthesis of AA-derived pro-inflammatory eicosanoids, which are associated with the increased risk of atherosclerotic vascular disease [44].

9.1.3 Research Findings

9.1.3.1 The Diet of Crete and Its Relation to Cardiovascular Disease and Cancer

Over the past 15 years, a number of animal experiments, epidemiological investigations, and double-blind controlled clinical trials have confirmed the hypotriglyceridemic, anti-inflammatory, and antithrombotic aspects of ω-3 fatty acids [45–50] as well as the essentiality of ω-3 fatty acids, particularly DHA, for the development of retina and brain in the premature infant. It therefore became important to investigate the ω-3 fatty acid composition of diets that have been shown to be associated with a decreased rate of cardiovascular disease and cancer. Such an opportunity then arose with the diet of Crete [51].

The population of Crete was one of the populations participating in the Seven Countries Study [52]. The others were the populations of the former Yugoslavia, Italy, the Netherlands (Zutphen), Finland, USA, and Japan. The results of the Seven Countries Study are interesting because they show that the population of Crete had the lowest rate of cardiovascular disease and cancer, followed by the population of Japan. The investigators concluded that the reason must be the high olive oil intake and the low saturated fat intake of the "Mediterranean diet." The fact that Crete had a high-fat diet, 37 % of energy from fat, and Japan had a low-fat diet, 11 % of energy from fat, was not very much discussed nor were any other fatty acids considered, despite the fact that the people of Crete ate 10 times more fish than the US population. Furthermore, the people of Crete ate plenty of vegetables, fruits, nuts, and legumes, all rich sources of folate, calcium, vitamins, and minerals. In addition, since the meat came from animals that grazed, rather than being grain-fed, it contained ω-3 PUFAs as did their milk and milk products, such as cheese. The population of Crete ate snails about three times per week throughout the year. Renaud has shown that the snails of Crete and Greece contain more ω-3 fatty acids and less ω-6 fatty acids than the snails of France (personal communication). Figs are the most popular fruit eaten throughout the year. Both fresh and dried figs contain equal amounts of LA + ALA and are rich in vitamins and minerals, especially calcium.

The traditional Greek diet, including the diet of Crete, includes wild plants. Wild plants are rich sources of ω-3 fatty acids and antioxidants [19, 22, 25]. A commonly eaten plant is purslane. Purslane is rich in ALA (400 mg/100 g) as well as in vitamin E (12 mg/100 g) [25]. In Crete and Greece, purslane is eaten fresh in salads, soups, and omelets, or cooked with poultry, and during the winter months, the dried purslane is used in soups, vegetable pies, and as a tea for sore throats and earaches. It is highly recommended for pregnant and lactating women and for patients with diabetes. The purslane study was just the beginning in our involvement in a series of studies that investigated the ω-3 fatty acids in the Greek diet under conditions similar to those prior to 1960 [19, 21, 22, 25]. In the Greek countryside, chickens wander in the farm, eat grass, purslane, insects, worms, and dried figs, all good sources [19, 21, 22, 25] of ω-3 fatty acids. As mentioned earlier, their eggs have a ratio of ω-6/ω-3 fatty acids of 1.3, whereas the USDA egg has a ratio of 19.4.

Similarly, Greek cheeses contain ω-3 fatty acids, whereas American cheeses do not. Noodles made with milk and eggs in Greece also contain ω-3 fatty acids. A pattern thus began to unfold. The diet of Greece, including Crete prior to 1960, contained ω-3 fatty acids in every meal—breakfast, lunch, dinner, and snacks. Figs stuffed with walnuts are a favorite snack. Both figs and walnuts contain ω-3 fatty acids. Contrast this snack with a chocolate chip cookie, which contains trans-fatty acids and ω-6 fatty acids from the partially hydrogenated oils used in preparation [28]. While these studies were carried out between 1984 and 1986, further analyses of blood specimens from the Seven Countries Study [52] were published in 1993 by Sandker et al. [53], indicating that the serum cholesteryl esters of the population in Crete had threefold the amount of ALA as compared to the population of Zutphen in the Netherlands. Similar data indicated that the Japanese population also had higher concentrations of ω-3 fatty acids than that of the Netherlands. Here was the missing link. It was the higher concentrations of ω-3 fatty acids and the lower concentration of ω-6 fatty acids leading to a balanced ω-6/ω-3 that added protection for cardiovascular disease. This is in addition to the olive oil, wine, fruits, and vegetables of the diet of Crete, which is superior to other Mediterranean diets due to a balanced ω-6 and ω-3 fatty acid intake and high antioxidants (Table 9.3). Participants from the two countries (Crete and Japan) with the lowest CHD in the Seven Countries Study had a higher intake of fish and ALA. The Japanese obtained ALA from perilla oil and soybean oil and the population of Crete from purslane, other wild plants, walnuts, and figs. Additional studies showed that the population of Crete not only had higher serum cholesteryl ester levels of ALA but also lower LA (18:2w-6) [53].

Renaud had been working with ALA and shown that it decreases platelet aggregation [54]. Everything seems now to fall into place in terms of defining the characteristics of the diet of the population of Crete. Their diet was very similar to the Paleolithic diets in composition. It was low in saturated fat, balanced in the essential fatty acids (ω-6 and ω-3) [40] (Table 9.3), very low in trans-fatty acids, and high in vitamins E and C. This diet formed the basis of the diet used by Renaud and de Lorgeril in their now famous Lyon Study [51, 54–56]. The Lyon study is a prospective randomized single-blinded secondary prevention trial that compared the effects of a modified Crete diet enriched with ALA to those of a Step I American Heart Association Diet. A total of 605 patients were divided into two groups, 302 in the experimental group were fed the modified diet of Crete, including 2 g of ALA per day, and 303 in the control group followed the Step I American Heart Association diet. Two months after randomization, plasma levels of vitamins C and E ($P < 0.5$) and ω-3 fatty acids ($P < 0.001$) were higher in experimental subjects while those of ω-6 fatty acids were lower ($P < 0.001$). The ratio of ω-6/ω-3 fatty acids was 4/1. After a mean follow-up of 27 months, there were 16 cardiac deaths in the control group and three in the experimental group, and 17 non-fatal myocardial infarctions in the control group and five in the experimental group. The combined risk ratio for these two main end points was 0.27 (95 % CI 0.12–0.59, $P = 0.001$)

after adjustment for prognostic variables. Overall, mortality was 20 in the control group, eight in the experimental group, and an adjusted risk ratio of 0.30 (95 % CI 0.11–0.82, $P = 0.02$).

This study showed a decrease in death rate by 70 % in the experimental group and clearly showed that a modified Crete diet low in butter and meats, such as deli products but high in fish and fruits and vegetables, and enriched with ALA, is more efficient than that of the American Heart Association or similar prudent diets in the secondary prevention of coronary events and death. The same subjects were followed for 5 years. At 4 years of follow-up, de Lorgeril et al. [57] reported that the reduction of risk in the experimental subjects compared with control subjects was 56 % for total deaths, and 61 % for cancers, indicating that a modified diet of Crete is associated with lower risk for CHD and cancer.

The diet of Crete or the traditional diet of Greece resembles hunter-gatherers' diets in terms of antioxidants, saturated fat, and monounsaturated fat and in the ω-6/ω-3 ratio. The Lyon Heart Study and subsequently the Singh et al. study [58] support the importance of having a diet consistent with human evolution. Western diets today deviate from Paleolithic diets and are associated with high rates of cardiovascular disease, diabetes, obesity, and cancer. In as much as the health of the individual and the population in general is the result of the interaction between the genetic profile and the environment, nutrition is one of the most important environmental factors [59].

9.1.3.2 Other Beneficial Effects of ω-3 Fatty Acids

In addition to the studies of CHD and Cancer, a number of clinical intervention studies have shown that the high ω-6/ω-3 ratio in Western diets has increased the risk of many of the chronic "diseases of civilization" as a result of a "mismatch" between our genes and the environment. Clinical intervention studies in patients with asthma [60], osteoporosis [61, 62], mental health and major depression [63], bipolar disorder [64, 65], and mood [66] have shown improvements upon treatment with EPA + DHA. A number of studies show that LA increases low-density lipoprotein oxidation and the severity of coronary atherosclerosis and inhibits EPA incorporation from dietary fish oil supplements in human subjects, whereas decreasing LA in the diet while maintaining constant ALA increases EPA in plasma phospholipids in healthy men. A lower ω-6/ω-3 ratio as part of a Mediterranean diet decreases vascular endothelial growth factor. As the ω-6/ω-3 ratio decreases, so does the platelet aggregation. The higher the ratio of ω-6/ω-3 fatty acids in platelet phospholipids, the higher the death rate from cardiovascular disease. A high plasma ω-6/ω-3 ratio increases inflammatory markers, thus increasing the risk of chronic diseases. EPA and DHA attenuate the rate of shortening of telomere length suggesting a decrease in the rate of the aging process [67], whereas LA intake is associated with shorter telomere length [68].

9.1.4 Implications for Policy and Practice

Studies on the evolutionary aspects of diet suggest that during evolution, ω-3 fatty acids were present in practically all foods that humans ate and in equal or higher amounts than the ω-6 fatty acids. Western diets are characterized by high ω-6 and low ω-3 fatty acid intake, whereas during most of the human evolutionary history, there was a balance between ω-6 and ω-3 fatty acids. Today, human beings live in a nutritional environment that differs from that for which their genetic constitution was selected. The balance of ω-6/ω-3 fatty acids is an important determinant in maintaining homeostasis, normal development, and mental health throughout the life cycle. Excessive amounts of ω-6 PUFA and a very high ω-6/ω-3 ratio, as is found in today's western diets, promote the pathogenesis of many diseases, including cardiovascular disease, cancer, and inflammatory and autoimmune diseases, and interfere with normal brain development.

Diets must be balanced regarding ω-6 and ω-3 fatty acids to be consistent with the evolutionary understanding of the human diet. This balance can best be accomplished by decreasing the intake of oils rich in ω-6 fatty acids (corn oil, sunflower, safflower, cottonseed, and soybean) and increasing the intake of oils rich in ω-3s (canola, flaxseed, perilla, and chia) and olive oil which is particularly low in ω-6 fatty acids and high in monounsaturated fatty acids.

The ratio of ω-6 and ω-3 fatty acids in the brain is between 1:1 and 2:1, which is in agreement with the data from the evolutionary aspects of diet, genetics, and the studies with the fat-1 transgenic animal model [69, 70]. Therefore, a ratio of 1:1 to 2:1 ω-6 to ω-3 fatty acids should be the target ratio for health [10, 71]. Because chronic diseases are multigenic and multifactorial, it is quite possible that the therapeutic dose of the ω-3 fatty acids will vary between individuals: It will depend on the degree of severity of diseases, which in turn results from the genetic predispositions, and notably the endogenous metabolism of LA and ALA.

In the Lyon Heart Study, a ratio of 4:1 LA:ALA decreased total mortality by 70 % in patients with one episode of myocardial infarction. Whether an ω-6/ω-3 ratio of 3:1 to 4:1 could prevent the pathogenesis of many diseases induced by today's Western diets, a target of 1:1 to 2:1 appears to be consistent with studies on evolutionary aspects of diet, neurodevelopment, and genetics. The ω-6/ω-3 fatty acid ratio in red cell membrane phospholipids could be used as a biomarker for dietary intake and endogenous metabolism, thus providing a more accurate nutritional status for dietary recommendations.

It is essential that food labels distinguish between ω-3 and ω-6 fatty acids instead of the current label that distinguishes only saturated fatty acids (SFAs), monounsaturated fatty acids (MONOs), and polyunsaturated fatty acids (PUFAs). Regulatory agencies should follow the scientific advances in forming dietary regulations and recommendations.

Clinical intervention studies should include complete fatty acid information of the background diet as well as fatty acid levels in red cell membrane phospholipids. Discrepancies in the results of clinical studies are often due to inadequate data on fatty acid levels before, during, and at the completion of the studies.

Glossary

Essential fatty acids (EFA)	Fatty acids that human and other animals must ingest because the body requires them for good health but cannot synthesize them
Polyunsaturated fatty acids (PUFAs)	Fatty acids that contain more than one double bond in their backbone
Linoleic acid (LA)	An EFA ω-6 fatty acid found in corn, sunflower, cottonseed, safflower, soybean, and other vegetable oils
Alpha-linolenic acid (ALA)	An EFA ω-3 fatty acid found in nuts, rapeseed, flaxseed, perilla, canola, and chia oils
Arachidonic acid (AA)	A PUFA ω-6 fatty acid
Eicosapentaenoic acid (EPA)	A PUFA ω-3 fatty acid
Docosahexaeboic acid (DHA)	A PUFA ω-3 fatty acid that is a primary structural component of the human brain, cerebral cortex, skin, sperm, testicles, and retina
Fatty acid desaturase 1 (FADS1)	An enzyme that in humans is encoded by the FADS1 gene
Fatty acid desaturase 2 (FADS2)	An enzyme that in humans is encoded by the FADS2 gene
Haplotype	The group of genes that a progeny inherits from one parent

References

1. Simopoulos AP (1991) Ω-3 fatty acids in health and disease and in growth and development. Am J Clin Nutr 54:438–463
2. de Gomez Dumm INT (1975) Brenner RR. Oxidative desaturation of alphalinolenic, linoleic, and stearic acids by human liver microsomes. Lipids 10:315–317
3. Emken EA, Adlof RO, Rakoff H, Rohwedder WK (1989) Metabolism of deuterium-labeled linolenic, linoleic, oleic, stearic and palmitic acid in human subjects. In: Baillie TA, Jones JR (eds) Synthesis and application of isotopically labeled compounds 1988. Elsevier Science Publishers, Amsterdam, pp 713–716

4. Hague TA, Christoffersen BO (1984) Effect of dietary fats on arachidonic acid and eicosapentaenoic acid biosynthesis and conversion to C22 fatty acids in isolated liver cells. Biochim Biophys Acta 796:205–217

5. Hague TA, Christoffersen BO (1986) Evidence for peroxisomal retroconversion of adrenic acid (22:4n6) and docosahexaenoic acid (22:6n3) in isolated liver cells. Biochim Biophys Acta 875:165–173

6. Indu M, Ghafoorunissa (1992) n-3 fatty acids in Indian diets—comparison of the effects of precursor (alpha-linolenic acid) vs product (long chain n-3 polyunsaturated fatty acids). Nutr Res 12:569–582

7. Carlson SE, Rhodes PG, Ferguson MG (1986) Docosahexaenoic acid status of preterm infants at birth and following feeding with human milk or formula. Am J Clin Nutr 44:798–804

8. Singer P, Jaeger W, Voigt S, Theil H (1984) Defective desaturation and elongation of n-6 and n-3 fatty acids in hypertensive patients. Prostaglandins Leukot Med 15:159–165

9. Honigmann G, Schimke E, Beitz J, Mest HJ, Schliack V (1982) Influence of a diet rich in linolenic acid on lipids, thrombocyte aggregation and prostaglandins in type I (insulin-dependent) diabetes. Diabetologia 23:175 (abstract)

10. Simopoulos AP (2008) The importance of the omega-6/omega-3 fatty acid ratio in cardiovascular disease and other chronic diseases. Exp Biol Med (Maywood). 233(6):674–688

11. Simopoulos AP (1999) Essential fatty acids in health and chronic disease. Am J Clin Nutr 70 (suppl):560S–569S

12. Simopoulos AP (2011) The importance of the Ω-6/Ω-3 Balance in health and disease: evolutionary aspects of diet. In: Simopoulos AP (ed) Healthy agriculture, healthy nutrition, healthy people, vol 102. World Rev Nutr Diet. Karger, Basel, pp 10–21

13. Eaton SB, Konner M (1985) Paleolithic nutrition. A consideration of its nature and current implications. New Engl J Med 312:283–289

14. Simopoulos AP (1998) Overview of evolutionary aspects of w3 fatty acids in the diet. In: Simopoulos AP (ed) The return of w-3 fatty acids into the food supply. I. Land-based animal food products and their health effects. World Rev Nutr Diet. Karger, Basel, pp 1–11

15. Crawford MA (1968) Fatty acid ratios in free-living and domestic animals. Lancet 1:1329–1333

16. Wo CKW, Draper HH (1975) Vitamin E status of Alaskan Eskimos. Am J Clin Nutr 28:808–813

17. Crawford MA, Gale MM, Woodford MH (1969) Linoleic acid and linoleic acid elongation products in muscle tissue of Syncerus caffer and other ruminant species. Biochem J 115:25–27

18. Raper NR, Cronin FJ, Exler J (1992) Ω-3 fatty acid content of the US food supply. J Am College Nutr 11:304

19. Simopoulos AP, Salem N Jr (1986) Purslane: a terrestrial source of ω-3 fatty acids. N Engl J Med 315:833

20. van Vliet T, Katan MB (1990) Lower ratio of n-3 to n-6 fatty acids in cultured than in wild fish. Am J Clin Nutr 51:1–2

21. Simopoulos AP, Salem N Jr (1992) Egg yolk as a source of long-chain polyunsaturated fatty acids in infant feeding. Am J Clin Nutr 55:411–414

22. Simopoulos AP, Norman HA, Gillapsy JE et al (1992) Common purslane: a source of ω-3 fatty acids and antioxidants. J Am College Nutr 11:374–382

23. Simopoulos AP (1989) Nutrition and fitness. JAMA 261:2862–2863

24. Simopoulos AP, Salem N Jr (1989) n-3 Fatty acid in eggs from range-fed Greek chickens. N Engl J Med 321:1412

25. Simopoulos AP, Norman HA, Gillapsy JE (1995) Purlsane in human nutrition and its potential for world agriculture. World Rev Nutr Diet 77:7–74

26. Simopoulos AP (1995) Trans-fatty acids. In: Spiller GA (ed) Handbook of lipids in human nutrition. CRC Press, Florida, pp 91–99

27. Dupont J, White PJ, Feldman EB (1991) Saturated and hydrogenated fats in food in relation to health. J Am College Nutr 10:577–592

28. Litin L, Sacks F (1993) Trans-fatty acid content of common foods. N Engl Med 329:1969–1970
29. Guil JL, Torija ME, Gimenez JJ et al (1996) Identification of fatty acids in edible wild plants by gas chromatography. J Chromatog A 719:229–235
30. Cordain L, Martin C, Florant G, Watkins BA (1998) The fatty acid composition of muscle, brain, marrow and adipose tissue in elk: evolutionary implications for human dietary lipid requirements. World Rev Nutr Diet 83:225–226
31. Sinclair AJ, Slattery WJ, O'Dea K (1982) The analysis of polyunsaturated fatty acids in meat by capillary gas-liquid chromatography. J Food Sci Agri 33:771–776
32. Simopoulos AP (1996) The role of fatty acids in gene expression: health implications. Ann Nutr Metab 40:303–311
33. Simopoulos AP (ed) (1995) Plants in human nutrition. World Rev Nutr Diet 77
34. Ledger HP (1968) Body composition as a basis for a comparative study of some East African animals. Symp Zool Soc London 21:289–310
35. Elton S (2008) Environments, adaptation, and evolutionary medicine: should we be eating a stone age diet? In: Elton S, O'Higgins P (eds) Medicine and evolution: current applications, future prospects. CRC Press, Boca Raton, pp 9–34
36. Eaton SB, Konner M, Shostak M (1988) Stone agers in the fast lane: chronic degenerative diseases in evolutionary perspective. Am J Med 84:739–749
37. Kuipers RS et al (2010) Estimated macronutrient and fatty acid intakes from an East African Paleolithic diet. Br J Nutr 104(11):1666–1687
38. Guil-Guerrero JL et al (2014) The fat from frozen mammals reveals sources of essential fatty acids suitable for Palaeolithic and Neolithic humans. PLoS ONE 9(1):e84480
39. Eaton SB, Eaton SB 3rd, Sinclair AJ, Cordain L, Mann NJ (1998) Dietary intake of long-chain polyunsaturated fatty acids during the Paleolithic. In: Simopoulos AP (ed) The return of w-3 fatty acids into the food supply. I. Land-based animal food products and their health effects, vol 83. World Rev Nutr Diet. Karger, Basel, pp 12–23
40. Simopoulos AP (2003) Importance of the ratio of ω-6/ω-3 essential fatty acids: evolutionary aspects. World Rev Nutr Diet 92:1–22
41. Ameur A et al (2012) Genetic adaptation of fatty-acid metabolism: a human-specific haplotype increasing the biosynthesis of long-chain Ω-3 and Ω-6 fatty acids. Am J Hum Genet 90:809–820
42. Li JZ, Absher DM, Tang H, Southwick AM, Casto AM, Ramachandran S, Cann HM, Barsh GS, Feldman M, Cavalli-Sforza LL, Myers RM (2008) Worldwide human relationships inferred from genome-wide patterns of variation. Science 319:1100–1104
43. Leonard WR, Snodgrass JJ, Robertson ML (2010) Evolutionary perspectives on fat ingestion and metabolism in humans. In: Montmighteur JP, Le Coutre J (eds) Fat detection: taste, texture, and post ingestive effects. CRC Press, Boca Raton, FL
44. Martinelli N, Consoli L, Olivieri O (2009) A 'desaturase hypothesis' for atherosclerosis: Janus-faced enzymes in ω-6 and ω-3 polyunsaturated fatty acid metabolism. J Nutrigenet Nutrigenomics 2:129–139
45. Simopoulos AP, Kifer RE, Martin RR (eds) (1986) Health effects of polyunsaturated fatty acids in seafoods, proceedings from the conference. Academic Press, Orlando
46. Galli C, Simopoulos AP (eds) (1989) Dietary w3 and w6 fatty acids: biological effeots and nutritional essentiality. Plenum Publishing Corporation, New York
47. Simopoulos AP, Kifer RR, Martin RE, Barlow SM (eds) Health effects of w3 polyunsaturated fatty acids in seafoods. World Rev Nutr Diet 66
48. Galli C, Simopoulos AP, Tremoli E (eds) Effects of fatty acids and lipids in health and disease. World Rev Nutr Diet 76
49. Galli C, Simopoulos AP, Tremoli E (eds) Fatty acids and lipids: biological aspects. World Rev Nutr Diet 75
50. Salem N Jr, Simopoulos AP, Galli C, Lagarde M, Knapp H (eds) Proceedings of the 2nd congress of ISSFAL on fatty acids and lipids from cell biology to human disease. Lipids 31 (Supplement):S-1–S-326

51. Simopoulos AP, Robinson J (1998) The Ω plan. The medically proven diet that gives you the essential nutrients you need. Macmillan, New York
52. Keys A (1970) Coronary heart disease in seven countries. Circulation 41(suppl):1–211
53. Sandker GW, Kromhout D, Aravanis C et al (1993) Serum lipids in elderly men in Crete and The Netherlands. Eur J Clin Nutr 47:201–208
54. Renaud S, Godsey F, Dumont F et al (1986) Influence of long-term diet modification on platelet function and composition in Moselle farmers. Am J Clin Nutr 43:136–150
55. De Lorgeril M, Renaud S, Mamelle N et al (1994) Mediterranean alpha-linolenic acid rich-diet in the secondary prevention of coronary heart disease. Lancet 343:1454–1459
56. Renaud S, de Lorgetil M, Delaye J et al (1995) Cretan Mediterranean diet for prevention of coronary heart disease. Am J Clin Nutr 6l(suppl):1360S–1367S
57. de Lorgeril M, Salen P, Martin J-L et al (1998) Mediterranean dietary pattern in a randomized trial. Prolonged survival and possible reduced cancer rate. Arch Intern Med 158:1181–1187
58. Singh RB, Niaz MA, Sharma JP et al (1997) Randomized, double-blind, placebo-controlled trial of fish oil and mustard oil in patients with suspected acute myocardial infarction: the Indian experiment of infarct survival 4. Cardivasc Drugs Ther 11:485–491
59. Simopoulos AP, Herbert V, Jacobson B (1995) The healing diet: how to reduce your risks and live a longer and healthier life if you have a family history of cancer, heart disease, hypertension, diabetes, alcoholism, obesity, food allergies. Macmillan Publishers, New York
60. Broughton KS, Johnson CS, Pace BK, Liebman M, Kleppinger KM (1997) Reduced asthma symptoms with n-3 fatty acid ingestion are related to 5-series leukotriene production. Am J Clin Nutr 65:1011–1017
61. Weiss LA, Barret-Connor E, von Muhlen D (2005) Ratio of n-6 to n-3 fatty acids and bone mineral density in older adults: the Rancho Bernardo Study. Am J Clin Nutr 81:934–938
62. Hogstrom M, Nordstrom P, Nordstrom A (2007) n-3 fatty acids are positively associated with peak bone mineral density and bone accrual in healthy men: the NO_2 study. Am J Clin Nutr 85:803–807
63. Maes M, Smith R, Christophe A, Cosyns P, Desnyder R, Meltzer H (1996) Fatty acid composition in major depression: decreased ω3 fractions in cholesteryl esters and increased C20:4 ω6/C20:5 ω3 ratio in cholesteryl esters and phospholipids. J Affect Disord 38:35–46
64. Locke CA, Stoll AL (2001) Ω-3 fatty acids in major depression. World Rev Nutr Diet 89(173–185):118
65. Stoll AL, Severus WE, Freeman MP, Rueter S, Zboyan HA, Diamond E, Cress KK, Marangell LB (1999) Ω3 fatty acids in bipolar disorder: a preliminary double-blind, placebo-controlled trial. Arch Gen Psychiatry 56:407–412
66. Fontani G et al (2005) Cognitive and physiological effects of Ω-3 polyunsaturated fatty acid supplementation in healthy subjects. Eur J Clin Invest 35(11):691–699
67. Kiecolt-Glaser JK, Epel ES, Belury MA, Andridge R, Lin J, Glaser R, Malarkey WB, Hwang BS, Blackburn E (2013) Ω-3 fatty acids, oxidative stress, and leukocyte telomere length: a randomized controlled trial. Brain Behav Immun 28:16–24
68. Cassidy A, De Vivo I, Liu Y, Han J, Prescott J, Hunter DJ, Rimm EB (2010) Associations between diet, lifestyle factors, and telomere length in women. Am J Clin Nutr 91(5):1273–1280
69. Kang JX, Liu A (2013) The role of the tissue ω-6/ω-3 fatty acid ratio in regulating tumor angiogenesis. Cancer Metastasis Rev 32(1–2):201–210
70. Kang JX (2011) The ω-6/ω-3 fatty acid ratio in chronic diseases: animal models and molecular aspects. World Rev Nutr Diet 102:22–29
71. Simopoulos AP (2010) Genetic variants in the metabolism of omega-6 and omega-3 fatty acids: their role in the determination of nutritional requirements and chronic disease risk. Exp Biol Med 235:785–795

Part IV
Cardiology

Chapter 10
Evolutionary Paradigms in Cardiology: The Case of Chronic Heart Failure

Prof. emeritus Bernard Swynghedauw, M.D.

Lay Summary In this chapter, attempts are made to use an evolutionary perspective for understanding heart failure (HF), a major issue in cardiology and the endpoint of most of cardiovascular diseases (CVDs) including myocardial infarction, arterial hypertension or valve diseases. Evolutionary medicine takes the view that illness is linked to incompatibilities between the environment in which humans currently live and their genome, which has been shaped by several environmental conditions during biological evolution. Chronic HF occurs after a long period of adjustment of the heart to the new working conditions imposed by CVD (e.g. atherosclerosis). Such an adjustment has been possible due to an ancient, widespread and evolved cellular response to physical forces, called mechanotransduction. Mechanotransduction renders the maximum cardiac contraction slower and more efficient under increased load. From an evolutionary perspective, the heart fails firstly because the adaptive process reaches its own limits. In addition, the anthropogenic increase in lifespan and the accompanying ageing have contributed a new dimension, cardiac fibrosis, that aggravates cardiac function and is one of main biological marker for HF.

B. Swynghedauw (✉)
3 rue Bossuet, Saint-Rémy-lès-Chevreuse, 78470 Paris, France
e-mail: Bernard.Swynghedauw@inserm.fr

B. Swynghedauw
U942-INSERM, Hôpital Lariboisière, 75475 Paris, France

© Springer International Publishing Switzerland 2016
A. Alvergne et al. (eds.), *Evolutionary Thinking in Medicine*,
Advances in the Evolutionary Analysis of Human Behaviour,
DOI 10.1007/978-3-319-29716-3_10

137

10.1 Introduction

Evolutionary medicine and evolutionary cardiology take the view that illness is linked to incompatibilities between the environment in which humans currently live and their genome which has been shaped by successive environmental conditions during biological evolution [1–6]. In this chapter, we described some conspicuous elements of evolutionary medicine in cardiology that concern the ultimate step of most of CV diseases, namely HF.

In developing countries, CVDs represent one of the leading causes of mortality and morbidity [7, 8]. In the past, CVD mainly originated from infections including valve heart diseases (also called Bouillaud disease), due to severe septic sore throat, endocarditis and myocarditis. In recent decades, both the increase in lifespan and the first epidemiological transition fully modified the medical landscape. The incidence of infections progressively disappeared; instead, within the field of CVD, the clinical consequences of both arterial hypertension and atherosclerosis became predominant.

The first epidemiological transition occurred at the beginning of the twentieth century and was mainly a consequence of a better control of infections. It resulted in a substantial drop of neonatal mortality and the beginning of the global increase of lifespan. A transition occurred in the latter half of the twentieth century and was characterized by an increase in non-transmissible age-related diseases [9, 10]. Presently, to prevent CVD, controlling infections have become less important than controlling new cardiac risk factors, such as sedentary behaviours, obesity, tobacco smoking, air pollution and diet (sugar, fat and salt). These new risks are strongly associated with low socio-economic status in high-income populations.

It is a common misunderstanding to pool all CVD together, and it is important to clarify this issue. From a clinical and aetiological point of view, CVD is a heterogeneous group. For example, clinical manifestations of atherosclerosis, such as myocardial infarction, have little in common with infection-related valve diseases, or congenital heart diseases. As a result, cardiology covers a heterogeneous physio-pathological field of investigations, from fully inherited monogenic diseases, such as hypertrophic cardiomyopathy, to the clinical manifestations of atherosclerosis, which largely depends on behaviour and environment. Consequently, it is impossible to provide the same evolutionary approach to such a wide variety of diverse conditions, and integrating the relevant evolutionary paradigm in cardiology training is far from being achieved, or, even, properly conceived, even by cardiologists!

In this chapter, we illustrate how cardiologists can make use of evolutionary thinking for improving the diagnostic and treatment of chronic HF. In most CVD, HF develops after a long period of compensatory adjustment, and the heart finally fails and does not perform a normal cardiac output and the corresponding normal oxygenation of peripheral tissues (this is the definition of heart failure, HF) [11]. HF is a syndrome, with several causes, which, we argue, indicates the endpoint of an adaptive process. Although we have restricted our approach to chronic HF, other cardiac conditions can be informed by evolutionary thinking, such as arterial hypertension [12, 13] or atherosclerosis ([11], See also Chap. 11).

10.2 The Myocardial Tissue Response to Cardiac Overload as an Adaptation: Research Findings

Cardiac and skeletal or smooth muscle functioning is based upon mechanics and creates movement against gravitational force. The muscle is a thermodynamic machine achieving maximum economy according to the thermodynamics law [14]. Not surprisingly, the permanent modifications of mechanical conditions by CVD, such as myocardial infarction or valve disease, will modify muscle economy (see Box 10.1). The myocardial tissue response to this change in economy is an activation of a very ancient adaptive process, called mechanotransduction, which is the cellular response to mechanical forces.

Box 10.1 Economy and Mechanotransduction in Muscle Physiology
According to the thermodynamic principles, economy is improved when more force is produced for less heat dissipated. In a muscle, when the heat produced per g of developed tension is reduced, the system becomes more efficient. In other words, a car is more economic when it dissipates less heat and utilizes less energy as gasoline, for the same distance, the same speed and the same load. With the same motor, for the same distance, an overloaded car will immediately have a worse economy. In the heart, the motor is the contractile apparatus, and when the load is increased, the contractile apparatus has to be modified by an appropriate gene reprogramming to recover a normal economy. Foetal gene reprogramming has been selected during evolution to modify the cardiac protein content and renders the contractile apparatus more efficient [29, 31].

From a thermodynamic point of view, living cells are dissipative systems with low entropy that funnel energy into their own production and reproduction [11], and any drop in economy imposed by an environmental pressure has to be readjusted to allow the cells to survive and reproduce. This is the basis of mechanotransduction in muscles that have to permanently produce force against gravitation. As shown in Fig. 10.1, mechanotransduction is mainly composed of two complexes that allow continuity in the transmission of the force signal to the chromosomes: (i) at the external membrane level, ion channels and the adhesion protein/integrin complex and (ii) at the nuclear membrane level, the linker of the nucleoskeleton and cytoskeleton complex.

Chronic HF indicates the limits of this adaptive process and occurs after a long period of adjustment. Acute HF is something different, and it frequently occurs as an exacerbating episode during the course of chronic HF, but can also be a totally unexpected event occurring on a normal heart, such as acute failure after a massive pulmonary embolism: the heart fails because it does not have enough time to develop any compensatory process [11]. From a therapeutic point of view, in acute HF, it is urgent to save the patient by a rapid activation of contractility, and several

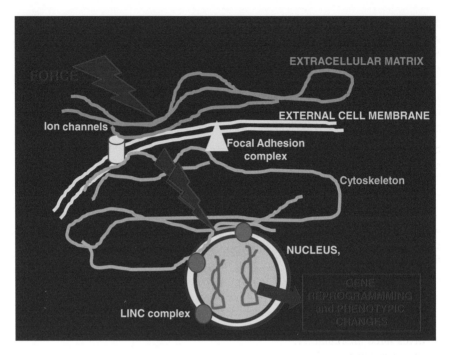

Fig. 10.1 Mechanotransduction. The general mechanism summarized above has been transmitted throughout evolution from the common ancestors of yeasts and humans [19–21]. The nucleus, which is the defining feature of eukaryotic cell, is tightly integrated into the structural network of the cell through a LINC complex. During evolution, LINC was essential for a broad range of cell functions, including meiosis and cell movements. Recent articles showed that the same LINC complex did also probe its mechanical environment, especially within the heart [22]. The mechanical continuity is assumed by several junctions from extracellular matrix to nucleus and allows forces to propagate relative long distances within the cells. (i) Physical forces act on the extracellular matrix and, by so doing, activate several ion channels and the adhesion proteins/integrin dimer complex, which transmits the signal to the cytoskeleton (actin and microtubules). (ii) Mechanosensing is enabled by protein conformations that accommodate the applied force. (iii) Finally, the transduction of the signal to chromatin through the nuclear membrane was performed by the LINC complex. Chromatin rearrangements result in a release of transcriptional factors and a gene reprogramming with specific phenotypic responses. For space limitations, the mechanism of the transmission of the force through the external membrane was not detailed (see [19])

inotropic drugs are fully indicated. By contrast, the adaptive process that occurs during chronic HF is characterized by a reduction of the contractile capacity of the heart, and inotropic drugs are usually not indicated; instead, the therapy is based on drugs which lower the cardiac load.

Several adaptations account for such a long-time preservation of the myocardial economy under mechanical overload. They are likely to be consequences of a foetal gene reprogramming and include cardiac hypertrophy, the reduction of V_{max} and the accompanying decrease in heat produced per g of tension that all allow the tissue to recover a normal economy. The failure of these adaptive processes to compensate anymore for the changes in working conditions constitutes a crucial determinant of HF. The same adaptive processes are also observed in exercise-induced cardiac and skeletal muscle hypertrophy. Nevertheless, in these physiological conditions, mechanical overload is not permanent (even professional athletes train only a few hours a day!) and failure does not happen.

Other components—the senescent process first, but also ischaemia, diabetes and obesity—have been superimposed on this basic process and render HF a more complicated issue. The most important of the biological determinants that limits the adaptive process is myocardial fibrosis, a multi-factorial marker of increasing electrical cardiac heterogeneity, diastolic stiffness and systolic dysfunction [11, 15]. In developing countries, the most important cause of HF is myocardial ischaemia due to coronary atherosclerosis. The wound healing response of the myocardium after myocardial infarction involves both the infarcted area and the non-infarcted ventricle. Reparative fibrosis is organized as a scar and is surrounded by reactive fibrosis. Consequently, ischaemia adds a new detrimental component to the general process of adjustment. The same is true for diabetes, which is equally fibrogenic (see details and references in [16]).

We will discuss below evolutionary paradigms that are essential for understanding the pathophysiology of HF and the limits of the adaptive process. HF involves three important and fully interrelated evolved processes or traits, namely mechanotransduction, the development of myocardial anatomical structures and ageing [15–17].

10.2.1 Mechanotransduction: An Evolutionary Legacy

Mechanotransduction is the cellular responses to physical forces (Fig. 10.1), and, in cardiovascular medicine, is involved at two levels: (i) in the heart, hypertrophy and the transcriptional modifications of the myocardium are the first responses of the tissue to mechanical overload and (ii) in the arterial wall, a transduction process is initiated in endothelial cells by the mechanical forces of the arterial lumen, the so-called shear stress, that will either contribute to atherogenesis (i.e. formation of abnormal fatty or lipid masses in arterial walls), or modify the arterial wall stiffness during hypertension, two major contributors of HF.

10.2.1.1 Mechanotransduction During Evolution: The General Process

The general process of mechanotransduction (Fig. 10.1) has been described in nearly every tissue, including skeletal muscle, lungs, ears, skin (touch), nerves, liver and kidney and in any eukaryotes, mammals and plants [18–22]. It has several basic features that are all linked to the evolution of life. The various genetic components and pathways involved in mechanotransduction have been favoured by natural selection during the evolutionary history of living species, as it enabled organisms to adjust to one of the most important variables of the environment, namely physical forces. Like every crucial biological pathway (another good example is circadian clocks), it is not surprising that many different components and sub-mechanisms involved with mechanotransduction involve load-bearing sub-cellular structures, such as plasma membrane itself (the phospholipid layer is sensitive to force), plasma membrane proteins (the stretch-sensitive ion channels or various cell adhesion complexes sense force), cytoskeleton (the widespread deformations of elastic cytoskeletal components are at the origin of several models of mechanotransduction), extracellular matrix components (as fibronectin or collagen) or the different constituents of the contractile apparatus itself [19–21]. Such a complexity is clearly a signature of the "tinkering" process during evolution [23].

Mechanotransduction can roughly be divided into three interrelated steps ([19, 20], Fig. 10.1): (i) the cellular response to forces is a rapid process enabled by mechanotransmission and consistent with the direct effects of mechanical overload in the heart [24]; (ii) mechanosensing generates protein folding that accommodates to physical forces, but the proteins involved are usually specific to a given tissue; (iii) the mechanoresponse influences general transcriptional networks that are not specifically force-dependent, and several different mechanisms have recently been documented [21, 22, 25]. The final result is a gene reprogramming and a modification of the myocardial phenotype with major physiological consequences.

Mechanotransduction itself may be also considered as one example, amongst several, of a broader phenomenon, called phenotypic plasticity, and, as such, is based on quantifiable reaction norms, i.e. on the relationships between phenotype and environmental factors for a given genotype [17, 26]. The genotype permitting such a "reaction norm" has been selected for by natural selection because it increased the ability of individuals to survive and reproduce in variable environments. Here, the CV diseases (e.g. atherosclerosis) create new environmental conditions imposed on the heart [16, 17].

10.2.1.2 Heart Failure: When Adaptive Plasticity Reaches Its Limits

Cardiac muscle sensors that have been selected for sense movements (cyclic as well as stable) and a special attention have recently been focused on the passive elastic elements of the sarcomere, the basic contractile unit of a muscle, such as the titin molecule which is a long "passive" molecule that runs from one end of the sarcomere to another and which may sense stress during diastole. Several mutations (especially on the titin kinase) have recently been discovered and shown to be involved in HF. Research suggests that the elastic protein titin kinase could play a decisive role in the deleterious process of cardiac dilatation, a crucial determinant of HF. Further, the same observation has been made concerning the active components of the sarcomere: it has been shown that in muscles, mechanotransduction is using a specific mechanotransmission pathway through its contractile apparatus. Then, the above-described general process of mechanotransduction has been modified and adapted to this particular tissue throughout evolution [21, 27].

The **mechanoresponse** to cardiac overload is both a quantitative and qualitative gene reprogramming. The final results are adaptive modifications of the cardiac structures, which allow the heart to function normally in these new working conditions [15–17]. The heart hypertrophies and the maximal contraction velocity of the muscle (V_{max}) becomes lower. Foetal gene reprogramming was first proposed in our group as a global molecular explanation of this mechanoresponse [16, 28, 29]. Foetal gene reprogramming was probably the only alternative available to deal with increased load during cardiac evolution [16, 30]. It simultaneously includes (i) a global activation of gene expression leading to cardiac hypertrophy, (ii) the re-expression or the increased expression of genes normally expressed during foetal life, such as those coding for the slow isoform of myosin [28] and for the brain natriuretic peptide (BNP) and (iii) the blunted expression of genes which are not expressed during the foetal life. The corresponding proteins include membrane components, such as enzymes regulating the calcium transient [29] and a potassium channel regulating the action potential duration. The latter accounts for the enlargement of the QT interval on the electrocardiogram (the time between the start of the Q wave and the end of the T wave in the heart's electrical cycle, representing electrical depolarization and repolarization of the ventricles), a marker of cardiac remodelling which is familiar to clinicians (references in [16]) (Table 10.1).

This foetal reprogramming is an adaptive process for several reasons [16]. First, cardiac hypertrophy normalizes the wall stress (according to the Laplace law) and multiplies the contractile units. The changes in contractile and membrane proteins account for a lower V_{max} that adapts the myocardial economy to the new loading conditions [15, 16, 31]. The same relationship does also exist in skeletal muscles and explains why red and white skeletal muscles have, in phylogeny, different shortening velocity [30]. Second, other changes in gene expression also contribute to adjustment. Three of them are widely utilized in clinical practice as biomarkers for HF: the plasma levels of the atrial natriuretic factor (ANF) and the brain natriuretic peptide (BNP), and the QT interval duration. In normal conditions, ANF forms small grains in atria and is physiologically regulated by water availability. Mechanical overload induces the ventricular expression of genes coding for ANF and is responsible for the enhanced plasma levels of ANF. BNP, which plays a similar role in homeostasis control, is normally expressed in both atria and ventricles, and its plasma level is enhanced by cardiac mechanical overload. In practice, BNP is a more sensitive biomarker for mechanical overload and is now widely adopted in clinical practice. ANF and BNP are both diuretic agents, and the enhancement of the production of urine is adaptive because it reduces the load of the heart. An enhanced level of BNP indicates cardiac overload but, sensu stricto, is not a marker of HF, as generally believed, because the definition of HF also includes functional signs. Another frequently forgotten direct marker of the adaptive process is the lengthening of the QT interval on the ECG and, its electrophysiological equivalent, the lengthening of the action potential duration. The increased QT interval and action potential durations are well-documented

Table 10.1 Foetal programme re-expression during cardiac remodelling

	Changes in gene expression	Phenotypic consequences Physiological and practical applications
Global increased expression	Collagen, contractile proteins, channels … with activation of pre-existing or imported stem cells	Cardiac hypertrophy, increased contractile units and improved wall stress
Genes re-expressed	Myosin isoform (slower)	Reduction of V_{max} and the accompanying normalization of myocardial economy
	General anaerobic switch	Better recovery period after the contractile event
	Ventricular expression of the atrial natriuretic factor (ANF) and increased brain natriuretic peptide (BNP)	Diuresis and reduction of the preload. BNP is the most widely utilized biomarker for cardiac mechanical overload and HF
Genes whose expression is blunted	Calcium ATPase of sarcoplasmic reticulum (SERCA2)	Increased relaxation time, participates in the slowing of contraction
	Early transient K^+ current, I_{tO}	Increased action potential duration, and QT duration on the ECG, a marker of cardiac remodelling, participates in the slowing of contraction
	Adrenergic and muscarinic receptors	Decreased heart rate variability and reduced response to exercising
	Myoglobin	Anaerobic switch and better recovery period

Phenotypic consequences and practical applications (from Refs. [16, 17])

characteristics of the hypertrophied heart, and, as components of the slowing of the contraction velocity, are adaptive.

Economy is indeed crucial to any tissue and especially to a tissue in charge of a mechanical function such as muscles (Box 10.1). To quantify the economy of this system, the energy flux and mechanical performances have been simultaneously measured on experimental models of cardiac overload in Alpert's group, and by so doing, they have demonstrated that the adaptive process mainly involves an improvement of energy utilization rather than one of energy production [31] (Fig. 10.2). In other words, the diminution of V_{max} is a beneficial event allowing the heart to contract at a normal energy cost. The same rationale also applies to other muscles [30].

The slowing of V_{max} is an evolutionary process by which a muscle can adapt economy to a wide range of load. Such a pheno-conversion is a non-specific response of the genome to any modification in the loading conditions. Depré et al. [32], for example, had developed a model of ventricular unloading in rats by

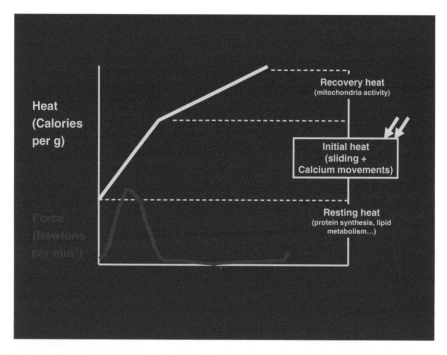

Fig. 10.2 Cardiac economy. For the cardiac muscle, economy is a central determinant of adaptation to new loading conditions. Heat production is measured on isolated papillary muscles of the heart during force development in normal conditions. Three types of heats are shown as follow: resting heat produced by the different cellular synthesis; initiation heat produced during the force development; and recovery heat, which indicates the process of recovery of energy. Initial heat is the only one that is reduced during chronic cardiac overload in hypertrophied heart (*arrows*). From a thermodynamic point of view, such a reduction in heat production indicates that an adaptive process has occurred in order to normalize muscle economy and that, in addition, the process results from changes in energy utilization at the levels of both the sliding mechanism (mainly the contractile proteins) and the calcium movements (enzymes and ion channels in charge of intracellular calcium movements) (see Box 10.1) (Adapted from Alpert and Mulieri [31])

heterotopic cardiac transplantation and showed that the foetal isoforms were all re-expressed, whereas the "adult" isoforms were downregulated.

From a physiological point of view, the degree of mechanical overload parallels both cardiac hypertrophy and myocardial economy. Such a relationship is familiar to cardiologists and constitutes the basis of a clinical diagnostic. From a therapeutic point of view, during chronic cardiac overload (the situation is different in acute HF, as explained previously), any inotropic intervention is basically deleterious since it goes against the adaptive process that has been selected for during thousands years of evolution. To conclude, at the beginning, cardiac remodelling is not a disease per se but the physiological response of the heart to a CVD. HF indicates the limits of this physiological adaptive process and is a disease. It is worth noting that mechanotransduction is also involved in two other CVDs fully related to HF, arterial hypertension [12, 13] and atherogenesis [33, 34].

10.2.2 Development and Myocardial Structure

The cardiac structure itself strongly depends on embryogenesis. The anatomical description, initially proposed by Torrent-Guasp et al. [35] for the heart, is crucial for understanding HF. The same helical cardiac structure is found in humans, horses, oxen, sheep, dogs, pigs, cats and rabbits and also chicken, lissamphibians, chelonians, fish and sharks. This means that at least tetrapod possesses the same basal cardiac morphology (we here used the systematic classification of Lecointre [36]). Data concerning cardiac morphology and the torsion process in clades living before tetrapod are not, for the moment, well documented [37–40].

The primitive heart evolves from a singular tube in chordate ancestor, into a dual pumping chamber with separate right and left sides [37, 40, 41]. The complex structure of the heart (a triple figure-eight spiral band with three S-shaped helixes) correlates with the conventional embryologic development [40, 42]. Recent investigations have indeed led to a better understanding of how the 3D force-producing cellular behaviours are regulated during bending [41, 43].

Cardiac morphology includes two simple loops that start at the pulmonary artery and end in the aorta. The so-called contractile band is responsible for a spiral horizontal basal loop that surrounds the two ventricular cavities, with a change in direction causing a second spiral and giving rise to the double helical structure of the ventricular mass. Such a "rope structure" does explain why cardiac contraction is indeed more analogous to a mop torsion than to a balloon contraction. Such a torsion mechanism of the ventricles has likely been selected for because it improves the efficiency of ejection and allows an active suction during cardiac filling [40, 41] (Fig. 10.3).

Such a complex helicity is an evolved feature that is significant for cardiac functioning. In turn, the impairment of systolic torsion observed during the early stages of cardiac overload is an important determinant of HF [42]. The detailed analysis of such a complexity through recent advance in cardiac imaging provides new prognostic indicators in cardiology [42], and cardiac torsion is now frequently used as a prognosis index. Age also induces modifications of the ventricular twist, which can be measured in vivo [44].

10.2.2.1 Ageing

One of the major new conditions accounting for the symptoms of HF is ageing and the accompanying fibrosis [45, 46]. The senescent processes are responsible for the recent emergence of age-related non-transmissible chronic diseases including the clinical manifestations of atherosclerosis, as well as many cancers, and neuro-degenerative diseases [8]. Ageing are new partners in the present medical landscape, and the role of the senescent cell is determinant in the development of age-associated diseases [46]. Ageing results from the improvement of hygiene,

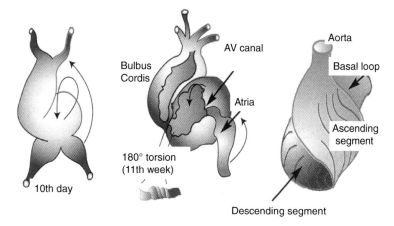

Fig. 10.3 Embryologic development of cardiac ventricles. At ten days, the torsion of the tubular heart begins. At eleven weeks, a 180° torsion happens which finally results in the final two loop rope structures of the adult heart (rearranged using data from [35]). The process exists in every tetrapod, and the torsion structure improves the efficiency of cardiac contraction, which becomes analogous to the torsion of a mop. Measuring the cardiac torsion provides information of cardiac contractility and the prognosis of HF

nutritional status and medical sciences, which are all the consequences of human activity; ageing in humans has clearly an anthropogenic origin. At this scale, such a process has never been experienced in any living species during the history of life and has nothing to do with the variations in ageing rates across species, which have been reported so far.

In 2016, HF represents the endpoint of most CVD. Average lifespan reaches 80 years in most developed countries, and because the main cause of the CV manifestations is atherosclerosis, HF is mainly observed in aged persons with ischaemic heart disease. Both ischaemia and HF generate myocardial fibrosis, thereby increasing myocardium heterogeneity and stiffness [16, 45]. Fibrosis is actually the main biomarker for HF [16, 46]. From an evolutionary perspective, the heart fails while attempting to maintain a normal cardiac output firstly because the adaptive process has its own limits.

10.3 Implications for Policy and Practice

Several CVDs can be understood within an evolutionary framework. Here, we focuses on chronic HF, a leading cause of morbi-mortality and the endpoint of most CVD including ischaemic heart diseases, arterial hypertension and valve diseases.

Chronic HF occurs after a long period of adjustment of the heart to the new working conditions imposed by CVD (e.g. atherosclerosis). Cardiac hypertrophy and the slowing of the maximum shortening velocity (Vmax) are the main components of this adaptive process. Hypertrophy is adaptive by both increasing the number of contractile units and reducing the wall stress. The slowing of V_{max} is adaptive by normalizing cardiac muscle economy. Such an adjustment has been made possible by an ancient, widespread and evolved cellular response to physical forces, called mechanotransduction. Mechanotransduction renders the cardiac contractions slower and more efficient under increased load. From an evolutionary perspective, the heart fails firstly because the adaptive process reaches its own limits. In addition, the anthropogenic increase in lifespan and the accompanying ageing have contributed to a new dimension, cardiac fibrosis that aggravates cardiac function and is one of the main biological markers for HF. To conclude, at the beginning, cardiac remodelling is not a disease per se but the physiological response of the heart to a CV disease. HF indicates the limits of this physiological adaptive process and is a disease.

Evolutionary thinking has practical consequences for the diagnostic and treatment of HF.

New diagnostic biomarkers cardiac hypertrophy, the lengthening of the QT interval duration on ECG and the slowing of Vmax are clinical markers for adaptation. In addition, BNP plasma level is a routine biomarker for HF in clinical practice. Its increase is the result of foetal reprogramming, and as a diuretic factor, BNP participates to the adaptive process by reducing cardiac load.

The helical cardiac structure has been shaped by embryo development in every tetrapod, and systolic torsion impairment observed during the early stages of cardiac overload is an important prognostic factor for HF. Attempts to quantify cardiac torsion were proposed to measure cardiac performances instead of the ejection fraction that is currently utilized by clinicians.

As a new perspective on the treatment and from an evolutionary perspective, the major goal of the treatment of chronic HF is to improve muscle economy and a positive inotropic effect is not indicated in chronic HF (as opposed to acute HF). The major goal of therapy is then to reduce the load either by suppressing the cause (e.g. by treating a valve disease or hypertension), or by reducing the loading conditions (e.g. with anti-aldosterone drugs, converting enzyme inhibitors or diuretics).

The present modifications of the medical landscape render HF a more complicated disease. Any evolutionary thinking has to include a new crucial factor, namely the age-associated cardiac fibrosis, which renders the heart stiffer both mechanically and electrically heterogeneous. Cardiac fibrosis is a consequence of the activity of the aged cell, and ageing itself results from the recent anthropogenic enhancement of lifespan. Cardiac fibrosis is presently an important determinant of HF both in clinical and in experimental condition.

Acknowledgments Many thanks to Pr. JF Toussaint (IRMES Paris) who kindly accepted to reread this chapter.

Glossary

Atheroma and Atherosclerosis	Lipid accumulation in arterial wall intima, media and sometimes adventitia, leading to foam cells and an important inflammatory reaction. The process leads to arterial stenosis and angina pectoris. Ultimately, atherosclerotic plaques are formed, and plaque rupture within arterial lumen causes thrombosis and myocardial infarction (or brain stroke). The disease is multi-factorial and involves numerous genetic and environmental factors
Cardiac remodelling	Qualifies changes that result in the rearrangement of normally existing structures. Cardiac remodelling includes cardiac hypertrophy, foetal gene reprogramming and fibrosis
Fibrosis	An increased concentration and mass of extracellular matrix components, mainly collagen
Foetal gene reprogramming	Gene reprogramming is essential for any organ facing new environmental conditions, such as pressure or volume overload in the heart. For the heart, the only alternative programme available is the foetal programme (see Table 10.1). For the skeletal muscle, an additional programme is also possible and re-expressed, the embryonic programme. Most elements of these two reprogramming phenomena are adaptive ... by chance (see discussion in Ref. [30])
Heart failure	The heart fails when it cannot assume anymore the normal oxygenation of peripheral tissues by a normal output [11, 14, 15]
Inotrope	Agent that increases cardiac force and contractility. Calcium itself is the physiological inotrope, but is not utilized as a drug. Major drugs having an inotrope effect include digitalis, digoxin, several

150   B. Swynghedauw

adrenergic agents and calcium sensitizers.
Most of them have additional pharmacological effects. Digitalis, for example, is
also a diuretic, and this is why digitalis is
still prescribed in certain conditions of
chronic HF by clinicians

Maximum contraction velocity of the   See V_{max}
unloaded cardiac muscle

Mechanotransduction   The cellular responses to mechanical forces. The process is essential in physiologic
homeostasis and for embryonic development and is present in nearly every cell
and living species. One essential piece of
the transduction is the linker of nucleoskeleton ad cytoskeleton (LINC) complex, which exists from yeasts to humans
(Fig. 10.1). LINC was initially a determinant of several basic cellular processes
such as meiosis, nuclear shaping and
chromosome organization. As such, the
complex has preserved its general architecture throughout evolution before being
able to act as an essential component of
mechanotransduction [21]. In the
switch-like models of mechanotransduction, applied forces are instantaneously
transmitted to load-bearing structure and
induce conformational changes in
mechanosensitive proteins, including
those from the LINC complex.
Mechanotransduction is sensitive to cyclic
and steady stretch, vibration, stress and
pressure through different pathways and
plays, for example, a role in hearing, the
inflation/deflation of the lungs and touch
sensation, and in many diseases including
cancer, osteoporosis, myopathies and
muscular dystrophies [19, 20]

Sarcomere   Contractile unit of the myofibril. It
includes two main filaments (thin and
thick). The sliding of these filaments is the
basis of muscle contraction. Sarcomere
proteins include actin, myosin, titin and

	the troponin components and participate in mechanotransduction both as a sensor and as an actor
Shear stress	Force component applied tangentially to the surface of a material (the luminal side of a vessel wall), which tends to cause deformation. In vascular physiology, shear stress depends on blood content, pressure and turbulences. It is a major atherogenic component. Arterial hypertension also increases shear stress
Stiffness	Pressure per volume change, a physical measurement of the rigidity of vessels or of the cardiac cavity
V_{max} or maximum contraction velocity of the unloaded cardiac muscle	Physiological measurement commonly made on isolated papillary muscle indicating the maximal contractile capacity of the heart. In phylogeny, V_{max} correlates with the ATPase activity of the main contractile protein, myosin. Such a correlation is usually considered as the biochemical explanation to explain the differences between fast (white) and slow (red) muscles [30]

References

1. Nesse RM, Williams G (eds) (1994) Why we get sick: the new science of Darwinian medicine. Times Books, New York
2. Nesse RM, Bergstrom CT, Ellison PT, Flier JS, Gluckman P et al (2010) Making evolutionary biology a basic science for medicine. Proc Nat Acad Sci USA 107:1800–1816
3. Trevathan WR, Smith EO, McKenna JJ (eds) (2007) Evolutionary medicine and health. Oxford University Press, Oxford
4. Stearns SC, Koella JC (eds) (2007) Evolution in health and disease, 2nd edn. Oxford University Press, Oxford
5. Swynghedauw B (ed) (2009) Quand le gène est en conflit avec son environnement. Une introduction à la médecine darwinienne. De Boeck, Bruxelles/Paris
6. Frelin C, Swynghedauw B (eds) (2011) Biologie de l'évolution et médecine. Lavoisier, Paris
7. de Peretti C, Chin F, Tuppin P et al (2012) Personnes hospitalisées pour infarctus du myocarde en France: tendances 2002-2008. Bul Epid Hebdomadaire 41:459–465
8. Lozano R, Naghavi M, Foreman K et al (2012) Global and regional mortality from 235 causes of death from 20 age groups in 1990 and 2010: a systematic analysis for the global burden of disease study 2010. Lancet 380:2095–2128

9. MacMichael AJ (2013) Globalization, climate change, and human health. N Engl J Med 368:1335–1343
10. Labonté R, Mohindra K, Schrecker T (2011) The growing impact of globalization for health and public health practice. Annu Rev Public Health 32:263–283
11. Poole-Wilson PA, Colucci WS, Massie BM et al (1997) Heart failure. Scientific principles and clinical practice. Churchill Livingstone, New York
12. Danzinger RS (2001) Hypertension in an anthropological and evolutionary paradigm. Hypertension 38:19–22
13. Weder AB (2007) Evolution and hypertension. Hypertension 49:260–265
14. Schneider ED, Kay JJ (1994) Life as a manifestation of the second law of thermodynamics. Mathl Comput Model 19:25–48
15. Swynghedauw B (ed) (1990) Hypertrophy and heart failure. INSERM/J. LIBBEY pub, Paris Londres
16. Swynghedauw B (1999) Molecular mechanisms of myocardial remodeling. Physiol Rev 79:215–262
17. Swynghedauw B (2006) Phenotypic plasticity of adult myocardium. Molecular mechanisms. J Exp Biol 209:2320–2327
18. Erdös T, Butler Browne GS, Rappaport L (1991) Mechanogenetic regulation of transcription. Biochimie (Paris) 73:1219–1231
19. Hoffman BD, Grashoff C, Schwartz MA (2011) Dynamic molecular process mediate cellular mechanotransduction. Nature 475:316–323
20. Orr AW, Helmke BP, Blackman BR et al (2006) Mechanisms of mechanotransduction. Develop Cell 10:11–20
21. Rothballer A, Kutay U (2013) The diverse functional LINCs of the nuclear envelope to the cytoskeleton and chromatin. Chromosoma 122: 415–429
22. Buyandelger B, Mansfield C, Knöll R (2014) Mechanosignaling in heart failure. Pflugers Arch-Eur J Physiol 466:1093–1099
23. Jacob F (1977) Evolution and tinkering. Science 196:1161–1166
24. Hatt PY, Ledoux C, Bonvalet JP (1965) Lyse et synthèse des protéines myocardiques au cours de l'insuffisance cardiaque expérimentale. Arch Mal Coeur Vx 12:1703–1720
25. Guilluy C, Swaminathan V, Garcia-Mata R et al (2011) The Rho GEFs LARG and GEF-H1 regulate the mechanical response to force on integrins. Nat Cell Biol 13:722–727
26. Dewitt TJ, Schneider SM (eds) (2004) Phenotypic plasticity: functional and conceptual approaches. Oxford University Press, New York
27. Takahashi K, Kakimoto Y, Toda K et al (2013) Mechanobiology in cardiac physiology and diseases. J Cell Mol Med 17:225–232
28. Lompré AM, Schwartz K, d'Albis A et al (1979) Myosin isoenzyme redistribution in chronic heart overloading. Nature 282:105–107
29. Lompré AM, Lambert F, Lakatta EG et al (1991) Expression of sarcoplasmic reticulum $Ca2^{+ATPase}$ and calsequestrine genes in rat heart during ontogenic development and aging. Circ Res 69:1380–1388
30. Swynghedauw B (1986) Developmental and functional adaptation of contractile proteins in cardiac and skeletal muscle. Physiol Rev 66:710–771
31. Alpert NR, Mulieri LA (1982) Increased myothermal economy of isometric force generation in compensated cardiac hypertrophy induced by pulmonary artery constriction in the rabbit. Circ Res 5:491–500
32. Depré C, Shipley GL, Chen W et al (1998) Unloaded heart in vivo replicates fetal gene expression of cardiac hypertrophy. Nat Med 4:1269–1275
33. Libby P, Ridder PM, Hansson GK (2011) Progress and challenges in translating the biology of atherosclerosis. Nature 473:317–325
34. Conway DE, Schwartz MA (2013) Flow-dependent cellular mechanotransduction in atherosclerosis. J Cell Sci 126:5101–5109

35. Torrent-Guasp F, Buckberg GD, Clemente C (2001) The structure and function of the helical heart and its buttress wrapping. I. The normal macroscopic structure of the heart. Semin Thorac Cardiovasc Surg 13:301–319

36. Lecointre G, Le Guyader H (eds) (2007) The tree of life. A phylogenetic classification. Belknap Press of Harvard University Press, Cambridge

37. Burggren WW, Christoffels VM, Crossley DA II et al (2014) Comparative cardiovascular physiology: future trends, opportunities and challenges. Acta Physiol 210:257–276

38. Moorman AFM, Christoffels VM (2003) Cardiac chamber formation: development, genes, and evolution. Physiol Rev 83:1223–1267

39. Torrent-Guasp F, Kocica MJ, Corno A et al (2004) Systolic ventricular filling. Eur J Cardio-thoracic Surg 25:376–386

40. Buckberg GD (2001) The structure and function of the helical heart and its buttress wrapping. II. Interface between unfolded myocardial band and evolution of primitive heart. Semin Thorac Cardiovasc Surg 13:320–332

41. Keller R, Shook D (2011) The bending of cell sheets—from folding to rolling. BMC Biol 9:90–94

42. Buckberg GD, Weisfeldt ML, Ballester M et al (2004) Left ventricular form and function. Circulation 110:e333–e336

43. Kanzaki H, Nakatani S, Yamada N et al (2006) Impaired systolic torsion in dilated cardiomyopathy: reversal of apical rotation at mid-systole characterized with magnetic resonance tagging method. Basic Res Cardiol 101:465–470

44. Dalen van BM, Soliman OII, Vietter WB et al (2008) Age-related changes in the biomechanics of left ventricular twist measured by speckle tracking echocardiography. Am J Physiol Heart Circ Physiol 295:H1705–H1711

45. Weber KT, Sun Y, Bhattachyara SK, Ahokas RA et al (2013) Myofibroblast-mediated mechanisms of pathological remodeling of the heart. Nature Rev Cardiol 10:15–26

46. van Deursen J (2014) The role of senescent cell. Nature 509:439–446

Chapter 11
Evolutionary Imprints on Cardiovascular Physiology and Pathophysiology

Robert S. Danziger, M.D.

Lay summary Prominent cardiovascular pathologies, including hypertension and atherosclerosis, may have evolutionary underpinnings. For example, selection for survival with a low-salt diet in early man may underlie salt-sensitivity and hypertension in modern civilization with high dietary salt consumption. Similarly, the evolutionary process may not have had sufficient time to adapt to a shift to high-fat and meat diets in contemporary society. Thinking in these terms results in an approach to these diseases focused on changes in environmental factors.

11.1 Introduction

The standard approach to the treatment of cardiovascular disease is based on empirical studies and does not consider the implications or imprint of evolution, which may provide insight into the pathogenesis of heart and vascular disease in contemporary times.

The cardiovascular system has evolutionary roots in invertebrate animals when simple absorption of nutrients, as by cells, is gradually replaced by an open circulatory system in which blood, or more appropriately termed hemolymph, is not contained in vessels (or a very few) and flows freely through the organism making direct contact with organs and cells. Subsequently, vessels with focal areas of contracting muscle to move the blood as seen in the earthworm. A two-chamber heart consisting of a single atrium and ventricle with a closed circulatory system, which exposes the blood to oxygen in the gills and then transport it throughout the

R.S. Danziger (✉)
University of Illinois, 840 S Wood Street, 60612 Chicago, IL, USA
e-mail: RDanziger@aol.com

© Springer International Publishing Switzerland 2016   155
A. Alvergne et al. (eds.), *Evolutionary Thinking in Medicine*,
Advances in the Evolutionary Analysis of Human Behaviour,
DOI 10.1007/978-3-319-29716-3_11

organs, emerged in fish. With the evolution of amphibians, e.g., frog, which exist in both water and land, a three-chamber heart developed with two atria and a single ventricle allowing oxygenated and deoxygenated blood to be separated as they enter the heart. With the advent of reptiles, e.g., turtles, a partial ventricular septum occurs to form a "three and one-half"-chamber heart. The human or mammalian heart and cardiovascular system are the most advanced, consisting of four chambers and highly developed arterial and venous circulations (reviewed in [1]). Thus, the cardiovascular system has evolved over approximately 4 billion years (see also Chap. 10).

The classical Darwinian concept of evolution is that environmental selection, along with other evolutionary forces (i.e., genetic drift, gene flow, mutation) has shaped what we are today and will, presumably, shape what we will become… slowly reducing the frequency of maladaptive traits to give way to one's that confer a reproductive advantage. However, cardiovascular disease is the number one cause of mortality in the world today [2] The question is why, if the system is constantly being optimized by selection.

A central theory is that many of the selective forces that have acted on the cardiovascular system are now being replaced by new forces to which the cardiovascular system is not adapted [3, 4]. Human agriculture is about 12,000 years old,

> only a minimal part throughout the history of human development, but its occurrence turns the human world upside down… [and] there has not been enough time for natural selection to change this design of the body to adapt us to the lifestyle of modern society [5].

Additionally, in general, cardiovascular diseases (CD) progress with aging and post-reproductive years so that the force of selection is weaker (see also Chap. 21). In this chapter, two leading causes of cardiovascular mortality, i.e., hypertension and atherosclerosis, are analyzed from this evolutionary perspective.

11.2 Research Findings

11.2.1 Hypertension

Arterial hypertension is a leading cause of heart failure, stroke, and renal failure.

It is reasoned that Darwinian selection has led to a highly regulated and complex system to maintain blood pressure for an optimal perfusion of organs and tissues, with delivery of nutrients and oxygen under all varieties of situations, such that there is scant evidence of hypertension in non-human organisms.

Blood pressure is a complex quantitative trait with both environmental and genetic components. Genome-wide associations and targeted gene studies have generated an expanding list of common and uncommon genetic variants linked to blood pressure (reviewed in [6]). Biometric strategies in the past have suggested 20–30 % of inter-individual variation in blood pressure is attributed to genes [7]. However, more recent and refined phenotyping from family studies suggest that

15–40 % of clinical systolic blood pressure and 50–60 % of ambulatory blood pressure are heritable [8–10]. Although blood pressure is heritable, "essential" hypertension does not follow a clear pattern of inheritance and is assumed to, in part, be due to the numerous interacting networks of molecular pathways, genes/protein modifications, and environmental confounders. Thus, it is reasoned that most of the genes associated with blood pressure were adaptive across human evolutionary history and may have had little phenotypic detriment until changes in human civilization occurred.

A key environmental factor in hypertension has been assigned to an increase in dietary sodium [11, 12]. The epidemiologic, clinical, and experimental support for this is overwhelming. First, in the INTERSALT Study conducted in 32 countries, the risk of developing hypertension over three decades of adult life was linearly and very tightly related to 24-hour urine sodium excretion, the best measure of dietary sodium intake. Second, reduced dietary sodium intake and diuretics have proven to be among the most effective treatments for primary hypertension. However, both normotensive and hypertensive persons show tremendous inter-individual variability in their blood pressure responses to dietary sodium loading and sodium restriction. This variability indicates a strong genetic underpinning. Third, the handful of rare Mendelian forms of human hypertension all involve excessive renal retention of salt and water, leading to severe salt-dependent hypertension.

A prevailing theory is that hypertension in human society, especially Western civilization, is the by-product of the selection for salt retention (reviewed in [13]). It has been speculated that the human diet, up until 10,000–25,000 years ago, consisted of 80 % meat with the rest being wild vegetables and fruits for an estimated daily intake of 600–770 mg sodium. With this diet, genes were selected for salt and water retention to the challenges of volume-depleting illnesses. Moreover, recent data show that dietary salt increases arterial stiffness, suggesting that the vasculature also has evolved in the context of a low-salt diet [14–16].

These genes are hypothesized to have become maladaptive when dietary salt intake increased with the agricultural revolution, harvested by solar evaporation and boiling, used to preserve and cure meats or used as a commodity of trade; and when an acquired taste for salt developed, e.g., salted fish [13, 17]. Now the average daily consumption of salt is 10–12 g/day in the USA and 24 g/day in Japan, representing 10–20 times the consumption estimated prior to the agricultural revolution [12]. Even a low-salt diet today is 6 g/day, representing a 350 % increase in about 10,000 years.

11.2.2 Atherosclerosis

Atherosclerosis, frequently referred to as "hardening of the arteries" is the principal cause of heart attack, stroke, and peripheral vascular disease. Heritability of atherosclerotic disease is well established and a family history of coronary artery disease is a risk factor included in established criteria for preventative treatment for cardiovascular disease [18–20]. The range of genetic variance in coronary artery disease is between 40 and 60 % based on family pedigree in twins [21, 22].

Rodents and other lower organisms do not normally develop and are resistant to developing atherosclerotic lesions even when subjected to pro-atherogenic interventions. Induction of atherosclerotic changes in vessels has not been reported in amphibians or lower species. Under non-experimental conditions, atherosclerosis is not observed in rodents on "normal" rat diets and severe interventions are required even to evoke mild atherosclerosis in rodents. In order to induce significant atherosclerosis, genetic manipulations, e.g., ApoE knockout [23, 24] are required to use these for research studies. Providing Western-style diets with high levels of saturated fats (approx. 35 % kcal %fat), cholesterol (0.5–1 % w/w), and cholic acid (0.1–5 % w/w) induces mild atherosclerosis in some mouse strains [25]. Guinea pigs, unlike other rodents, have a cholesterol profile similar to humans and develop diet-induced atherosclerotic lesions [26].

Like hypertension, atherosclerosis is considered as a function of environmental and genetic components. The difficulty in discerning these two components has been presented as evidence that the risk of the genetic variants is dependent on environmental influences [27]. Genome-wide association studies (GWAS) have identified at least 150 suggestive loci associated with coronary artery disease [28]. However, over 50 % of the associated variants occur in half of the population, and a quarter occur in 75 % of the population [29]. Thus, it is likely that on the average, each variant confers a minimal to modest risk and it has been estimated that the contribution of these to coronary artery disease and similar complex diseases is less than 10 % [27]. Many of these genes implicated in coronary artery disease are involved in inflammation and stem cell biology, and a lesser number are associated with known pathways for lipid variants [30].

Atherosclerotic disease is widely believed to be a disease of Western societies and changes in lifestyle brought about in the post-agricultural era. Major established environmental risk factors include high cholesterol, cigarette smoking, obesity, physical inactivity, and diabetes (review [31]).

A primary genetic focus of the link between increased cholesterol and genetics in the pathogenesis of atherosclerosis is the gene coding for apolipoprotein (ApoE) [32–35]. Among the principal variants, the alleles epsilon 2 (E2), epsilon 3 (E3), and epsilon 4 (E4) are the most common and have been shown to affect lipoproteins through regulation of hepatic binding, chylomicron catabolism, and uptake. Epidemiological studies have demonstrated that the epsilon 4 allele (coding for the protein ApoE4) most predisposes to atherosclerosis [36]. This allele is an "evolutionary relic from the pre-agricultural history of Homo sapiens and has not adapted to a nutrient-rich culture" [35, 37] (see also Chaps. 19 and 21). It remains at a high frequency; however, populations living in regions where agricultural economies have first been established, e.g., in the Mediterranean basin, have the lowest frequencies (0.05–0.13), while the frequency of this allele remains high among foragers, e.g., Pygmies (0.41), aborigines of Malaysia (0.24), and Papuans (0.37) [38].

A central question in evolutionary terms is whether ApoE and other proteins/gene variants linked to the lipid handling and the inflammatory response have persisted simply because they were neutral, or are under bidirectional selection (positive and negative). Both lipid handling and the inflammatory response are complex processes of interacting molecular signaling pathways and are central to cell and organ survival. The handling of fats and lipids, which include sterols, vitamins, phospholipids, and triglycerides, is key to dietary emulsification, digestion, absorption of nutrients, cell membranes, and metabolism required for cell and organ survival. The evolutionary paradigm suggests that the constellation of genes/proteins controlling lipid metabolism was selected to handle a pre-agricultural diet [39]. Inflammation plays a pathogenic role in a variety of other modern human diseases, including hypertension [40]. Inflammation has classically been defined as an evolutionary response to injury and infection; however, it is now associated with many post-agricultural human diseases, not including cardiovascular ones but also articular, inflammatory digestive, degenerative, and oncological disorders (see Chap. 18). A detailed analysis of the relevant theories has recently been address by Okin et al. [41].

A new twist in the atherosclerotic story and its link with a post-agricultural meat diet is the role of gut flora-mediated formation of pro-atherogenic compounds from meat, e.g., trimethylamine-N-oxide (TMAO) [42]. This is especially intriguing in an evolutionary context since vegetarian diets shift the microbiome to produce less TMAO, suggesting that dietary meat may have altered the gut flora to make it more pro-atherogenic.

11.3 Implications for Policy and Practice

Evolutionary underpinnings of contemporary CD in the post-agricultural period are based on selective pressures that have shifted from adaptive to maladaptive, or possess both adaptive and maladaptive features. In other words, "created by evolutionary hangovers… [and] biological evolution to cure hangovers can be very slow" [43]. Additionally, most CD are, in large part, age-related, occurring mostly in post-productive life, i.e., when the effect of selective pressures is markedly reduced. Thus, we cannot depend on natural selection to correct these over any predictable, albeit long time period. Policies and practices must be directed both (1) toward reducing or preventing the adverse effects of the cardiovascular risk factors derived from and associated with the post-agricultural period and (2) to use our understanding of the specific evolved genetic underpinning to develop molecular and genetics-targeted therapeutics. These strategies have already been engaged for hypertension and atherosclerosis.

For hypertension, current recommendations are to reduce salt intake to 1500–2000 mg/day [44, 45]. It should be noted that there is significant inter-individual variability and the generalized recommendation for indiscriminate reduction of salt in all populations has been a subject of recent debate [46]. Drugs targeted to reducing sodium retention and promoting salt/water excretion, including thiazides and furosemide, are among the most effective in reducing blood pressure and associated cardiovascular morbidity and mortality.

For atherosclerotic disease, there is convincing data showing that targeting both inflammatory process and genes/proteins in cholesterol/lipid metabolism associated with atherosclerosis is highly effective in reducing the incidence of coronary artery disease and stroke. Low-dose aspirin is anti-inflammatory and has proven to be one of the most efficacious treatments for coronary artery disease. Prominent among proven agents targeted to lipid metabolism are those which target HMG CoA reductase, e.g., statins, linked to atherosclerosis (reviewed in [47]). Agents are being developed to target ApoE, which is also believed to play a role in Alzheimer's disease [48] (see Chaps. 19 and 21). Again, the situation appears much more complex. Common clinical guidelines are to reduce red meat and saturated fat consumption; however, the empirical data indicate that the relationship is "complicated" [49]. Cardiovascular risk can be reduced by decreasing saturated fats and replacing them with polyunsaturated and monounsaturated fats. Although most doctors also recommend a diet rich in plant proteins rather than in meat and poultry, the evidence for a benefit is scant [47]. Reduction of total dietary protein itself has not been shown to be of significant benefit in the prevention of coronary heart disease [50].

Glossary

Atherosclerosis	Atherosclerotic plaques are aggregates of plasma lipids (especially cholesterol) cells (smooth muscle cells and monocytes/macrophages), and connective tissue matrix (collagen fibers and proteoglycans). Inflammation is the "dominant process" with atherosclerotic plaques characterized by increased cellular proliferation, lipids accumulation, calcification, ulceration, hemorrhage, and thrombosis. Typically, a major acute coronary syndrome occurs when an atherosclerotic plaque in a coronary artery ruptures with subsequent thrombosis. Chronic ischemia caused by reduced blood flow in the coronary artery due to narrowing of the vessels by atherosclerosis triggers chronic stable angina and may cause heart failure. The most common form of stroke arises when blood clots form on atherosclerotic plaques in carotid and cerebral arteries and blocks flow (reviewed in [30]).
Hypertension	Over 500 million people experience hypertension worldwide and its prevalence increases with age especially in Western civilization. There is overwhelming evidence that reducing hypertension reduces cardiovascular morbidity and mortality. Antihypertensive therapy has been associated with reductions in stroke incidence averaging 35–40 %, myocardial infarction 20–25 %, and heart failure >50 %. It is estimated that in patients with stage 1 hypertension and additional cardiovascular risk

factors, achieving a sustained 12 mmHg reduction in systolic blood pressure over 10 years will prevent 1 death for every 11 patients treated [51]. The origins of current standard and effective treatment of hypertension can be derived from evolutionary insights.

References

1. Bishopric NH (2005) Evolution of the heart from bacteria to man. Ann N Y Acad Sci 1047:13–29
2. www.who.int/mediacentre/factsheets/fs317/en/
3. Archer E, Blair SN (2011) Physical activity and the prevention of cardiovascular disease: from evolution to epidemiology. Prog Cardiovasc Dis 53:387–396
4. deGoma EM, Knowles JW, Angeli F, Budoff MJ, Rader DJ (2012) The evolution and refinement of traditional risk factors for cardiovascular disease. Cardiol Rev 20:118–129
5. Yang D, Liu Z (2013) An evolutionary perspective on cardiovascular disease. Phylogenet Evol Biol 1:1–2
6. Ai X, Jiang A, Ke Y, Solaro RJ, Pogwizd SM (2011) Enhanced activation of p21-activated kinase 1 in heart failure contributes to dephosphorylation of connexin 43. Cardiovasc Res 92:106–114
7. Sing CF, Boerwinkle E, Turner ST (1986) Genetics of primary hypertension. Clin Exp Hypertens A 8:623–651
8. Bochud M, Bovet P, Elston RC, Paccaud F, Falconnet C, Maillard M, Shamlaye C, Burnier M (2005) High heritability of ambulatory blood pressure in families of East African descent. Hypertension 45:445–450
9. Hottenga JJ, Boomsma DI, Kupper N, Posthuma D, Snieder H, Willemsen G, de Geus EJ (2005) Heritability and stability of resting blood pressure. Twin Res Hum Genet 8:499–508
10. Kupper N, Willemsen G, Riese H, Posthuma D, Boomsma DI, de Geus EJ (2005) Heritability of daytime ambulatory blood pressure in an extended twin design. Hypertension 45:80–85
11. Elliott P, Marmot M, Dyer A, Joossens J, Kesteloot H, Stamler R, Stamler J, Rose G (1989) The INTERSALT study: main results, conclusions and some implications. Clin Exp Hypertens A 11:1025–1034
12. Stamler J (1997) The INTERSALT study: background, methods, findings, and implications. Am J Clin Nutr 65:626S–642S
13. Lev-Ran A, Porta M (2005) Salt and hypertension: a phylogenetic perspective. Diabetes Metab Res Rev 21:118–131
14. Kusche-Vihrog K, Schmitz B, Brand E (2015) Salt controls endothelial and vascular phenotype. Pflugers Arch 467:499–512
15. Edwards DG, Farquhar WB (2015) Vascular effects of dietary salt. Curr Opin Nephrol Hypertens 24:8–13
16. Kullo IJ, Leeper NJ (2015) The genetic basis of peripheral arterial disease: current knowledge, challenges, and future directions. Circ Res 116:1551–1560
17. Eaton SB, Konner M (1985) Paleolithic nutrition. A consideration of its nature and current implications. N Engl J Med 312:283–289
18. Kullo IJ, Trejo-Gutierrez JF, Lopez-Jimenez F, Thomas RJ, Allison TG, Mulvagh SL, Arruda-Olson AM, Hayes SN, Pollak AW, Kopecky SL, Hurst RT (2014) A perspective on the New American College of Cardiology/American Heart Association guidelines for cardiovascular risk assessment. Mayo Clin Proc 89:1244–1256

19. Nasir K, Budoff MJ, Wong ND, Scheuner M, Herrington D, Arnett DK, Szklo M, Greenland P, Blumenthal RS (2007) Family history of premature coronary heart disease and coronary artery calcification: Multi-Ethnic Study of Atherosclerosis (MESA). Circulation 116:619–626
20. Marenberg ME, Risch N, Berkman LF, Floderus B (1994) de FU: Genetic susceptibility to death from coronary heart disease in a study of twins. N Engl J Med 330:1041–1046
21. Marenberg ME, Risch N, Berkman LF, Floderus B (1994) de FU: Genetic susceptibility to death from coronary heart disease in a study of twins. N Engl J Med 330:1041–1046
22. Bjorkegren JL, Kovacic JC, Dudley JT, Schadt EE (2015) Genome-wide significant loci: how important are they? Systems genetics to understand heritability of coronary artery disease and other common complex disorders. J Am Coll Cardiol 65:830–845
23. Jawien J (2012) The role of an experimental model of atherosclerosis: apoE-knockout mice in developing new drugs against atherogenesis. Curr Pharm Biotechnol 13:2435–2439
24. Imaizumi K (2011) Diet and atherosclerosis in apolipoprotein E-deficient mice. Biosci Biotechnol Biochem 75:1023–1035
25. Nishina PM, Verstuyft J, Paigen B (1990) Synthetic low and high fat diets for the study of atherosclerosis in the mouse. J Lipid Res 31:859–869
26. Fernandez ML, Volek JS (2006) Guinea pigs: a suitable animal model to study lipoprotein metabolism, atherosclerosis and inflammation. Nutr Metab (Lond) 3:17
27. Schadt EE, Bjorkegren JL: NEW: network-enabled wisdom in biology, medicine, and health care. Sci Transl Med 2012;4:115rv1
28. Bjorkegren JL, Kovacic JC, Dudley JT, Schadt EE (2015) Genome-wide significant loci: how important are they? Systems genetics to understand heritability of coronary artery disease and other common complex disorders. J Am Coll Cardiol 65:830–845
29. Roberts R (2015) A genetic basis for coronary artery disease. Trends Cardiovasc Med 25:171–178
30. Goldschmidt-Clermont PJ, Dong C, Seo DM, Velazquez OC (2012) Atherosclerosis, inflammation, genetics, and stem cells: 2012 update. Curr Atheroscler Rep 14:201–210
31. McGill HC Jr (1978) Risk factors for atherosclerosis. Adv Exp Med Biol 104:273–280
32. Davignon J, Gregg RE, Sing CF (1988) Apolipoprotein E polymorphism and atherosclerosis. Arteriosclerosis 8:1–21
33. Phillips MC (2014) Apolipoprotein E isoforms and lipoprotein metabolism. IUBMB Life 66:616–623
34. Davignon J, Cohn JS, Mabile L, Bernier L (1999) Apolipoprotein E and atherosclerosis: insight from animal and human studies. Clin Chim Acta 286:115–143
35. Mertens G (2010) Gene/environment interaction in atherosclerosis: an example of clinical medicine as seen from the evolutionary perspective. Int J Hypertens 2010:654078
36. Eichner JE, Dunn ST, Perveen G, Thompson DM, Stewart KE, Stroehla BC (2002) Apolipoprotein E polymorphism and cardiovascular disease: a HuGE review. Am J Epidemiol 155:487–495
37. Mahley RW, Rall SC Jr (1999) Is epsilon4 the ancestral human apoE allele? Neurobiol Aging 20:429–430
38. Corbo RM, Scacchi R (1999) Apolipoprotein E (APOE) allele distribution in the world. Is APOE*4 a 'thrifty' allele? Ann Hum Genet 63:301–310
39. Kuipers RS, Luxwolda MF, Dijck-Brouwer DA, Eaton SB, Crawford MA, Cordain L, Muskiet FA (2010) Estimated macronutrient and fatty acid intakes from an East African Paleolithic diet. Br J Nutr 104:1666–1687
40. De MC, Rudemiller NP, Abais JM, Mattson DL (2015) Inflammation and hypertension: new understandings and potential therapeutic targets. Curr Hypertens Rep 17:507
41. Okin D, Medzhitov R (2012) Evolution of inflammatory diseases. Curr Biol 22:R733–R740
42. Tang WH, Hazen SL (2014) The contributory role of gut microbiota in cardiovascular disease. J Clin Invest 124:4204–4211
43. Ehrlich P (2001) The human natures: genes, cultures, and the human prospect. Penguin Books, New York, p 36

44. Kotchen TA, Cowley AW Jr, Frohlich ED (2013) Salt in health and disease–a delicate balance. N Engl J Med 368:1229–1237
45. Whelton PK, Appel LJ, Sacco RL, Anderson CA, Antman EM, Campbell N, Dunbar SB, Frohlich ED, Hall JE, Jessup M, Labarthe DR, MacGregor GA, Sacks FM, Stamler J, Vafiadis DK, Van Horn LV (2012) Sodium, blood pressure, and cardiovascular disease: further evidence supporting the American Heart Association sodium reduction recommendations. Circulation 126:2880–2889
46. Stolarz-Skrzypek K, Staessen JA (2015) Reducing salt intake for prevention of cardiovascular disease–times are changing. Adv Chronic Kidney Dis 22:108–115
47. Darioli R (2011) Dietary proteins and atherosclerosis. Int J Vitam Nutr Res 81:153–161
48. Liao F, Hori Y, Hudry E, Bauer AQ, Jiang H, Mahan TE, Lefton KB, Zhang TJ, Dearborn JT, Kim J, Culver JP, Betensky R, Wozniak DF, Hyman BT, Holtzman DM (2014) Anti-ApoE antibody given after plaque onset decreases Abeta accumulation and improves brain function in a mouse model of Abeta amyloidosis. J Neurosci 34:7281–7292
49. Michas G, Micha R, Zampelas A (2014) Dietary fats and cardiovascular disease: putting together the pieces of a complicated puzzle. Atherosclerosis 234:320–328
50. Haring B, Gronroos N, Nettleton JA, von Ballmoos MC, Selvin E, Alonso A (2014) Dietary protein intake and coronary heart disease in a large community based cohort: results from the Atherosclerosis Risk in Communities (ARIC) Study. PLoS ONE 9:e109552
51. Chobanian AV, Bakris GL, Black HR, Cushman WC, Green LA, Izzo JL Jr, Jones DW, Materson BJ, Oparil S, Wright JT Jr, Roccella EJ (2003) The Seventh report of the joint national committee on prevention, detection, evaluation, and treatment of high blood pressure: the JNC 7 report. JAMA 289:2560–2572

Part V
Oncology

Chapter 12
Darwinian Strategies to Avoid the Evolution of Drug Resistance During Cancer Treatment

John W. Pepper

Lay Summary A major reason cancer is so difficult to cure is that it changes during treatment, such that drugs that were initially effective soon stop working. After a cancer acquires resistance to the drug that was used to treat it, the patient often relapses and dies. This accounts for many cancer deaths. Acquired drug resistance happens because a cancer is an evolving population of abnormal (mutated) human cells. Evolutionary theory has been used to understand exactly why and how cancers and other disease-causing cells acquire the ability to resist the drugs used against them. Evolutionary theory also suggests ways to prevent the problem with a different kind of therapeutic drugs. 'Antisocial' therapies do not selectively kill individual cells that are sensitive to them, but instead disrupt the ability of cancer cells to thrive collectively by providing benefits to each other. The approach has been shown to work on other kinds of disease-causing cells, without driving the evolution of drug resistance. This has not been tested yet for cancer, and it needs to be.

12.1 Introduction: Acquired Drug Resistance is a Central Problem in Cancer Biology

The acquisition of drug resistance by cancer during treatment is a central problem in cancer biology. Acquired drug resistance results from Darwinian selection and evolution among cancer cells, fuelled by their heritable heterogeneity. The outcome is grimly consistent:

J.W. Pepper (✉)
Biometry Research Group, Division of Cancer Prevention,
National Cancer Institute, Bethesda, MD, USA
e-mail: pepperjw@mail.nih.gov

© Springer International Publishing Switzerland 2016   167
A. Alvergne et al. (eds.), *Evolutionary Thinking in Medicine*,
Advances in the Evolutionary Analysis of Human Behaviour,
DOI 10.1007/978-3-319-29716-3_12

The reality is that targeted therapies are generally not curative or even enduringly effective, because of the adaptive and evasive resistance strategies developed by cancers under attack [1].

The outlook is even worse for disseminated cancers:

Patients with complete responses to targeted therapies invariably relapse. Most of the initial lesions generally recur, and the time frame at which they recur is notably similar. This time course can be explained by the presence of resistance mutations that existed within each metastasis before the onset of the targeted therapy [2].

It is now clear that the problem of acquired drug resistance usually results directly from cancer's evolutionary response to cytotoxins, drugs that are designed to kill cancer cells. A heterogeneous population of cancer cells varies in susceptibility or resistance to virtually any cytotoxin, including modern drugs that specifically target cancer cells. Cancer cell heterogeneity arises through somatic mutations, epigenetic changes and sporadic somatic aneuploidy (see Chap. 13). All of these are inherited across cell divisions, and therefore, any variants that increase cell survival by resisting cytotoxins cause clonal expansion of their cell lineage, gradually replacing the drug-sensitive cell lineages that are removed by the cytotoxin.

Avoiding acquired resistance during therapy will require alternatives to the standard drug classes, such as cytotoxins, that are known to drive cancer's evolutionary response towards acquired resistance. Several approaches have been proposed, and they all take some emphasis off trying to completely eradicate cancer cells, which so far has not been successful. Instead, they focus on providing benefit to patients by reducing the damage done by the cancer [3–6]. Among these alternative approaches, one with a solid theoretical foundation and some encouraging empirical results exploits the fact that cancer cells rely on cooperatively building a cancer-supportive micro-environment within the body. This approach has been called 'antisocial' therapy because it targets the social cooperation among cancer cells and the supportive micro-environment they collectively build for themselves. In this chapter, I review the successful application of this approach against other cellular pathogens and discuss its application to cancer.

12.2   Research Findings

12.2.1   Cancer Cells Rely on Cooperative Niche Building to Survive and Proliferate

It has long been recognized that important characteristics of cancer cannot be attributed solely to the individual cells composing it, but instead emerge as collective properties of the population of cancer cells [7]. Such collective properties do not vary among individual cells and thus do not provide fuel for cellular selection and resulting cancer evolution. Consequently, therapies targeting such collective properties cannot drive the cellular evolution of drug resistance. Many such collective properties arise from molecular cooperation among genetically distinct cells or clones [8, 9]. Such molecular cooperation often involves producing and secreting diffusible 'public good' molecules that increase the survival or proliferation of all nearby cells, collectively creating a niche that cancer cells thrive in [10, 11]. Therapies that target the social cooperation among pathogen cells instead of targeting the cells themselves have been called 'antisocial' therapies [12].

12.2.2   Theory Predicts that 'AntiSocial' Drugs Targeting Cancer Cell Cooperation will Retain Effectiveness Better than Cytotoxins can

The applicable aspects of evolutionary theory have been formalized mathematically (Fig. 12.1), and the resulting population dynamics have been analysed in computational models (Fig. 12.2), but the underlying logic is quite simple for why antisocial

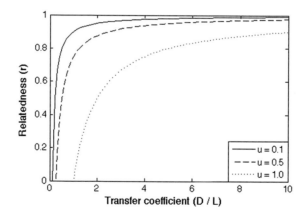

Fig. 12.1 Graphic depiction of a mathematical model of the conditions for the evolution of traits (such as drug resistance) that permit the production of secreted molecules providing a 'public good' to a cell population: Only in the parameter space above the *curve* does selection favour pathogen cells that secrete public good molecules. If an antisocial drug interferes with the secreted molecule, cells that resist its effect are positively selected only above the *curve*. With high relatedness *r* (trait similarity of producer to recipient), the targeted molecule and its benefits are mostly transferred to neighbouring cells that are similar to the producer cell, in also being effective (drug-resistant) producers of the shared benefit. A higher value of *u* (uptake rate of shared cell product) indicates that producer cells take up more of the secreted molecule themselves instead of allowing it to transfer to other cells. Figure re-drawn from [10]. The parameter space above the curve is smallest, indicating the most restrictive conditions for evolution of drug resistance, when the targeted molecule is the most 'public' (easily transferred between cells, as indicated by a high transfer coefficient

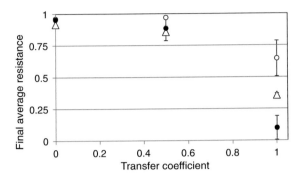

Fig. 12.2 Computer simulation results re-drawn from [11]: Frequency of drug-resistant cells after 1000 cell generations of mutation and evolution. High transferability of cell product (transfer coefficient *T*) corresponds to diffusible drug targets that mostly transfer benefits to neighbouring cells other than the producer. A transfer coefficient of zero corresponds to a cell-intrinsic drug target that is not shared with neighbouring cells, as in targeted cytotoxins. Each marker represents the mean for a different drug dosage: (*filled circle* = 100 % of full dose). *Error bars* show standard error across 10 independent simulation runs. When a simulated drug targeted a non-transferable molecule that was private to the cell producing it, evolution produced a population of cells with a high level of drug resistance. In contrast, when a simulated drug targeted a highly transferred 'public good' molecule, evolution produced a population of cells with a much lower level of drug resistance

therapies targeting shared 'public good' molecules retain their effectiveness longer than do cytotoxins that directly target cancer cells: in contrast to cytotoxins, which stop cells from surviving and proliferating, these antisocial drugs stop their targets from helping other cells to survive and proliferate. When a mutant cell resists a cytotoxin, it provides a competitive advantage to itself and its mutant progeny, so that its lineage increases over time and comes to dominate the tumour. In contrast, when a mutant cell resists an antisocial drug, it provides a competitive advantage to other cells and lineages. Thus, the mutant lineage of drug-resistant cells does not out-compete other lineages and does not expand to dominate the tumour (Fig. 12.3).

The crucial implication is that under antisocial therapy, reduced selective increase per cell generation results in slower evolution of drug resistance and greater likelihood of tumour eradication before evolution of complete drug resistance.

12.2.3 Empirical Results Bear Out the Theoretical Promise of AntiSocial Therapies Against Cellular Pathogens Including Cancer

The theoretical prediction that antisocial drugs would mitigate evolved drug resistance was first supported in the realm of infectious diseases through an accidental rather than a planned experiment. A contrast has been observed to vaccines that directly target the cells of pathogenic bacteria. These have often driven the

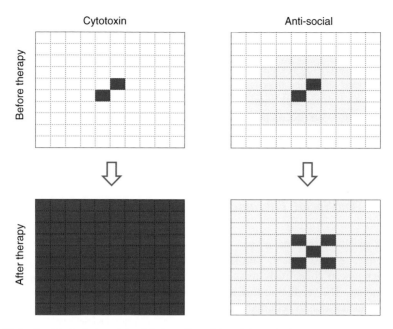

Fig. 12.3 Graphic depiction of the differing effect of Darwinian selection on mutant cells resistant to a standard cytotoxin (*left*) versus an antisocial drug (*right*). Each grid depicts a fixed total capacity to support 100 cancer cells. In each scenario, before therapy, only two (2 %) of these cells (*shaded red*) are mutants that resist the effect of the drug in question. Left panel: Cytotoxin scenario: a cytotoxin kills all cancer cells except the two cytotoxin-resistant mutants. After tumour re-growth from these surviving cells, the next cell generation will go from containing 2 % to 100 % drug-resistant cells. Right panel: Antisocial drug scenario: all cancer cells produce a diffusible public good necessary for cancer cell survival, and an antisocial drug blocks the ability of susceptible cells to provide this shared benefit. Only the two mutant cells resist this effect and continue to produce the shared public good benefit. This allows the 40 cells within the diffusion neighbourhood of the mutants (*shaded yellow*) to receive the required benefit and therefore survive and contribute to the next cell generation. After tumour re-growth from these surviving cells, the next cell generation will go from containing 2 % drug-resistant cells to containing 2/40 = 5 %. The other 95 % will be descended not from the mutants, but from the drug-sensitive neighbours they protected by continuing to provide the shared benefit they needed to survive

evolution of vaccine resistance or 'vaccine escape', rendering the vaccine ineffective. In contrast, other 'antisocial' vaccines instead prevent illness by targeting secreted toxins that function as bacterial public goods. These have retained effectiveness worldwide, over much longer time spans of many decades [12].

The first controlled experiment designed to test the expected advantage of an antisocial therapy against an evolving pathogen was conducted only last year [13]. In a laboratory study using the bacterial pathogen *Pseudomonas aeruginosa* and an insect host, the conventional cytotoxic antibiotic gentamicin drove the evolution of antibiotic resistance within just 12 days. In the same system, gallium was used as an antisocial therapy to inactivate the siderophores secreted by bacteria to make iron available as a public good. This drug was also effective in suppressing bacterial growth, but unlike the cytotoxic antibiotic, it remained effective throughout the experiment, with no evidence of evolved drug resistance.

Even before this test, both theoreticians and practitioners have noted the parallels between evolved drug resistance in bacteria and in cancer cells and have suggested that antisocial strategies may succeed against both [4, 11, 14, 15]. Both cytotoxic and antisocial drugs fall short of total eradication of every last pathogen cell. However, cytotoxins act as powerful drivers of Darwinian evolution by 'selecting' only the most drug-resistant cells to survive and proliferate in the absence of competitors, thereby founding an entirely drug-resistant new cell population.

12.2.4 AntiSocial Therapies Against Cancer

Several specific targets have been proposed and in some cases tried for antisocial therapeutics against cancer. The first target was the signalling of cancer cells in solid tumours to collectively recruit more blood vessels to support continued cell proliferation. The prediction that suppressors of these angiogenesis factors could forestall relapse due to acquired drug resistance [16] soon found empirical support [17], although anti-angiogenic drugs are not perfect in this regard [18]. The growth of a solid tumour is also limited by its ability to destroy the normal tissues that stand in its way. Cancerous tumours typically overcome this limit by secreting acids that break down surrounding tissues. In animal studies, those destructive acids have been successfully targeted by buffers such as bicarbonate that neutralize the secreted acid in the tumour's environment [19, 20]. Cancer cells also use secreted proteases to break through barriers to their growth, and in animal studies, those shared 'public good' proteases have been successfully targeted with the drug verapamil to suppress cancer cell proliferation, invasion and metastasis without killing cancer cells [21]. Finally, many cancer cells are limited in proliferation by the availability of diffusible growth factors, which they sometimes secrete themselves, as another public good. These cancer-produced growth factors present further potential targets for antisocial therapies [22].

12.2.5 Future Directions

The most immediate and promising future direction is to test, specifically in cancer, the theoretical prediction that some antisocial therapies already developed and tested for short-term efficacy will also retain effectiveness better than do cytotoxins.

While some specific antisocial cancer therapies listed above have been shown to be effective against cancer, none has yet been directly compared with targeted cytotoxins in their ability to provide a sustained benefit. Such tests are a needed next step in this promising direction. It will also be possible to systematically search for further targets for antisocial cancer therapies, as diffusible public good secretions are likely involved wherever there are cooperative effects between genetically different cancer cell lineages, or clones. Such cooperation between cells with different genetic lesions has been observed in cell line models of transformation and in mouse models of several cancers [9, 23].

12.3 Implications for Policy and Practice

To date, the approach of 'personalized' or 'precision' cancer medicine is largely based on molecular targeting and does not address the issues of cancer's cellular genetic heterogeneity and consequent acquired drug resistance. Any policies that promote continued development primarily of targeted cytotoxins will continue to face the limitation of temporary effectiveness described here. Developing and testing potentially sustainable treatments such as antisocial therapies should also be pursued.

Because their origins and molecular biology are human, and because of their inherent adaptability, endogenous and heterogeneous human malignancies will never yield easily to the kind of targeted cytotoxins that have succeeded against infectious diseases with stable non-human molecular biology. This makes primary prevention the necessary top priority for cancer, with treatment as a sometimes necessary, but better avoided, 'plan B' [2].

Glossary

Acquired drug resistance	A situation where a cancer that is initially susceptible to a therapeutic drug becomes resistant to it during therapy often leading to tumour re-growth and patient relapse
Tumour heterogeneity	The observation that cells within the same tumour differs in their genes and their traits
Antisocial therapies	A class of therapeutic drugs that block the ability of cancer cells to provide mutual benefits to each other rather than killing susceptible cancer cells directly (and selectively)
Darwinian evolution	Population change by natural selection in which only those individuals (or cells) best able to survive and reproduce in a given environment will pass on their genes to the next generation, with the result that only those genes and traits allowing the most survival and reproduction will persist over time

Somatic cells	In contrast to germ cells (eggs and sperm) that create offspring somatic (within-body) cells make up the tissues of an individual during its single lifetime. Mutations in these cells can change cell traits and be inherited by daughter somatic cells, but are not passed on to offspring
Cytotoxin	A drug to kill cells often of a single targeted type ('targeted cytotoxins')

References

1. Hanahan D (2014) Rethinking the war on cancer. Lancet 383(9916):558–563. doi:http://dx. doi.org/10.1016/S0140-6736(13)62226-6
2. Vogelstein B, Papadopoulos N, Velculescu VE, Zhou S, Diaz LA, Kinzler KW (2013) Cancer genome landscapes. Science 339(6127):1546–1558. doi:10.1126/science.1235122
3. Gatenby RA (2009) A change of strategy in the war on cancer. Nature 459(7246):508–509. doi:10.1038/459508a
4. Pepper JW (2011) Somatic evolution of acquired drug resistance in cancer. In: Gioeli (ed) Targeted therapies: Mechanisms of resistance, vol 7, Springer, New York, pp 127–134
5. Jansen G, Gatenby R, Aktipis CA (2015) Opinion: Control versus eradication: Applying infectious disease treatment strategies to cancer. Proc Natl Acad Sci USA 112(4):937–938. doi:10.1073/pnas.1420297111
6. Oronsky B, Carter CA, Scicinski J, Oronsky N, Caroen S, Parker C, Lybeck M, Reid T (2015) The war on cancer: A military perspective. Front Oncol 4. doi:10.3389/fonc.2014.00387
7. Deisboeck TS, Couzin ID (2009) Collective behavior in cancer cell populations. BioEssays 31(2):190–197. doi:10.1002/bies.200800084
8. Ohsawa S, Takemoto D, Igaki T (2014) Dissecting tumour heterogeneity in flies: Genetic basis of interclonal oncogenic cooperation. J Biochem 156(3):129–136. doi:10.1093/jb/mvu045
9. Polyak K, Marusyk A (2014) Cancer: Clonal cooperation. Nature 508(7494):52–53. doi:10.1038/508052a
10. Driscoll WW, Pepper JW (2010) Theory for the evolution of diffusible external goods. Evolution 64(9):2682–2687. doi:10.1111/j.1558-5646.2010.01002.x
11. Pepper JW (2012) Drugs that target pathogen public goods are robust against evolved drug resistance. Evol Appl 5(7):757–761. doi:10.1111/j.1752-4571.2012.00254.x
12. Pepper JW (2014) The evolution of bacterial social life: From the ivory tower to the front lines of public health. Evol Med Public Health 2014(1):65–68. doi:10.1093/emph/eou010
13. Ross-Gillespie A, Weigert M, Brown SP, Kummerli R (2014) Gallium-mediated siderophore quenching as an evolutionarily robust antibacterial treatment. Evol Med Public Health 1:18–29. doi:10.1093/emph/eou003
14. Lambert G, Estevez-Salmeron L, Oh S, Liao D, Emerson BM, Tlsty TD, Austin RH (2011) An analogy between the evolution of drug resistance in bacterial communities and malignant tissues. Nat Rev Cancer 11(5):375–382. doi:10.1038/nrc3039
15. Jansen G, Gatenby R, Aktipis CA (2015) Opinion: Control versus eradication: Applying infectious disease treatment strategies to cancer. In: Proceedings of the national academy of sciences 112(4):937–938. doi:10.1073/pnas.1420297111
16. Kerbel RS (1991) Inhibition of tumor angiogenesis as a strategy to circumvent acquired-resistance to anticancer therapeutic agents. BioEssays 13(1):31–36. doi:10.1002/bies.950130106

17. Boehm T, Folkman J, Browder T, Oreilly MS (1997) Antiangiogenic therapy of experimental cancer does not induce acquired drug resistance. Nature 390(6658):404–407. doi:10.1038/37126
18. Bergers G, Hanahan D (2008) Modes of resistance to anti-angiogenic therapy. Nat Rev Cancer 8(8):592–603. doi:10.1038/nrc2442
19. Robey IF, Baggett BK, Kirkpatrick ND, Roe DJ, Dosescu J, Sloane BF, Hashim AI, Morse DL, Raghunand N, Gatenby RA, Gillies RJ (2009) Bicarbonate increases tumor pH and inhibits spontaneous metastases. Cancer Res 69(6):2260–2268. doi:10.1158/0008-5472.can-07-5575
20. Fais S, Venturi G, Gatenby B (2014) Microenvironmental acidosis in carcinogenesis and metastases: New strategies in prevention and therapy. Cancer Metastasis Rev 33(4):1095–1108. doi:10.1007/s10555-014-9531-3
21. Farias EF, Ghiso JAA, Ladeda V, Joffe EBD (1998) Verapamil inhibits tumor protease production, local invasion and metastasis development in murine carcinoma cells. Int J Cancer 78(6):727–734. doi:10.1002/(sici)1097-0215(19981209)78:6<727:aid-ijc10>3.0.co;2-a
22. Archetti M, Ferraro DA, Christofori G (2015) Heterogeneity for IGF-II production maintained by public goods dynamics in neuroendocrine pancreatic cancer. In: Proceedings of the National Academy of Sciences. doi:10.1073/pnas.1414653112
23. Ortmann CA, Kent DG, Nangalia J, Silber Y, Wedge DC, Grinfeld J, Baxter EJ, Massie CE, Papaemmanuil E, Menon S, Godfrey AL, Dimitropoulou D, Guglielmelli P, Bellosillo B, Besses C, Döhner K, Harrison CN, Vassiliou GS, Vannucchi A, Campbell PJ, Green AR (2015) Effect of mutation order on myeloproliferative neoplasms. N Engl J Med 372(7):601–612. doi:10.1056/NEJMoa1412098

Chapter 13
Why Chemotherapy Does Not Work: Cancer Genome Evolution and the Illusion of Oncogene Addiction

Aleksei Stepanenko, Ph.D. and Prof. Vadym Kavsan, Ph.D.

Lay Summary To shift from largely palliative chemotherapy to tumour control and cure, a tumour should be considered through the prism of evolutionary biology. In a tumour, the genome level alterations produce profound phenotype leaps and fast adaptations to micro-environmental stresses that are not achievable through simple gene mutations. Constant chromosome changes create striking intercellular genomic heterogeneity, drive dynamic expression changes of hundreds of genes, rewire metabolic and signalling pathways, and give rise to genetic and phenotypic diversification of tumour cells, which is a basis for cancer evolutionary selection. Overall, genomic heterogeneity can be used as a predictor of tumour evolutionary potential; treatment applied at different phases of genome evolution may have differential impact on tumour evolution and patient survival; efforts directed at pushing the genome of cancer cells towards a stable phase may lower the evolutionary potential of a tumour due to reducing population genetic/phenotypic diversity.

Vadym Kavsan—Deceased

I dedicate this chapter to the memory of my tutor, professor Vadym Kavsan, a prominent Ukrainian scientist (1939–2014). His patience, advice, guidance, and attention to detail were invaluable. His original ideas and striking personality will inspire me in future.

A. Stepanenko (✉) · V. Kavsan
Department of Biosynthesis of Nucleic Acids, Institute of Molecular Biology
and Genetics, National Academy of Sciences of Ukraine,
Zabolotnogo Str. 150, Kiev 03680, Ukraine
e-mail: a.a.stepanenko@gmail.com

© Springer International Publishing Switzerland 2016
A. Alvergne et al. (eds.), *Evolutionary Thinking in Medicine*,
Advances in the Evolutionary Analysis of Human Behaviour,
DOI 10.1007/978-3-319-29716-3_13

13.1 Introduction

The long-held and dominating view that cancer is a genetic disease caused by the deterministic sequential mutations in a restricted number of cancer-driver genes (oncogenes and tumour suppressor genes), occurring in a continuous linear pattern of tumour progression (linear clonal model), and directly determining the hallmarks of cancer, has now been disproved. The cancer sequencing studies revealed a large number of stochastic gene mutations and multiple genetic subclones evolving in parallel (branching clonal model) (Fig. 13.1). The majority of somatic mutations are not detectable in all tumour regions; some genes undergo multiple distinct and spatially separated mutations within a single tumour, and diverse mutations emerge

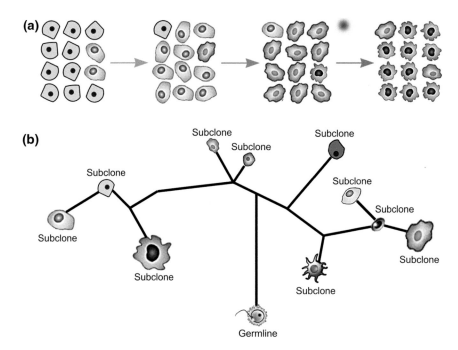

Fig. 13.1 Linear *versus* branching clonal evolution. **a** The linear clonal model of cancer evolution is based on assumption that tumour progression occurs in a continuous linear pattern due to sequential dominant clones (clonal sweep), which accumulate mutations in the key cancer "driver" genes in a deterministic stepwise manner and overwhelm earlier clones carrying only some of the mutations. **b** The branching structure of the phylogenetic tree demonstrates the number of simultaneous subclonal populations (represented by different coloured cell cartoons) within the cancer samples and their genetic relationships. The length of the trunk (represents the complement of mutations shared by all malignant cells within the cancer) and individual branches (represents the complement of mutations shared only by subclones on a branch) is proportional to the number of non-synonymous mutations separating the branching points. The size of a cell cartoon denotes the size/frequency of a subclone. Unique subclones emerge as a consequence of random genome alterations and gene mutations and may individually be capable of giving rise to disease relapse and metastasis. The dynamic clonal architecture is shaped by environmental selection pressures, including cancer treatments. The complexity of branching is underestimated on this figure for simplicity

and disappear during tumour progression to metastases and recurrences. There is a substantial evidence for mutation and amplification heterogeneity of the key cancer genes within individual tumour specimens and among multiple specimens from individual patients. The unprecedented level of inter- and intra-tumour heterogeneity is reflected in the statistics of the COSMIC database (http://www.sanger.ac.uk/genetics/CGP/cosmic/) [1]. The list of cancer genes and mutations within them is constantly growing in the Cancer Gene Census (http://cancer.sanger.ac.uk/census) and the Network of Cancer Genes databases (http://ncg.kcl.ac.uk/; http://bio.ieo.eu/ncg2/) [2]. Thereby, Heng [3] concludes:

> Now it is clear that cancer progression is a stochastic process both at the genome and gene levels, and is not a stepwise process defined by sequential genetic aberrations.

Similarly, Salk et al. [4] deduce:

> The large number and breadth of diversity in genes mutated among individual tumor specimens emphasize the fundamentally stochastic nature of cancer evolution.

Both cancer sequencing and genome instability studies strongly support the view that cancer evolution is not the sequential order of genetic alterations (specific cancer genes and common chromosome alterations) but, instead, represents multiple cycles of punctuated/discontinuous and gradual/step-wise evolution where stochastic (random, non-clonal) genome-level alterations are the primary and most important driving force of genetic heterogeneity and phenotype diversity. Todorovic-Rakovic [5] makes a conceptual deduction:

> oncogene mutation profiling now reveals all the complexity of cancer and provide the final explanation of the oncogenic pathways, based on stochastic (onco)genomic variation rather than (onco)genic concepts

and Brosnan and Iacobuzio-Donahue [6] emphasize:

> The evolution of cancer is not as straightforward as a stepwise series of mutations. As a result of genetic instability… cancers are often a heterogeneous mix of genomes.

Thereby, there is an appeal to shift "from the generally accepted view of cancer as a disease of a gene into that of a genome-based disease" [5], "from cancer gene-focused research to genome-focused research" [7], and from "analysing tumours not just as a ground-up bulk tissue, but as a population of individual tumour cells" [8].

13.2   Research Findings

13.2.1   The Genome Theory of Cancer: Building a Framework for Cancer Biology

There are two levels of genetic information organization: gene and genome. The gene mutation theory of cancer states that cancer somatic evolution is stepwise, clonal occurring by progressively acquiring more and more genetic and epigenetic abnormalities, whereas the hallmarks of cancer (e.g. self-sufficiency in growth signals, replicative immortality, resistance to growth suppressors and apoptosis, sustained angiogenesis, invasion, and metastasis) are driven by the several key oncogenes and tumour suppressor genes. The advocates of this concept disregard genome instability and judge it largely as a by-product of tumourigenesis. They focus on the identification of the shared gene mutations in cancer cells and on the individual molecular pathways responsible for the initiation and progression of disease. In contrast, the genome theory of cancer is based on the concept that genome-level instability (structural and numerical chromosome aberrations) together with random gene mutations serves as a driving force of cancer somatic evolution by increasing the cell population diversity, which is the raw material for evolutionary selection (reviewed in [9–15]). Genome is not just total DNA sequence (all genes and regulatory elements). Instead, genome is a self-organizing three-dimensional chromatin structure that governs the physical relationships between thousands of genes and regulatory elements through both *cis* and *trans* mechanisms along and between chromosomes (genome topology). It means that the loss of a chromosome not only results in reduction of expression of majority of genes located on that chromosome by half, but also leads to the disruption of the regulatory interactions within the nucleus that were established by that chromosome. Similarly, the chromosome translocation may or may not affect the gene structure at the break point but it inevitably entails changes in the expression pattern of genes located not only on the translocated chromosomes but also on non-translocated chromosomes, by *in trans* mechanism. An altered distribution of chromatin (changes in genome topology) and an aberrant expression of transcription factors favour the deregulation of transcriptome and cellular functions. Genome-level alterations rewire protein interaction network, alter timing and amplitude of signalling response, and change the functions of signal transduction pathways by multiple mechanisms. Altogether, these modifications affect cellular growth, division, migration, death, and other cellular processes. Moreover, transcriptome and proteome including protein–protein interactome are under constant change in cells with unstable genome. Therefore, the genome-level alterations play a key role in cancer evolution [9–15]. Single-cell analysis studies confirmed biochemical individuality of each cancer cell (different types and strengths of protein–protein interactions, proliferation rate, variation in responsiveness to stimuli and drugs, and potential to invade) and showed that a cancer cell relies on a unique series of pathways that ensure survival upon drug treatment [16, 17].

Heng et al. [10] describe the evolutionary mechanism of cancer as encompassing three components: (1) diverse stresses induce genome instability (e.g. mutagenic and non-mutagenic chemical and physical carcinogens, viral/bacterial infection, inflammation, ageing), (2) genome instability produces genetic and epigenetic heterogeneity, which is the raw material for evolution, and (3) cancer somatic evolution, based largely on random genome alterations, results in overcoming the system constraints such as cell death, tissue architecture, or immune system surveillance (Fig. 13.2). There are two phases of somatic genome evolution. One is

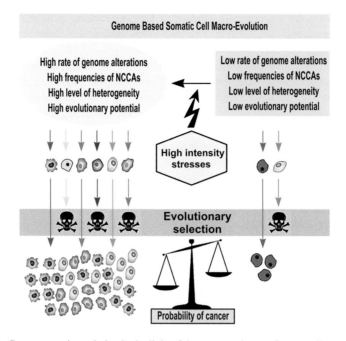

Fig. 13.2 Cancer somatic evolution in the light of the genome theory of cancer. Cancer evolution is based primary on genome-level alterations rather than gene mutations. A cell population without/low genome instability has a low diversity and, therefore, low evolutionary potential to overcome system constraints such as cell death, tissue architecture, or immune system surveillance. However, exposure of cells to high-level stresses (internal: e.g. oxidative stress, replication stress, endoplasmic reticulum stress, metabolic stress; external/environmental: e.g. drug treatment, radiation, viral/bacterial infection, metabolic, mechanical; experimental manipulations: e.g. gene overexpression, gene knock out/down, chemical inhibitors, culture conditions) promotes genome instability and results in an increase in population genetic, epigenetic and non-genetic heterogeneity and, therefore, high evolutionary potential and probability of cancer formation. Due to stress-induced genome chaos, many cells are not viable and undergo mitotic cell death or apoptosis. Eventually, more stable rare cancer-causing karyotypes are selected. NCCAs: non-clonal chromosome aberrations

the discontinuous phase characterized by heterogeneous karyotypes between cells and ongoing progressive karyotype changes; the other phase is the stepwise continuous phase within which the majority of cells share similar karyotypes for a long time. These two phases represent punctuated (or macro-) evolution and Darwinian (or micro-) evolution, respectively. Relationship between these two phases and genome system stability, measured by the level of stochastic genome alterations, showed that the punctuated phase is characterized by genome system instability (high frequencies of non-clonal chromosome aberrations (NCCAs), whereas the Darwinian stepwise phase demonstrates relative genome system stability (dominant clonal chromosome aberrations (CCAs) and low frequencies of NCCAs). Extremely high genome-level heterogeneity in the punctuated phase provides the genetic underpinning of the high degree of heterogeneity universally detected in cancers. Follow-up experiments evidenced that all factors, genetic/non-genetic, internal/external, functioning as a stress to a given system, can contribute to cancer evolution, either through micro- or macroevolution [10, 17–22].

System stress (CIN-promoting factors such as drug treatment, radiation exposure, metabolic perturbations, oxidative stress, infections/inflammation or experimental manipulations) during the different phases may lead to very different responses [12]. During the stepwise phase, acute stress may increase instability, destabilize CCAs, and abolish growth advantage that those CCAs endowed. While stabilizing genome, selection and spread throughout the population of chromosome aberrations result eventually in the fixation of newly formed advantageous dominant CCAs able to continue tumour progression. During the punctuated phase, acute stress may result in such instability when most of unstable cells are not viable [12]. Actually, high-level aneuploidy has a negative impact on cellular fitness and generates non-neoplastic and nonviable cells [23, 24]. However, genome chaos significantly increases the population genome diversity and the evolutionary potential of tumour. These phases of cancer cell evolution may shed light on the tumour-promoting and suppressing effects of CIN in tumourigenesis as well as on the "paradoxical" relationship between excessive CIN and improved survival outcome in cancer [25, 26].

Karyotypic heterogeneity has been linked to tumourigenicity. All cell lines displaying high tumourigenicity were characterized by high levels of genome heterogeneity (the high frequencies of NCCAs), regardless of which molecular mechanisms were deregulated. In contrast, all cell lines with low tumourigenicity displayed distinctly lower frequencies of NCCAs.

Genomic instability drives resistance in two ways. First, it generates population heterogeneity which increases the probability to survive drug treatment. The more different combinations of molecular mechanisms exist within a cancer cell population, the more likely a population survives. Second, drug treatment-mediated stress may induce genome chaos, which is accompanied by large-scale genome changes and increased population heterogeneity. This favours the emergence of resistant cells. Altogether, cancer genome evolution is the key event in cancer initiation-progression and drug resistance (reviewed in [10–15, 20, 27], see also Chap. 12).

In building a framework for cancer biology, Nicholson and Duesberg [28] conceptualized:

> Neoplastic transformation occurs because carcinogens, including conventional mutagenic and nonmutagenic carcinogens, or activated oncogenes destabilize the karyotype by inducing random aneuploidy. Aneuploidy destabilizes the karyotype by unbalancing teams of proteins that segregate, synthesize and repair chromosomes - in proportion to the degree of aneuploidy. Aneuploidy initiates and maintains karyotypic evolutions automatically because of the inherent instability of aneuploidy... Occasionally, rare cancer-causing karyotypes evolve stochastically... Flexibility and heterogeneity of cancer karyotypes is also the basis for the further, spontaneous evolutions that are known as tumor progression, such as metastasis and drug resistance.

In support of this argument, induction of aneuploidy by carcinogens and activated oncogenes is well documented (reviewed in [29–34]).

Dr. Peter Duesberg and colleagues formulated questions of great experimental and clinic importance that the gene mutation theory of cancer founds difficult answering: why do transgenes produce conditionally reversible hyperplasias and dysplasias early *versus* irreversible cancers late in conditional transgenic mice models? Why do a single transgene or a group of the same transgenes induce diverse cancers with different karyotypes, phenotypes, and transcriptomes in mice models? Why do some cancers of transgenic mice continue growing despite loss of or failure to express transgenic oncogenes? What does maintain the transformed phenotype of transgene-negative cancers? Why is cancer caused by non-mutagenic carcinogens? Why do cancers develop years to decades after the initiation by carcinogens (long latent periods) and follow pre-neoplastic aneuploidy? Why is cancer chromosomally and phenotypically unstable and generates much more complex phenotypes than conventional mutation as well as non-selective phenotypes (i.e. phenotypes that are unnecessary for tumour formation at the site of its origin), such as metastasis or multidrug resistance? Another set of questions referred to (multi-)drug resistance, in particular: why are cancer cells intrinsically resistant or rapidly acquire resistance against numerous drugs, and sometimes loose drug resistance in the absence of drugs? How do cancer cells generate complex resistance phenotypes against dozens of drugs? Why are drug treatment response and the acquisition of drug resistance accompanied by alterations of the genome, DNA methylation, transcriptome/proteome, metabolome, morphology, etc.? These fundamental questions remain unanswered by the advocates of the gene mutation theory of cancer. By contrast, these issues are faithfully explained by the genome theory of cancer (reviewed in [28, 35–42]).

13.2.2 The Illusion of Oncogene Addiction: Why Cancer Models Do Not Recapitulate a Natural Tumour

The term "oncogene addiction" was introduced by Bernard Weinstein to describe the dependency of tumour cells on a single activated oncogenic protein or pathway to maintain their malignant properties [43]. The oncogene addiction concept is based on results derived from tumour culture cell lines and conditional transgenic animal models in which acute inactivation of the overexpressed wild-type or mutated oncogenes resulted in rapid apoptosis or growth arrest and consequent tumour regression. However, many research groups monitoring long-term tumour response in diverse mouse models after oncoprotein withdrawal repeatedly observed tumour relapses. Tumour escape from oncogene addiction upon the primary oncogene inactivation was attributed to the acquisition of the diverse novel genetic lesions through CIN (reviewed in [34]).

The advocates of the concept of oncogene addiction name chronic myeloid leukaemia (CML) and the drug imatinib as the most successful example of this concept application in clinic. However, imatinib is an exception in cancer research and its success does not carry over to most solid tumours (reviewed in [27]). Actually, CML patients in the chronic phase of disease (characterized by a relatively stable genome and comparable to the benign phase of solid tumours), but not in the late phases (accelerated or blast crisis characterized by highly dynamic genome instability), perfectly respond to imatinib treatment. Imatinib was originally designed to target BCR-ABL fusion tyrosine kinase and a "magic bullet" effect of imatinib was ascribed due to the specific inhibition of BCR-ABL tumour "driver". However, it is now well documented that imatinib also inhibits the receptor tyrosine kinases and non-receptor tyrosine kinases, as well as RAF kinase family members, and the oxidoreductase NQO2. Thus, imatinib should be referred to as a multi-targeted cytotoxic drug (reviewed in Stepanenko and Dmitrenko, 2015). Moreover, recent studies revealed that imatinib fails to eradicate BCR-ABL-positive CML stem cells even at high concentrations. Authors concluded that "cancers are never truly oncogene addicted" [44].

The models, which afforded grounds for establishing the oncogene addiction concept, have obvious shortcomings and pitfalls. Cell lines display the markers of genetic drift and are characterized by low genomic heterogeneity as a consequence of selection and adaptation for cell culture conditions [45]. Furthermore, pure cultures of cells ignores the fact that tumours are surrounded by stromal cells that can provide nutrients and additional signals needed for cell growth [46].

In addition, other simplicities significantly reduce the generalization potential of *in vitro* tests: lack of cells of the immune system; lack of a complex of vessels supplying and removing fluids with many different parameters of flow, pressure, and nutrient levels; lack of spatial and temporal variations in external drivers of cancer evolution and growth; and lack of gradients in oxygen tension, CO_2 levels, applied drugs, etc. Gillet et al. [47] investigated the multidrug-resistant transcriptome of six cancer types in established cancer cell lines (grown in monolayer, 3D scaffold, or in xenograft) and clinical samples and revealed that:

> All of the cell lines, grown either *in vitro* or *in vivo*, bear more resemblance to each other, regardless of the tissue of origin, than to the clinical samples they are supposed to model.

Authors [47, 48] invoke investigators "to be aware of the associated caveats and temper their extrapolations so as not to infer direct applicability to clinical medicine". Furthermore, many anticancer agents have dose-limiting organ toxicities that are not represented in model systems such as cultured cancer cells [49].

Advantages and disadvantages of orthotropic xenografts of human tumours and genetically engineered mouse cancer models are thoroughly considered elsewhere (reviewed in [50–53]). Due to a limited number of initiating genetic alterations, transgenic mouse tumours are typically more homogeneous than human tumours [54]. In transgenic models, the initial conditions (such as the overexpression of specific oncogenes) form a dominant pathway through artificial selection that drastically reduces genome heterogeneity and artificially favours cancer progression [11].

Artificially activated oncogenes, benign levels of CIN, intra-tumour genetic homogeneity, and fostered evolution of tumour cells and their microenvironment make mouse tumour model inappropriate for the targeted treatment of human cancers. As Richmond and Su [53] sharply notice, "If one wants to know whether a patient's tumor will respond to a specific therapeutic regime, one must examine the response of that human tumor, not a mouse tumor, to the therapy" and further "We can cure many mouse tumors, but there is not a direct correlation between response in the mouse and response in the clinic".

Actually, cancer therapy based on the oncogene addiction concept is palliative rather than curative. Targeted drugs are highly successful from a financial perspective but have little curative impact/beneficial effect on patient survival. The unprecedented level of intra-tumour heterogeneity, the existence of myriads of genetic networks, and the complex protective phenotype response to chronic drug treatment make solid tumours independent from any particular oncogene and clinical utility of many cancer genes as targets for cancer treatment uncertain and challengeable. A failure of targeted therapies in clinical trials and the limited relevance of the oncogene addiction concept for the majority of tumours should lead researchers to abandon the idea of seeking for and targeting the putative addictive oncogene that maintains one's cancer.

13.3 Implications for Policy and Practice

Firstly, genomic heterogeneity may be used as a predictor of evolutionary potential as heterogeneity provides a greater chance to adapt and survive. Indeed, the overall genomic heterogeneity significantly correlates with tumourigenic potential of cells, tumour disease progression, patient survival, intrinsic and acquired (multi)drug resistance, and radio resistance [55, 56].

Secondly, the phase of cancer evolution should be monitored. Treatment applied at the different phases of genome evolution may have differential impact on tumour aggressiveness and patient survival [12, 57].

Thirdly, we should not aim at killing tumour cells with the highest tolerable drug concentrations. High-dose chemotherapeutic-mediated stress may significantly increase tumour evolution by generating novel phenotypes through induction of genome chaos. Therapy-induced genome instability should be avoided; instead, efforts should be directed at pushing the genome of cancer cells towards a stable phase and supporting the immunological system as well as homeostasis of the individual. These should lower the evolutionary potential of a tumour and constrain its dynamics due to reducing population diversity [14]. By contrast, the therapeutic promotion of excessive instability of the genome in tumour cells is a double-edged sword: while the primary objective response in the form of reduced cell viability will be positive, the price for a moderate inhibition of tumour growth will be changes in the genomic landscape, tumour subclonal architecture, and, eventually, promotion of cancer evolution that impacts both the patient survival and the therapeutic management of recurrence [13, 57, 58].

Fourthly, we should stop cataloguing putative cancer genes and classifying them on tumour suppressor genes and oncogenes. Context-dependent antagonistic functional duality of cancer genes is widely documented (reviewed in [16, 59]). The genome changes rewire the genetic network and may result in alterations of the role and function of the same genes and pathways within different genetic background. Cells within the same tumour may differentially respond to a drug, succumbing to senescence and death or on the contrary demonstrating enhanced growth and invasion. These are the instructive examples of "paradoxical" effects of anticancer drugs depending on the cellular genetic background/signalling network [16].

Acknowledgments This work was supported in frames of the programs "Fundamental grounds of molecular and cell biotechnologies" and "Nanotechnologies and nanomaterials for 2010–2014 years" by the National Academy of Sciences of Ukraine (NASU).

Glossary

Aneuploidy	Refers to a state of karyotype when whole chromosome(s) or parts of chromosome(s) are lost or supernumerary
Chromosome instability (CIN)	Is a high rate of genome changes of a given cell population (cell to cell variation). It implies a constant process of generation of numerical (loss or gain of whole chromosomes) and structural (loss of chromosome arm, translocations, amplifications, deletions, insertions) aneuploidy variants
Clonal chromosome aberration	Is an aberration found in two cells or more among at least 20 examined metaphases
Dysplasia	Is appearance of the tissue as disordered, with the increased number of immature cells, and great variability between cells
Evolutionary potential	Is a probability of cell population to persist, adapt, and survive the harsh microenvironment, intrinsic or extrinsic stresses
Genetic network of a cell	Includes the whole gene content, RNA, and protein expression and their interaction in space and time
Genetically engineered mouse cancer model	Is a model when a mouse genetic profile is altered such that one or several genes thought to be involved in tumourigenesis are mutated, deleted, or overexpressed

Hyperplasia	Is a condition when cell number increases due to hyperproliferation unbalanced by cell elimination that results to an increase in the amount/volume of a tissue/organ
Non-clonal chromosome aberration	Is an aberration found in only one cell among at least 20–50 examined metaphases
Orthotopic xenograft	Is the transplantation of a primary human tumour mass or the injection of human tumour cell line into a mouse tissue from which a tumour mass/cell line naturally originated
Transcriptome	Is the complete set of mRNA, rRNA, tRNA, and other non-coding RNA transcripts produced by the genome of a cell or a population of cells at any given time
Transgene	Is a foreign gene that has been deliberately transferred into genome of a cell/an organism by the genetic engineering techniques
Transgene-negative tumours	Are tumours formed by cells, which lost a transgene that endowed advantageous traits and accelerated propagation
Somatic evolution	Is the accumulation of heritable (through mitosis) variations such as mutations, epigenetic changes, and sporadic aneuploidy in somatic cells within a body during a lifetime
Stochastic nature of cancer evolution	Implies that cancer is mainly driven by random, non-clonal, and transitional genome alterations, and these dynamic genome changes are neither shared by cells of the same tumour nor by the different tumours

References

1. Forbes SA, Beare D, Gunasekaran P et al (2014) COSMIC: exploring the world's knowledge of somatic mutations in human cancer. Nucleic Acids Res 43:D805–D811. doi:10.1093/nar/gku1075
2. An O, Pendino V, D'Antonio M et al (2014) NCG 4.0: the network of cancer genes in the era of massive mutational screenings of cancer genomes. Database (Oxford) 2014. doi:10.1093/database/bau015
3. Heng HHQ (2007) Cancer genome sequencing: the challenges ahead. BioEssays 29:783–794. doi:10.1002/bies.20610

4. Salk JJ, Fox EJ, Loeb LA (2010) Mutational heterogeneity in human cancers: origin and consequences. Annu Rev Pathol 5:51–75. doi:10.1146/annurev-pathol-121808-102113

5. Todorovic-Rakovic N (2011) Genome-based versus gene-based theory of cancer: possible implications for clinical practice. J Biosci 36:719–724

6. Brosnan JA, Iacobuzio-Donahue CA (2012) A new branch on the tree: next-generation sequencing in the study of cancer evolution. Semin Cell Dev Biol 23:237–242. doi:10.1016/j.semcdb.2011.12.008

7. Heng HHQ, Liu G, Stevens JB et al (2011) Decoding the genome beyond sequencing: the new phase of genomic research. Genomics 98:242–252. doi:10.1016/j.ygeno.2011.05.008

8. Polyak K (2008) Is breast tumor progression really linear? Clin Cancer Res 14:339–341. doi:10.1158/1078-0432.CCR-07-2188

9. Heng HHQ, Liu G, Bremer S et al (2006) Clonal and non-clonal chromosome aberrations and genome variation and aberration. Genome 204:195–204. doi:10.1139/G06-023

10. Heng HHQ, Stevens JB, Bremer SW et al (2010) The evolutionary mechanism of cancer. J Cell Biochem 109:1072–1084. doi:10.1002/jcb.22497

11. Heng HHQ, Stevens JB, Bremer SW et al (2011) Evolutionary mechanisms and diversity in cancer. Adv Cancer Res. doi:10.1016/B978-0-12-387688-1.00008-9

12. Heng HH, Bremer SW, Stevens JB et al (2013) Chromosomal instability (CIN): what it is and why it is crucial to cancer evolution. Cancer Metastasis Rev 32:325–340. doi:10.1007/s10555-013-9427-7

13. Horne SD, Pollick SA, Heng HHQ (2015) Evolutionary mechanism unifies the hallmarks of cancer. Int J Cancer 136:2012–2021. doi:10.1002/ijc.29031

14. Heng HHQ, Bremer SW, Stevens JB et al (2009) Genetic and epigenetic heterogeneity in cancer: a genome-centric perspective. J Cell Physiol 220:538–547. doi:10.1002/jcp.21799

15. Heng HHQ (2009) The genome-centric concept: resynthesis of evolutionary theory. BioEssays 31:512–525. doi:10.1002/bies.200800182

16. Stepanenko AA, Vassetzky YS, Kavsan VM (2013) Antagonistic functional duality of cancer genes. Gene 529:199–207. doi:10.1016/j.gene.2013.07.047

17. Abdallah BY, Horne SD, Stevens JB et al (2013) Why unstable genomes are incompatible with average profiles single cell heterogeneity. Cell cycle 12:3640–3649

18. Liu G, Stevens JB, Horne SD et al (2014) Genome chaos: survival strategy during crisis. Cell Cycle 13:528–537. doi:10.4161/cc.27378

19. Stevens JB, Abdallah BY, Liu G et al (2011) Diverse system stresses: common mechanisms of chromosome fragmentation. Cell Death Dis 2:e178. doi:10.1038/cddis.2011.60

20. Stevens JB, Horne SD, Abdallah BY et al (2013) Chromosomal instability and transcriptome dynamics in cancer. Cancer Metastasis Rev 32:391–402. doi:10.1007/s10555-013-9428-6

21. Ye CJ, Liu G, Bremer SW, Heng HHQ (2007) The dynamics of cancer chromosomes and genomes. Cytogenet Genome Res 118:237–246. doi:10.1159/000108306

22. Ye CJ, Stevens JB, Liu G et al (2009) Genome based cell population heterogeneity promotes tumorigenicity: the evolutionary mechanism of cancer. J Cell Physiol 219:288–300. doi:10.1002/jcp.21663

23. Li L, McCormack AA, Nicholson JM et al (2009) Cancer-causing karyotypes: chromosomal equilibria between destabilizing aneuploidy and stabilizing selection for oncogenic function. Cancer Genet Cytogenet 188:1–25. doi:10.1016/j.cancergencyto.2008.08.016

24. Silk AD, Zasadil LM, Holland AJ et al (2013) Chromosome missegregation rate predicts whether aneuploidy will promote or suppress tumors. Proc Natl Acad Sci USA 110:E4134–E4141. doi:10.1073/pnas.1317042110

25. Birkbak NJ, Eklund AC, Li Q et al (2011) Paradoxical relationship between chromosomal instability and survival outcome in cancer. Cancer Res 71:3447–3452. doi:10.1158/0008-5472.CAN-10-3667

26. Roylance R, Endesfelder D, Gorman P et al (2011) Relationship of extreme chromosomal instability with long-term survival in a retrospective analysis of primary breast cancer. Cancer Epidemiol Biomarkers Prev 20:2183–2194. doi:10.1158/1055-9965.EPI-11-0343

27. Horne SD, Stevens JB, Abdallah BY et al (2013) Why imatinib remains an exception of cancer research. J Cell Physiol 228:665–670. doi:10.1002/jcp.24233
28. Nicholson JM, Duesberg P (2009) On the karyotypic origin and evolution of cancer cells. Cancer Genet Cytogenet 194:96–110. doi:10.1016/j.cancergencyto.2009.06.008
29. Fukasawa K (2007) Oncogenes and tumour suppressors take on centrosomes. Nat Rev Cancer 7:911–924. doi:10.1038/nrc2249
30. Harrison MK, Adon AM, Saavedra HI (2011) The G1 phase Cdks regulate the centrosome cycle and mediate oncogene-dependent centrosome amplification. Cell Div 6:2. doi:10.1186/1747-1028-6-2
31. Janssen A, Medema RH (2013) Genetic instability: tipping the balance. Oncogene 32:4459–4470. doi:10.1038/onc.2012.576
32. Lecona E, Fernández-Capetillo O (2014) Replication stress and cancer: it takes two to tango. Exp Cell Res 329:26–34. doi:10.1016/j.yexcr.2014.09.019
33. Thompson SL, Bakhoum SF, Compton DA (2010) Mechanisms of chromosomal instability. Curr Biol 20:R285–R295. doi:10.1016/j.cub.2010.01.034
34. Stepanenko A, Kavsan V (2013) Cancer genes and chromosome instability. In Oncogene and Cancer – Fom Bench to Clinic (Siregar Y ed.), pp. 151–182. InTech Publisher, Rijeka, Croatia.
35. Duesberg P, Li R, Fabarius A, Hehlmann R (2005) The chromosomal basis of cancer. Cell Oncol 27:293–318
36. Duesberg P, Li R, Sachs R et al (2007) Cancer drug resistance: the central role of the karyotype. Drug Resist Updat 10:51–58. doi:10.1016/j.drup.2007.02.003
37. Fabarius A, Li R, Yerganian G et al (2008) Specific clones of spontaneously evolving karyotypes generate individuality of cancers. Cancer Genet Cytogenet 180:89–99. doi:10.1016/j.cancergencyto.2007.10.006
38. Li L, McCormack AA, Nicholson JM et al (2009) Cancer-causing karyotypes: chromosomal equilibria between destabilizing aneuploidy and stabilizing selection for oncogenic function. Cancer Genet Cytogenet 188:1–25. doi:10.1016/j.cancergencyto.2008.08.016
39. Klein A, Li N, Nicholson JM et al (2010) Transgenic oncogenes induce oncogene-independent cancers with individual karyotypes and phenotypes. Cancer Genet Cytogenet 200:79–99. doi:10.1016/j.cancergencyto.2010.04.008
40. McCormack A, Fan JL, Duesberg M et al (2013) Individual karyotypes at the origins of cervical carcinomas. Mol Cytogenet 6:44. doi:10.1186/1755-8166-6-44
41. Duesberg P, McCormack A (2013) Immortality of cancers: a consequence of inherent karyotypic variations and selections for autonomy. Cell Cycle 12:783–802. doi:10.4161/cc.23720
42. Duesberg P, Mandrioli D, McCormack A, Nicholson JM (2011) Is carcinogenesis a form of speciation? Cell Cycle 10:2100–2114
43. Weinstein IB (2002) Cancer. Addiction to oncogenes–the Achilles heal of cancer. Science 297:63–64. doi:10.1126/science.1073096
44. Pellicano F, Mukherjee L, Holyoake TL (2014) Concise review: cancer cells escape from oncogene addiction: understanding the mechanisms behind treatment failure for more effective targeting. Stem Cells 32:1373–1379. doi:10.1002/stem.1678
45. Liedtke C, Wang J, Tordai A et al (2010) Clinical evaluation of chemotherapy response predictors developed from breast cancer cell lines. Breast Cancer Res Treat 121:301–309. doi:10.1007/s10549-009-0445-7
46. Kessler DA, Austin RH, Levine H (2014) Resistance to chemotherapy: patient variability and cellular heterogeneity. Cancer Res 74:4663–4670. doi:10.1158/0008-5472.CAN-14-0118
47. Gillet J-P, Calcagno AM, Varma S et al (2011) Redefining the relevance of established cancer cell lines to the study of mechanisms of clinical anti-cancer drug resistance. Proc Natl Acad Sci USA 108:18708–18713. doi:10.1073/pnas.1111840108
48. Gillet J-P, Varma S, Gottesman MM (2013) The clinical relevance of cancer cell lines. J Natl Cancer Inst 105:452–458. doi:10.1093/jnci/djt007
49. Weinstein JN (2012) Drug discovery: cell lines battle cancer. Nature 483:544–545. doi:10.1038/483544a

50. Taneja P, Zhu S, Maglic D et al (2011) Transgenic and knockout mice models to reveal the functions of tumor suppressor genes. Clin Med Insights Oncol 5:235–257. doi:10.4137/CMO. S7516
51. Herter-Sprie GS, Kung AL, Wong K-K (2013) New cast for a new era: preclinical cancer drug development revisited. J Clin Invest 123:3639–3645. doi:10.1172/JCI68340
52. Politi K, Pao W (2011) How genetically engineered mouse tumor models provide insights into human cancers. J Clin Oncol 29:2273–2281. doi:10.1200/JCO.2010.30.8304
53. Richmond A, Su Y (2008) Mouse xenograft models vs GEM models for human cancer therapeutics. Dis Model Mech 1:78–82. doi:10.1242/dmm.000976
54. Cheon D-J, Orsulic S (2011) Mouse models of cancer. Annu Rev Pathol 6:95–119. doi:10. 1146/annurev.pathol.3.121806.154244
55. Stepanenko AA, Kavsan VM (2012) Immortalization and malignant transformation of Eukaryotic cells. Cytol Genet 46:96–129. doi:10.3103/S0095452712020041
56. Stepanenko AA, Kavsan VM (2012) Evolutionary karyotypic theory of cancer versus conventional cancer gene mutation theory. Biopolym Cell 28:267–280. doi:10.7124/bc. 000059
57. Stepanenko A, Andreieva S, Korets K, Mykytenko D (2015) Step-wise and punctuated genome evolution drive phenotype changes of tumor cells. Mutat Res—Fundam Mol Mech Mutagen 771:56–69. doi:10.1016/j.mrfmmm.2014.12.006
58. Huang S (2013) Genetic and non-genetic instability in tumor progression: link between the fitness landscape and the epigenetic landscape of cancer cells. Cancer Metastasis Rev 32:423–448. doi:10.1007/s10555-013-9435-7
59. Lou X, Zhang J, Liu S et al (2014) The other side of the coin: the tumor-suppressive aspect of oncogenes and the oncogenic aspect of tumor-suppressive genes, such as those along the CCND-CDK4/6-RB axis. Cell Cycle 13:1677–1693. doi:10.4161/cc.29082

Chapter 14
Evolution, Infection, and Cancer

Prof. Paul W. Ewald, Ph.D. and Holly A. Swain Ewald, Ph.D.

Lay Summary The occurrence of cancer depends on three evolutionary processes: normal cells evolve into cancer cells, humans and other species have evolved biological protections against cancer, and disease organisms have evolved countermeasures that subvert these protective mechanisms. Evolutionary thinking has led not only to the recognition of these three processes, but also to the emergence of a more balanced and comprehensive understanding of cancer, which emphasizes that causes of cancer need to be understood through the interplay among genes, germs, and the environment. This understanding is framed by a focus on the function of genes that promote oncogenesis and the effects of these genes on the environment both within and outside of the cells; cancer-promoting genes may belong to cancer cells, parasites, or both. An evolutionary perspective helps to identify the processes that most importantly contribute to cancer—and are therefore the most important to prevent—even though innumerable processes are altered during oncogenesis. It suggests that generation of cancer solely by mutations is difficult because several critical barriers must be abrogated in succession without the occurrence of other mutations that make the cell non-functional. On the other hand, although natural selection commonly moulds parasites to compromise simultaneously the critical protections against cancer, infection alone is also insufficient to bring about human cancer.

Parasites are known to contribute about 20 % of all human cancer and may play a role in much, if not most, of the remaining 80 %. Evolutionary considerations and epidemiological evidence suggest that pathogens transmitted

P.W. Ewald (✉) · H.A. Swain Ewald
Department of Biology, University of Louisville, Louisville, KY 40292, USA
e-mail: pw.ewald@louisville.edu

H.A. Swain Ewald
e-mail: h.swainewald@louisville.edu

© Springer International Publishing Switzerland 2016
A. Alvergne et al. (eds.), *Evolutionary Thinking in Medicine*,
Advances in the Evolutionary Analysis of Human Behaviour,
DOI 10.1007/978-3-319-29716-3_14

by sexual contact, saliva, and hypodermic needles are disproportionately important infectious causes of human cancer. The possible influence of reducing these routes of transmission can be evaluated by comparing cancer incidences among populations in the context of the full spectrum of risk factors, as illustrated by comparisons among residents of Utah. Overall, evolutionary considerations suggest that inadequate attention has been given to possible infectious causes of cancer, and that control of infection may prove to be one of the most effective ways to control cancer.

14.1 Introduction to Evolution and Selection in Oncogenesis

Three evolutionary actions of selection are important in oncogenesis, the process by which normal cells acquire the characteristics of cancer cells. First, oncogenic selection acts through the increased survival and reproduction of cells that are genetically modified relative to normal cells in the body [1]. Second, natural selection generates adaptations in multicellular organisms that reduce their risks of cancer [2–6]. Third, natural selection also moulds infectious agents in ways that may compromise the adaptations that multicellular hosts have evolved to reduce cancer risk [1, 7, 8]. We consider each of these selective processes below.

14.1.1 Oncogenic Selection

Normal cells evolve into cancer cells in part through selection acting on genetic variation that arises from genetic mutations. This process is similar to natural selection in that genetic composition of a population changes over time as a result of differences in survival and reproduction. It is referred to as oncogenic selection [1], however, because it differs from natural selection in two distinct ways.

The most basic difference is that oncogenic selection involves the differential survival and reproduction of cells within the organism rather than that of the organism itself. Oncogenic selection results in changes in genetic composition of a population of cells within an organism rather than changes in the genetic composition of a population of organisms. Natural selection has moulded normal cells to restrict their own survival and reproduction when this regulation increases the survival and reproduction of the multicellular organism to which they belong. In contrast, oncogenic selection favours cells that lose such regulatory mechanisms when this loss increases their number relative to other cells in the body. Oncogenic selection therefore tends to involve the breaking rather than the refinement of regulatory adaptations.

The second major difference between oncogenic and natural selection involves long-term opportunities for evolutionary adaptation. Natural selection acting on an organism is open-ended, whereas selection of somatic cells within an organism is truncated by the death of the organism. Oncogenic selection therefore cannot generate the unending cumulative change and sophistication of adaptations that arises from natural selection. The reason is that cancer cells cannot, as a rule, be transmitted from one individual to another. In the rare exceptions to this rule, opportunities for future evolution become open-ended, and the cancer cells go through the transition from being cells of a multicellular organism to becoming a parasitic organism. When this transition occurs, oncogenic selection ends and natural selection begins. Two cancers that have passed through this transition have been well studied: transmissible venereal tumour of dogs and the facial tumours of Tasmanian devils [9, 10].

14.1.2 Natural Selection on Multicellular Organisms

Natural selection has led to adaptations that guard against oncogenesis. Barriers are defined as adaptations that block oncogenesis when they are in place [1]. The four main barriers are cell cycle arrest, apoptosis, telomerase regulation, and cell adhesion [1]. The presence of barriers may vary according to cell type, resulting in different vulnerabilities to oncogenesis [1]. Restraints inhibit but do not block oncogenesis [1]. Regulation of division rate of a dividing cell, for example, is a restraint, which retards rather than prevents oncogenesis, because even a slowly replicating cell can proceed down the path of oncogenesis. Alterations that compromise barriers are defined as essential causes of cancer; those that interfere with restraints are exacerbating causes [1]. Understanding whether a target is part of an essential or exacerbating cause is important for determining whether an intervention could be preventive, curative, or just ameliorative.

14.1.3 Natural Selection on Infectious Agents of Cancer

Approximately 20 % of human cancers are caused by parasites [11], here defined as self-replicating entities (e.g. viruses, bacteria, protozoa, or multicellular organisms) that live in or on a host organism and harm it. The extent to which infection contributes to the remaining 80 % of human cancer is not known, because a causal role for parasites can be ruled out for very few of these cancers.

Natural selection moulds infectious agents to exploit their hosts in ways that increase their own evolutionary fitness. When an infectious agent contributes to oncogenesis, it is important to determine the extent to which this contribution compromises restraints or barriers. If a parasite compromises one or more barriers, then prevention or cure of the infection may prevent oncogenesis.

Barriers to cancer can also be barriers to persistence within a host, particularly for intracellular parasites such as viruses. Breaking cell cycle arrest allows the viral genome within the cell to replicate in concert with cellular replication. By relaxing control of the synthesis of telomerase, the viral genome removes the cap on the total number of divisions a cell can undergo. By inhibiting apoptosis, a virus can reduce the chance that it will be destroyed by cellular self-destruction. By altering cell adhesion, viruses allow infected cells to disperse to new locations in the body. Together, these compromises of barriers to cancer enhance persistence because they allow a virus to replicate its genome with less exposure to the immune system than would occur through the release of virions from cells. By favouring viruses that subvert the host barriers to persistence, natural selection may lead to the evolution of increased oncogenicity in viruses.

Several cancer-causing viruses of humans have been sufficiently well investigated to evaluate whether they compromise these barriers: Epstein Barr virus (EBV), Kaposi's sarcoma-associated herpes virus (KSHV), hepatitis B virus (HBV), hepatitis C virus (HCV), human papillomavirus (HPV), and human T-cell lymphotropic virus type 1 (HTLV-1). Each of these viruses compromises all four barriers (reviewed by [8]). Each virus therefore contributes to four essential causes of cancer. This simultaneous compromising of four barriers to cancer is important for oncogenesis because it can generate large populations of infected, dividing cells, within which a relatively small number of additional mutations are needed to complete the progression to cancer.

This emerging understanding of oncogenic viruses contrasts markedly with earlier presumptions about the ways in which infectious agents contributed to oncogenesis. When parasites were first associated with oncogenesis, it was generally presumed that they exacerbated the mutation-driven process of cancer. Infection results in inflammation, which can increase rates of cellular proliferation [12] and generate reactive compounds that cause mutations. Through these effects, infection would generally contribute exacerbating rather than essential causes of oncogenesis, because most mutations would tend to occur in genes other than the few that maintain barriers. In contrast, oncogenic viruses contribute to multiple essential causes of oncogenesis, because from the onset of infection they are abrogating multiple barriers to cancer.

14.2   Research Findings

14.2.1   The Triad of Disease Causation

There are three general categories of disease causation: genetic, parasitic, and non-parasitic environmental causes. Diseases are often referred to using other adjectives, such as developmental, endocrinological, or neurological, but to explain aetiology more deeply, one must invoke at least one of the three general categories of causation. The triad of disease causation (Fig. 14.1) emphasizes that different aetiologies

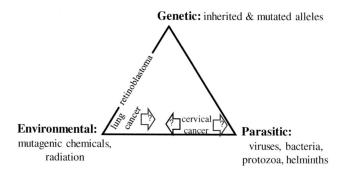

Fig. 14.1 The triad of disease causation: a visual aid for thinking about the spectrum of causation and the interplay between different causes. Placement of a cancer corresponds to the degree to which the cause compromises barriers (i.e. adaptations that block oncogenesis) as opposed to restraints (i.e. adaptations that hinder oncogenesis)

co-occur and interact. Genetic causes of cancer can be inherited or can arise because of mutations that are generated by parasitic or nonparasitic environmental causes. Parasitism can contribute to oncogenesis by compromising barriers and restraints but rarely if ever generate cancers without contributions from mutations. In addition, genetic variation in resistance to parasitism is ubiquitous and therefore must also influence contributions of parasites to oncogenesis.

The precise placement of a cancer within the triangle is almost always tentative and corresponds to the extent to which each of the three categories encompasses essential causes. Retinoblastoma is the best example of a cancer for which evidence implicates inherited alleles and additional mutations without contributions from an infectious agent [1]; it is therefore placed directly on the genetic-environmental axis in Fig. 14.1. Cervical cancer is placed close to the parasitic vertex because the causal agent, HPV, encodes proteins that compromise four barriers to cancer. Current evidence indicates that the frequencies of mutations in genes that maintain barriers tend to be relatively low in cervical cancer; these mutations therefore are not necessary for oncogenesis when viruses are present [13]. The arrows containing a question mark in Fig. 14.1 are inserted to acknowledge the remaining uncertainty about the overall contribution of environmentally induced mutations (e.g. by compounds in tobacco smoke and the overall net effect of mutations).

Inherited vulnerability to cervical cancer contributes to the placement of cervical cancer above the environmental/parasitic axis. Much of this contribution involves variation in immunological defences against viral infection [14–16] and therefore represents exacerbating causes of cervical cancer.

Consideration of the triad of causation guards against the error of overextending one category of explanation. The strong association between tobacco smoke and lung cancer, for example, has led to the sense that infection does not contribute to lung cancer, even among experts who recognize a broad role for infectious causation of cancer (e.g. [17]). It is well known that smoking can increase the probability of pulmonary infections; yet until recently there has been little investigation of the possible involvement of infectious agents in lung cancer. The few studies that have addressed this issue have reported associations with JC virus, merkel cell polyomavirus, EBV, and HPV [18–20]. Each of these viruses compromises three or more barriers to cancer [8, 19].

14.2.2   The Extended Phenotype in Oncogenesis

Oncogenesis involves not only the evolution of cancer cells, but also modifications of their microenvironments, including alterations of extracellular molecules and effects on non-cancerous cells [21, 22]. Although these modifications may be complex and diffuse, the role of the microenvironment can be grasped by applying the concept of the extended phenotype, which is defined as the effects of a genetic variant on its environment [23]. The extended phenotype of a cancer cell, for example, includes elevated metalloproteases, which may degrade cell adhesion molecules and influence proliferation, angiogenesis, and metastasis [24, 25]. Viral effects on barriers to oncogenesis are part of the extended phenotype of the virus.

An important aspect of the extended phenotype of oncogenic viruses involves epigenetic changes: alterations in gene expression that are not associated with changes in DNA sequence but can be stable from one cellular generation to the next. These modifications can be associated with tumour initiation and progression. Tumours infected with oncogenic viruses have shown distinct epigenetic alterations relative to non-cancerous patient tissue [26] and to uninfected cancers that are of the same type [27]. Methylation of gene promoter regions has been shown to silence genes involved in barriers to cancer; investigators have found this category of epigenetic alteration associated, for example, with HPV-positive cervical cancer [13], HPV-positive oropharyngeal carcinoma, [28], and EBV-positive cancers [29]. Telomerase expression is often up-regulated in virally associated cancers. HPV type 16 can increase telomerase expression in infected cervical carcinoma cell lines by decreasing methylation of the promoter for the hTERT subunit of telomerase [30]. In addition, epigenetic modifications at gene loci associated with cell adhesion have been identified in cervical cancer, HPV-positive oropharyngeal carcinomas, and EBV-related nasopharyngeal carcinoma [27, 31, 32]. Importantly, because gene expression abnormalities associated with epigenetic modifications are not muta-tions, they are potentially reversible; for example, experimental knock down of the E6 protein of HPV types 16 and 18 in cervical cancer cell lines restored silencing of telomerase [30].

Without the conceptual structure provided by the extended phenotype, the vast collection of microenvironmental and intracellular alterations during oncogenesis could be overwhelming. Together with the concepts of barriers and essential causes, the extended phenotype maintains the emphasis on a relatively small number of processes that result directly or indirectly from oncogenic genes in the genomes of the cancer cell and oncogenic parasites.

14.2.3 Transmission of Oncogenic Parasites

An evolutionary perspective suggests that selection for persistence of viruses, and hence potential oncogenicity, will be especially strong when opportunities for transmission are widely spaced over time. This condition applies to sexual trans- mission because opportunities for sexual transmission depend on changes in sexual partnerships, which tend to occur less frequently than, for example, opportunities for transmission by sneezing or coughing. Similarly, pathogens transmitted by intimate kissing should be subject to strong selective pressure for persistence because new intimate kissing partnerships tend to be temporally spaced. Selection for persistence will also be strong when pathogens are maintained across genera- tions by transmission through milk, because such transmission requires persistence within a host from infancy until adulthood and is augmented by extended persis- tence over periods that encompass successive births. Opportunities for needle-borne transmission through intravenous drug use or blood donation are also relatively infrequent. Because molecular mechanisms for persistence often compromise crit- ical barriers to oncogenesis (as discussed in Sect. 14.1.3), evolutionary consider- ations lead to the expectation that oncogenic capabilities should occur disproportionately among pathogens transmitted by sex, saliva, needles, and milk. This prediction is particularly applicable to viral pathogens because viruses infect intracellularly.

This evolutionary logic accords with the transmission routes of viruses that are accepted causes of human cancer. Among oncogenic viruses, the predominant route of transmission is sexual, with needle-borne and salivary transmission being present to a lesser extent (Table 14.1). In contrast, these routes apply to only about one-quarter of all known human viruses. HTLV-1 is the only accepted tumour virus that is known to be maintained substantially in humans across generations through transmission by milk (Table 14.1). Candidate oncogenic viruses are transmitted largely by these routes (Tables 14.1 and 14.2).

Two unicellular pathogens are also accepted infectious causes of cancer: *Plasmodium falciparum* and *Helicobacter pylori* (Table 14.1). Oncogenic mecha- nisms are less well understood for these pathogens than for oncogenic viruses. Burkitt's lymphoma can be caused by EBV in the absence of *P. falciparum*, which apparently contributes to this cancer by enhancing EBV [33–36]. *P. falciparum* therefore appears to be an exacerbating rather than an essential cause. *H. pylori* enhances telomerase activity [37–39] but has complex and sometimes contradictory effects on other barriers [40, 41]. One of its proteins exerts anti-apoptotic effects that counter the host cell's apoptotic responses to the bacterium [42]. *H. pylori* also stimulates pro-inflammatory and growth factor signalling [43, 44], and is associated with increased telomerase expression [45] as well as with reduced adhesion and abrogation of cell cycle arrest [46]. *H. pylori* therefore can compromise the four barriers to oncogenesis that are abrogated by oncogenic viruses. *H. pylori* can infect intracellularly [47] and thus may benefit directly from the replication of its host cell.

Table 14.1 Parasites for which a causal role in human cancer has been generally accepted

Virus[a]	Mode of transmission	Cancers for which the parasite is	
		An accepted cause	A candidate cause
EBV	Saliva, sex	Burkitt's lymphoma, Hodgkin's lymphoma, gastric carcinoma, post-transplant proliferative disease, nasopharyngeal carcinoma	Breast, acute lymphoblastic leukaemia, ovarian, lung
HPV	Sex	Cervical, oropharyngeal, penile, anal, vulval, vaginal cancers	Breast, bladder, oesophagus, prostate, lung, skin
HTLV-1	Sex, needle, milk	Adult T-cell leukaemia and lymphoma	None
KSHV	Saliva, sex	Kaposi's sarcoma	Lung
HBV	Sex, needle, milk	Hepatocellular carcinoma	Cholangiocarcinoma, pancreas
HCV	Sex, needle	Hepatocellular carcinoma	Cholangiocarcinoma
MCPyV	Saliva	Merkel cell carcinoma	Lung
Helicobacter pylori	Saliva? Diarrhoea? Vomit?	Gastric carcinoma; Mucosa-associated lymphoid tissue MALT lymphoma	None
Plasmodium falciparum	Mosquitoes	Endemic Burkitt's lymphoma	None
Schistosomal and opisthorchid trematodes	Water contact, fish consumption	Cholangiocarcinoma, bladder	Colorectal

[a]*EBV* Epstein Barr virus = Human Herpes virus 4; *HPV* Human papilloma virus; *HTLV-1* Human T lymphotropic virus type 1 = human T-cell leukaemia/lymphoma virus type 1; *KSHV* Kaposi's sarcoma-associated herpes virus = human herpes virus 8; *HBV* Hepatitis B virus; *HCV* Hepatitis C virus; *MCPyV* Merkel cell polyomavirus. For references, see [1, 7, 8, 20, 67, 68]

When infecting extracellularly, it may benefit from stimulating the replication of host cells in its immediate vicinity if they provide some protection against stomach acidity.

14.2.4 Environmental and Infectious Risk Factors: An Illustration

The predominance of sexual and salivary transmission among tumour viruses suggests that reduction in the numbers of intimate partnerships would reduce the prevalence of a variety of cancers. Evaluation of this prediction is difficult because epidemiological studies of cancer generally must rely on correlation rather than experimentation.

Table 14.2 Parasites that have been associated with cancers but are not yet accepted causes of any cancer

Pathogen[a]	Mode of transmission	Cancers for which virus is a candidate cause	References
HCMV	Saliva, sex, milk	Brain (glioblastoma), prostate, breast	[69–72]
HHSV-2	Sex	Melanoma, prostate	[73]
JCV	Unknown	Brain, colorectal, oesophageal, lung, gastric	[18, 19, 74]
BKV	Unknown	Brain, bladder, kidney, ovary, prostate	[72]
SV40	Unknown	Brain, mesothelioma	[75]
MMTV	Milk in mice	Breast	[76]
BLV	Milk in cows	Breast	[77]
XMRV	Unknown	Prostate	[72]
Mycoplasma hominis	Sex	Prostate	[72]
Propionobacterium acnes	Skin contact	Prostate	[72]
Trichomonas vaginalis	Sex	Cervical	[78, 79]

[a]*HCMV* Human cytomegalovirus = Human Herpes virus 5; *HHSV-2* Human herpes simplex virus 2 = human herpes virus 2; *JCV* JC virus; *BKV* BK virus; *SV40* Simian virus 40; *MMTV* Mouse mammary tumour virus = human mammary tumour virus; *BLV* Bovine leukaemia virus; *XMRV* Xenotropic murine leukaemia virus-like retrovirus

Measured variables may be correlated with cancer but not causally involved, and important correlates may have been unmeasured. Quantification of the number of sexual partners, for example, does not incorporate the amount of sexual contact per partner and whether the sexual partners are themselves at high risk for infection. Moreover, measured variables could be correlated with unmeasured variables in unobvious ways. Smoking, for example, could be correlated with a tendency to be more risk prone and hence with unmeasured aspects of risky sexual behaviour.

In spite of these drawbacks, comparisons of cancer in populations characterized by different risk factors may provide a sense of the extent to which changes in behaviour might reduce cancer incidence. They may also illustrate the importance of considering alternative combinations of risk factors and unmeasured variables when attempting to determine causes of cancers.

Among the best-studied subjects for such assessments are members of the Church of the Latter Day Saints (LDS). LDS members have fewer sexual partners than non-LDS comparison populations [48]. Accordingly, cervical cancer among LDS women was about half as frequent as among non-LDS women [49–52]. Multivariate analyses indicated that about 40 % of this reduction was associated with fewer sexual partners [51]. Among LDS women who regularly attended church, the prevalence of cervical cancer was about 80 % lower, with just over one-third of the reduction being attributable to fewer sexual partners [51]. About 10 % of the lower prevalence of cervical cancer among LDS women was independently correlated with lower rates of smoking. Tobacco smoke contains carcinogenic compounds and is immunosuppressive [53, 54], and might therefore increase cancer rates by increasing mutation rates or reducing immunological control of precancerous lesions. Alternatively, as suggested above, smoking could be correlated with unmeasured variables that cause the differences in cancer rates.

The difficulties in interpreting smoking-associated risk of cervical cancer were addressed a quarter century ago. Two of the researchers involved with this debate [55] wrote:

> Definitive clarification of whether this association is causal will likely have to await definitive identification of the sexually transmitted agent which is probably the most important cause of cervical cancer. Only then will it be possible to clarify the contributions of risk factors with weaker associations with cervical cancer, such as cigarette smoking and socioeconomic status.

HPV was in the process of being recognized as the main cause of cervical cancer at about the time their paper was published. A few years later, a multifactorial analysis of the associations of smoking with cervical HPV infection showed that smoking was not significantly associated with HPV infection once sexual behaviour and other life-style variables were accounted for, leading the authors to conclude that smoking, alcohol, and drug use were correlates but not causes of HPV infection [56].

Most cancers are less prevalent among LDS members [49–52]. The lower rates of smoking-associated cancers (e.g. lung, cervical, bladder, colon, and laryngeal) among LDS members could be interpreted as a direct effect of lower exposure to tobacco smoke. Indeed when these reductions are discussed, the smoking-associated cancers are often grouped together implying that their lower rates result from lower exposure to tobacco smoke [50, 52]. An association with smoking, however, does not weaken the hypothesis that infectious causes are also involved, as illustrated by the associations of sexual behaviour with cervical cancer. The associations of sexually transmitted pathogens with smoking-associated cancers (e.g. lung, bladder, colon, and laryngeal in Tables 14.1 and 14.2) raise the possibility that infectious agents may be causally involved in more cancers than previously thought.

14.3 Implications for Policy and Practice

The control of infectious diseases through the use of vaccines, anti-infective agents, and interventions that block transmission has been among the most significant accomplishments in the history of medicine. Although the control of cancer remains largely an unfulfilled goal, control of infectious causes rank among the most successful interventions against cancer. These interventions include vaccination against

HPV and HBV (for cervical and liver cancers), prevention of transmission of HBV and HCV (for liver cancer), and control of *H. pylori* by improvements in hygiene and antibiotic treatment (for stomach cancers) [57–62].

Evolutionary considerations suggest that the relative importance of infectious causation is being underestimated, in part because pathogens evolve to compromise multiple barriers to cancer. To induce cancer without infection, multiple mutations (or epigenetic changes) must compromise several critical barriers in succession without making the cell non-functional [1].

Infectious causation has been accepted for about 20 % of all human cancer, and associations with infectious agents have been reported for most of the remaining 80 %. It is critically important to determine whether pathogens cause these cancers by compromising barriers. If so, their prevention should prevent the cancers they cause.

Many cancers may be controllable with the same interventions that are already in place but are being restricted to a particular cancer. The recent prophylactic vaccines against cervical cancer, for example, probably provide protection against other cancers induced by HPV, such as oropharyngeal, penile, and rectal cancers, and may provide protection against other cancers for which HPV is at present just a candidate cause, such as bladder cancer (Table 14.1).

Standard approaches of vaccination and prevention of transmission will undoubtedly contribute much to the future control of pathogen-induced cancers, once the causal pathogens are identified. Because pathogens differ from human cells in their biochemical make-up, discovery of infectious causes of cancer also offers new approaches to cancer prevention and control. Antivirals are becoming more effective and may soon provide anti-cancer benefits that mirror the effects of antibiotic treatment of stomach cancers. For example, researchers have identified cytomegalovirus (CMV) in a majority of glioblastoma (brain tumours) samples, and adjunctive treatment with antiviral therapy has shown a significantly extended patient survival rate relative to non-treated individuals [63].

Some possibilities capitalize on the sophistication of immune control of foreign organisms. Therapeutic vaccines based on oncogenic HPV proteins show efficacy for treatment of cervical cancer [64, 65]. EBV-specific T-cell therapy has shown promising results in the early phase clinical trials of recurrent and metastatic nasopharyngeal carcinoma, and efforts are underway to develop effective therapeutic vaccines, anti-EBV antibodies, and therapies that target viral-associated epigenetic changes [66]. Determining whether a therapeutic target is part of an essential or exacerbating cause is crucial because interference with essential causes offer promise for cures.

Concerted interventions on interacting causes of cancer have been enacted to reduce incidence of hepatocellular cancer by vaccination against HBV and reduction in exposure to aflatoxin [61]. Similar concerted efforts may help to control cancers caused by joint infections. A two-pronged attack on opisthorchid trematodes and hepatitis viruses is a promising example for the control of cholangiocarcinoma [67].

The progress and promise for controlling cancers by controlling their infectious causes warrants attention from individuals working across the spectrum of health sciences. Scientific policymakers need to weigh the benefits of funding research that attempts to identify infectious causes of cancers and development of interventions against known and candidate pathogens. Medical policymakers need to assess the appropriateness of alternative guidelines for interventions, such as vaccines, when the interventions have likely protective benefits against cancer in addition to documented benefits against other cancers (e.g. protection against oropharyngeal cancer in addition to cervical cancer for HPV vaccines) or against other diseases (e.g. vaccination against HBV for protection against hepatocellular carcinoma in addition to liver cirrhosis, or antibiotic treatment of *H. pylori* for protection against stomach cancer in addition to peptic ulcers). Similarly, practicing physicians need to be able to advise patients about possible benefits of particular interventions (e.g. protection against oropharyngeal cancer afforded by HPV vaccination in addition to protection against cervical cancer). As this research is continually developing, experts in each of these areas need to keep abreast of ongoing developments to enhance the accuracy of their decisions.

Acknowledgments This study was supported by a fellowship from the Wissenschaftskolleg zu Berlin and a grant from the Rena Shulsky Foundation to P.W.E. Sarah E. Ewald suggested that we look at LDS populations as to assess the possibility that populational differences in sexual behaviour are related to cancer risks.

Glossary

Aflatoxin	A toxin produced by a *Aspergillus* fungi; can damage liver cells and contribute to liver cancer
Angiogenesis	Generation of new blood vessels; can contribute to oncogenesis by increasing supply of resources to a tumour
Apoptosis	Programmed cell death; acts as a barrier to oncogenesis by terminating precancerous lineages of cells
Barrier to oncogenesis	A process that blocks oncogenesis
Cancer	A tumour with cells that are invasive or metastatic
Cell cycle arrest	The blocking of cellular replication by enforcement at a checkpoint in the cell cycle; an important checkpoint for oncogenesis is at the transition to the phase in which DNA replication occurs
Cholangiocarcinoma	A liver cancer derived from cells of the gall bladder

Epigenetic changes	Modifications to DNA that turn gene expression on or off (e.g. through methylation) and may be inherited across cellular divisions, but do not alter the DNA sequence
Essential causes of cancer	Factors that abrogate barriers to oncogenesis
Exacerbating causes of cancer	Factors that abrogate restraints on oncogenesis
Glioblastoma	An aggressive tumour that forms from glial cells of the brain or spinal cord
hTERT	The catalytic subunit of telomerase
Oncogenesis	The evolution of cancer cells from normal cells
Oncogenic selection	The differential survival and reproduction of cells during oncogenesis
Parasite	A replicating agent that lives in or on a host organism, on which it has a harmful effect
Pathogen	A parasite at or below the single-cell level of organization
Restraint on oncogenesis	A process that inhibits but does not block oncogenesis
T-cell therapy for cancer	A process in which T-cells are activated and used clinically to attack a tumour
Telomerase	The enzyme that maintains telomere length through the addition of telomere units that would otherwise be lost during each cycle of DNA synthesis; maintenance of telomeres allows the number of future cellular divisions to be unlimited
Tumour	An abnormal mass of new tissue growth
Virion	A virus particle released from an infected cell

References

1. Ewald PW, Swain Ewald HA (2013) Toward a general evolutionary theory of oncogenesis. Evol Appl 6:70–81. doi:10.1007/s00109-012-0891-2
2. Nowell PC (1976) The clonal evolution of tumor cell populations. Science 194(4260):23–28
3. Heppner GH, Miller FR (1998) The cellular basis of tumor progression. Int Rev Cytol 177:1–56
4. Merlo LM, Pepper JW, Reid BJ, Maley CC (2006) Cancer as an evolutionary and ecological process. Nat Rev Cancer 6(12):924–935. doi:10.1038/nrc2013
5. Gillies RJ, Verduzco D, Gatenby RA (2012) Evolutionary dynamics of carcinogenesis and why targeted therapy does not work. Nat Rev Cancer. doi:10.1038/nrc3298

6. Crespi B, Summers K (2005) Evolutionary biology of cancer. Trends Ecol Evol 20(10):545–552

7. Ewald PW (2009) An evolutionary perspective on parasitism as a cause of cancer. Adv Parasitol 68:21–43. doi:10.1016/S0065-308X(08)00602-7

8. Ewald PW, Swain Ewald HA (2012) Infection, mutation, and cancer evolution. J Mol Med (Berl) 90(5):535–541. doi:10.1007/s00109-012-0891-2

9. Ujvari B, Pearse AM, Peck S, Harmsen C, Taylor R, Pyecroft S, Madsen T, Papenfuss AT, Belov K (2013) Evolution of a contagious cancer: epigenetic variation in devil facial tumour disease. Proc Biol Sci R Soc 280(1750):20121720. doi:10.1098/rspb.2012.1720

10. Murchison EP, Wedge DC, Alexandrov LB, Fu B, Martincorena I, Ning Z, Tubio JM, Werner EI, Allen J, De Nardi AB, Donelan EM, Marino G, Fassati A, Campbell PJ, Yang F, Burt A, Weiss RA, Stratton MR (2014) Transmissible [corrected] dog cancer genome reveals the origin and history of an ancient cell lineage. Science 343(6169):437–440. doi:10.1126/science.1247167

11. zur Hausen H (2010) Infections causing human cancer. Wiley-VCH, Weinheim

12. Moss SF, Blaser MJ (2005) Mechanisms of disease: inflammation and the origins of cancer. Nat Clin Pract Oncol 2 (2):90–97. doi:10.1038/ncponc0081

13. Steenbergen RD, Snijders PJ, Heideman DA, Meijer CJ (2014) Clinical implications of (epi)genetic changes in HPV-induced cervical precancerous lesions. Nat Rev Cancer 14 (6):395–405. doi:10.1038/nrc3728

14. Magnusson PK, Gyllensten UB (2000) Cervical cancer risk: is there a genetic component? Mol Med Today 6(4):145–148

15. Maley SN, Schwartz SM, Johnson LG, Malkki M, Du Q, Daling JR, Li SS, Zhao LP, Petersdorf EW, Madeleine MM (2009) Genetic variation in CXCL12 and risk of cervical carcinoma: a population-based case-control study. Int J Immunogenet 36(6):367–375. doi:10.1111/j.1744-313X.2009.00877.x

16. Bodelon C, Madeleine MM, Johnson LG, Du Q, Galloway DA, Malkki M, Petersdorf EW, Schwartz SM (2014) Genetic variation in the TLR and NF-kappaB pathways and cervical and vulvar cancer risk: a population-based case-control study. Int J Cancer J Int du Cancer 134 (2):437–444. doi:10.1002/ijc.28364

17. Parsonnet J (2001) The enemy within. Am Sci 89:275

18. Abdel-Aziz HO, Murai Y, Hong M, Kutsuna T, Takahashi H, Nomoto K, Murata S, Tsuneyama K, Takano Y (2007) Detection of the JC virus genome in lung cancers: possible role of the T-antigen in lung oncogenesis. Appl Immunohistochem Mol Morphol 15(4):394–400. doi:10.1097/01.pai.0000213126.96590.64

19. Zheng H, Abdel Aziz HO, Nakanishi Y, Masuda S, Saito H, Tsuneyama K, Takano Y (2007) Oncogenic role of JC virus in lung cancer. J Pathol 212(3):306–315. doi:10.1002/path.2188

20. De Paoli P, Carbone A (2013) Carcinogenic viruses and solid cancers without sufficient evidence of causal association. Int J Cancer J Int du Cancer 133(7):1517–1529. doi:10.1002/ijc.27995

21. Mueller MM, Fusenig NE (2004) Friends or foes—bipolar effects of the tumour stroma in cancer. Nat Rev Cancer 4(11):839–849. doi:10.1038/nrc1477 nrc1477

22. Kim Y, Stolarska MA, Othmer HG (2011) The role of the microenvironment in tumor growth and invasion. Prog Biophys Mol Biol 106 (2):353–379. doi:10.1016/j.pbiomolbio.2011.06.006, pii:S0079-6107(11)00053-8

23. Dawkins R (1983) The extended phenotype. The gene as the unit of selection. Oxford University Press, Oxford

24. Egeblad M, Werb Z (2002) New functions for the matrix metalloproteinases in cancer progression. Nat Rev Cancer 2(3):161–174. doi:10.1038/nrc745

25. Bourboulia D, Stetler-Stevenson WG (2010) Matrix metalloproteinases (MMPs) and tissue inhibitors of metalloproteinases (TIMPs): positive and negative regulators in tumor cell adhesion. Semin Cancer Biol 20 (3):161–168. doi:10.1016/j.semcancer.2010.05.002, pii:S1044-579X(10)00028-3

26. Bennett KL, Lee W, Lamarre E, Zhang X, Seth R, Scharpf J, Hunt J, Eng C (2010) HPV status-independent association of alcohol and tobacco exposure or prior radiation therapy with promoter methylation of FUSSEL18, EBF3, IRX1, and SEPT9, but not SLC5A8, in head and neck squamous cell carcinomas. Genes Chromosom Cancer 49(4):319–326. doi:10.1002/gcc. 20742

27. Lechner M, Fenton T, West J, Wilson G, Feber A, Henderson S, Thirlwell C, Dibra HK, Jay A, Butcher L, Chakravarthy AR, Gratrix F, Patel N, Vaz F, O'Flynn P, Kalavrezos N, Teschendorff AE, Boshoff C, Beck S (2013) Identification and functional validation of HPV-mediated hypermethylation in head and neck squamous cell carcinoma. Genome Med 5 (2):15. doi:10.1186/gm419

28. van Kempen PM, Noorlag R, Braunius WW, Stegeman I, Willems SM, Grolman W (2014) Differences in methylation profiles between HPV-positive and HPV-negative oropharynx squamous cell carcinoma: a systematic review. Epigenet: Off J DNA Methylation Soc 9 (2):194–203. doi:10.4161/epi.26881

29. Niller HH, Wolf H, Minarovits J (2009) Epigenetic dysregulation of the host cell genome in Epstein-Barr virus-associated neoplasia. Semin Cancer Biol 19(3):158–164. doi:10.1016/j. semcancer.2009.02.012

30. Jiang J, Zhao LJ, Zhao C, Zhang G, Zhao Y, Li JR, Li XP, Wei LH (2012) Hypomethylated CpG around the transcription start site enables TERT expression and HPV16 E6 regulates TERT methylation in cervical cancer cells. Gynecol Oncol 124(3):534–541. doi:10.1016/j. ygyno.2011.11.023

31. Overmeer RM, Henken FE, Snijders PJ, Claassen-Kramer D, Berkhof J, Helmerhorst TJ, Heideman DA, Wilting SM, Murakami Y, Ito A, Meijer CJ, Steenbergen RD (2008) Association between dense CADM1 promoter methylation and reduced protein expression in high-grade CIN and cervical SCC. J Pathol 215(4):388–397. doi:10.1002/path.2367

32. Niemhom S, Kitazawa S, Kitazawa R, Maeda S, Leopairat J (2008) Hypermethylation of epithelial-cadherin gene promoter is associated with Epstein-Barr virus in nasopharyngeal carcinoma. Cancer Detect Prev 32(2):127–134. doi:10.1016/j.cdp.2008.05.005

33. Rochford R, Cannon MJ, Moormann AM (2005) Endemic Burkitt's lymphoma: a polymicrobial disease? Nat Rev Microbiol 3(2):182–187. doi:10.1038/nrmicro1089

34. Chene A, Donati D, Guerreiro-Cacais AO, Levitsky V, Chen Q, Falk KI, Orem J, Kironde F, Wahlgren M, Bejarano MT (2007) A molecular link between malaria and Epstein-Barr virus reactivation. PLoS Pathog 3(6):e80. doi:10.1371/journal.ppat.0030080

35. Chene A, Donati D, Orem J, Mbidde ER, Kironde F, Wahlgren M, Bejarano MT (2009) Endemic Burkitt's lymphoma as a polymicrobial disease: new insights on the interaction between *Plasmodium falciparum* and Epstein-Barr virus. Semin Cancer Biol 19(6):411–420. doi:10.1016/j.semcancer.2009.10.002

36. Snider CJ, Cole SR, Chelimo K, Sumba PO, Macdonald PD, John CC, Meshnick SR, Moormann AM (2012) Recurrent *Plasmodium falciparum* malaria infections in Kenyan children diminish T-cell immunity to Epstein Barr virus lytic but not latent antigens. PLoS ONE 7(3):e31753. doi:10.1371/journal.pone.0031753PONE-D-11-23913

37. Hur K, Gazdar AF, Rathi A, Jang JJ, Choi JH, Kim DY (2000) Overexpression of human telomerase RNA in *Helicobacter pylori*-infected human gastric mucosa. Jpn J Cancer Res 91 (11):1148–1153

38. Kameshima H, Yagihashi A, Yajima T, Kobayashi D, Denno R, Hirata K, Watanabe N (2000) *Helicobacter pylori* infection: augmentation of telomerase activity in cancer and noncancerous tissues. World J Surg 24(10):1243–1249

39. Chung IK, Hwang KY, Kim IH, Kim HS, Park SH, Lee MH, Kim CJ, Kim SJ (2002) *Helicobacter pylori* and telomerase activity in intestinal metaplasia of the stomach. Korean J Intern Med 17(4):227–233

40. Cheng AS, Li MS, Kang W, Cheng VY, Chou H, Lau SS, Go MY, Lee CC, Ling TK, Ng EK, Yu J, Huang TH, To KF, Chan MW, Sung JJ, Chan FK (2012) *Helicobacter pylori* Causes epigenetic dysregulation of FOXD3 to promote gastric carcinogenesis. Gastroenterology 144:122–133. doi:10.1053/j.gastro.2012.10.002, pii:S0016-5085(12)01467-9

41. Wang P, Mei J, Tao J, Zhang N, Tian H, Fu GH (2012) Effects of *Helicobacter pylori* on biological characteristics of gastric epithelial cells. Histol Histopathol 27(8):1079–1091
42. Oldani A, Cormont M, Hofman V, Chiozzi V, Oregioni O, Canonici A, Sciullo A, Sommi P, Fabbri A, Ricci V, Boquet P (2009) *Helicobacter pylori* counteracts the apoptotic action of its VacA toxin by injecting the CagA protein into gastric epithelial cells. PLoS Pathog 5(10): e1000603. doi:10.1371/journal.ppat.1000603
43. Ashktorab H, Daremipouran M, Wilson M, Siddiqi S, Lee EL, Rakhshani N, Malekzadeh R, Johnson AC, Hewitt SM, Smoot DT (2007) Transactivation of the EGFR by AP-1 is induced by *Helicobacter pylori* in gastric cancer. Am J Gastroenterol 102 (10):2135–2146. doi:10. 1111/j.1572-0241.2007.01400.x, pii:AJG1400
44. Suzuki M, Mimuro H, Kiga K, Fukumatsu M, Ishijima N, Morikawa H, Nagai S, Koyasu S, Gilman RH, Kersulyte D, Berg DE, Sasakawa C (2009) *Helicobacter pylori* CagA phosphorylation-independent function in epithelial proliferation and inflammation. Cell Host Microbe 5(1):23–34. doi:10.1016/j.chom.2008.11.010, pii:S1931-3128(08)00402-2
45. Zhu Y, Shu X, Chen J, Xie Y, Xu P, Huang DQ, Lu NH (2008) Effect of *Helicobacter pylori* eradication on oncogenes and cell proliferation. Eur J Clin Invest 38(9):628–633. doi:10.1111/ j.1365-2362.2008.01987.x
46. Yoon JH, Choi YJ, Choi WS, Ashktorab H, Smoot DT, Nam SW, Lee JY, Park WS (2013) GKN1-miR-185-DNMT1 axis suppresses gastric carcinogenesis through regulation of epigenetic alteration and cell cycle. Clin Cancer Res: Off J Am Assoc Cancer Res 19 (17):4599–4610. doi:10.1158/1078-0432.CCR-12-3675
47. Deen NS, Huang SJ, Gong L, Kwok T, Devenish RJ (2013) The impact of autophagic processes on the intracellular fate of *Helicobacter pylori*: more tricks from an enigmatic pathogen? Autophagy 9(5):639–652. doi:10.4161/auto.23782
48. West DW, Lyon JL, Gardner JW (1980) Cancer risk factors: an analysis of utah Mormons and non-Mormons. J Natl Cancer Inst 65(5):1083–1095
49. Lyon JL, Gardner JW, West DW (1980) Cancer incidence in Mormons and non-Mormons in Utah during 1967–75. J Natl Cancer Inst 65(5):1055–1061
50. Lyon JL, Gardner K, Gress RE (1994) Cancer incidence among Mormons and non-MORMONs in Utah (United States) 1971–85. Cancer Causes Control 5(2):149–156
51. Gardner JW, Sanborn JS, Slattery ML (1995) Behavioral factors explaining the low risk for cervical carcinoma in Utah Mormon women. Epidemiology 6(2):187–189
52. Merrill RM, Lyon JL (2005) Cancer incidence among Mormons and non-Mormons in Utah (United States) 1995–1999. Prev Med 40(5):535–541. doi:10.1016/j.ypmed.2004.10.011
53. Sopori M (2002) Effects of cigarette smoke on the immune system. Nat Rev Immunol 2 (5):372–377. doi:10.1038/nri803
54. Stampfli MR, Anderson GP (2009) How cigarette smoke skews immune responses to promote infection, lung disease and cancer. Nat Rev Immunol 9(5):377–384. doi:10.1038/nri2530
55. Layde PM, Broste SK (1989) Carcinoma of the cervix and smoking. Biomed Pharmacother 43 (3):161–165
56. Sikstrom B, Hellberg D, Nilsson S, Mardh PA (1995) Smoking, alcohol, sexual behaviour and drug use in women with cervical human papillomavirus infection. Arch Gynecol Obstet 256 (3):131–137
57. Ni YH, Chen DS (2010) Hepatitis B vaccination in children: the Taiwan experience. Pathol-Biol 58(4):296–300. doi:10.1016/j.patbio.2009.11.002
58. Gwack J, Park SK, Lee EH, Park B, Choi Y, Yoo KY (2011) Hepatitis B vaccination and liver cancer mortality reduction in Korean children and adolescents. Asian Pac J Cancer Prev 12 (9):2205–2208
59. Lansdorp-Vogelaar I, Sharp L (2013) Cost-effectiveness of screening and treating helicobacter pylori for gastric cancer prevention. Best Pract Res Clin Gastroenterol 27(6):933–947. doi:10. 1016/j.bpg.2013.09.005
60. Nakamura S, Matsumoto T (2013) Helicobacter pylori and gastric mucosa-associated lymphoid tissue lymphoma: recent progress in pathogenesis and management. World J Gastroenterol 19(45):8181–8187. doi:10.3748/wjg.v19.i45.8181

61. Sun Z, Chen T, Thorgeirsson SS, Zhan Q, Chen J, Park JH, Lu P, Hsia CC, Wang N, Xu L, Lu L, Huang F, Zhu Y, Lu J, Ni Z, Zhang Q, Wu Y, Liu G, Wu Z, Qu C, Gail MH (2013) Dramatic reduction of liver cancer incidence in young adults: 28 year follow-up of etiological interventions in an endemic area of China. Carcinogenesis 34(8):1800–1805. doi:10.1093/carcin/bgt007

62. Li WQ, Ma JL, Zhang L, Brown LM, Li JY, Shen L, Pan KF, Liu WD, Hu Y, Han ZX, Crystal-Mansour S, Pee D, Blot WJ, Fraumeni JF, Jr, You WC, Gail MH (2014) Effects of Helicobacter pylori treatment on gastric cancer incidence and mortality in subgroups. J Nat Cancer Inst 106 (7):1–14. doi:10.1093/jnci/dju116

63. Soderberg-Naucler C, Rahbar A, Stragliotto G (2013) Survival in patients with glioblastoma receiving valganciclovir. New Engl J Med 369(10):985–986. doi:10.1056/NEJMc1302145

64. Govan VA (2005) Strategies for human papillomavirus therapeutic vaccines and other therapies based on the E6 and E7 oncogenes. Ann N Y Acad Sci 1056:328–343. doi:10.1196/annals.1352.016

65. Morrow MP, Yan J, Sardesai NY (2013) Human papillomavirus therapeutic vaccines: targeting viral antigens as immunotherapy for precancerous disease and cancer. Expert Rev Vaccines 12(3):271–283. doi:10.1586/erv.13.23

66. Hutajulu SH, Kurnianda J, Tan IB, Middeldorp JM (2014) Therapeutic implications of Epstein-Barr virus infection for the treatment of nasopharyngeal carcinoma. Ther Clin Risk Manag 10:721–736. doi:10.2147/TCRM.S47434

67. Ewald PW, Swain Ewald HA (2014) Joint infectious causation of human cancers. Adv Parasitol 84:1–26. doi:10.1016/B978-0-12-800099-1.00001-6

68. Iloeje UH, Yang HI, Jen CL, Su J, Wang LY, You SL, Lu SN, Chen CJ (2010) Risk of pancreatic cancer in chronic hepatitis B virus infection: data from the REVEAL-HBV cohort study. Liver Int: off J Int Assoc Study Liver 30(3):423–429. doi:10.1111/j.1478-3231.2009.02147.x

69. Straat K, Liu C, Rahbar A, Zhu Q, Liu L, Wolmer-Solberg N, Lou F, Liu Z, Shen J, Jia J, Kyo S, Bjorkholm M, Sjoberg J, Soderberg-Naucler C, Xu D (2009) Activation of telomerase by human cytomegalovirus. J Natl Cancer Inst 101(7):488–497. doi:10.1093/jnci/djp031

70. Cinatl J Jr, Nevels M, Paulus C, Michaelis M (2009) Activation of telomerase in glioma cells by human cytomegalovirus: another piece of the puzzle. J Natl Cancer Inst 101(7):441–443. doi:10.1093/jnci/djp047

71. Michaelis M, Doerr HW, Cinatl J (2009) The story of human cytomegalovirus and cancer: increasing evidence and open questions. Neoplasia 11(1):1–9

72. Hrbacek J, Urban M, Hamsikova E, Tachezy R, Heracek J (2013) Thirty years of research on infection and prostate cancer: no conclusive evidence for a link. Syst Rev. Urol Oncol 31 (7):951–965. doi:10.1016/j.urolonc.2012.01.013

73. Thomas F, Elguero E, Brodeur J, Le Goff J, Misse D (2011) Herpes simplex virus type 2 and cancer: a medical geography approach. Infect Genet Evol: J Mol Epidemiol Evol Genet Infect Dis 11(6):1239–1242. doi:10.1016/j.meegid.2011.04.009

74. Murai Y, Zheng HC, Abdel Aziz HO, Mei H, Kutsuna T, Nakanishi Y, Tsuneyama K, Takano Y (2007) High JC virus load in gastric cancer and adjacent non-cancerous mucosa. Cancer Sci 98(1):25–31. doi:10.1111/j.1349-7006.2006.00354.x

75. Qi F, Carbone M, Yang H, Gaudino G (2011) Simian virus 40 transformation, malignant mesothelioma and brain tumors. Expert Rev Respir Med 5(5):683–697. doi:10.1586/ers.11.51

76. Lawson JS, Heng B (2010) Viruses and breast cancer. Cancers 2(2):752–772. doi:10.3390/cancers2020752

77. Buehring GC, Shen HM, Jensen HM, Choi KY, Sun D, Nuovo G (2014) Bovine leukemia virus DNA in human breast tissue. Emerg Infect Dis 20(5):772–82. doi:10.3201/eid2005.131298

78. Zhang ZF, Begg CB (1994) Is trichomonas vaginalis a cause of cervical neoplasia? Results from a combined analysis of 24 studies. Int J Epidemiol 23(4):682–690

79. Afzan MY, Suresh K (2012) Pseudocyst forms of Trichomonas vaginalis from cervical neoplasia. Parasitol Res 111(1):371–381. doi:10.1007/s00436-012-2848-3

Part VI
Immunology

Chapter 15
Microbes, Parasites and Immune Diseases

Gabriele Sorci, Ph.D., Emanuel Guivier, Ph.D., Cédric Lippens, M.Sc. and Bruno Faivre, Ph.D.

Lay Summary In the last decades, post-industrial countries have experienced spectacular advances in sanitary conditions and lifespan, yet many debilitating diseases due to a dysfunctioning of the immune system have increased in frequency. Even though the risk of becoming affected by immune diseases is influenced by many factors, the rapid changes in their frequency strongly suggest a major role of environmental factors. The rise in frequency of many immune disorders has paralleled improved hygienic condition, reduced exposure to environmental commensals (commensal bacteria) and parasites that regulate the host immune response (helminthic parasites). This "hygiene" hypothesis has received considerable attention and has been backed up by both epidemiological and experimental studies. Successive refinements of the hypothesis have shown that the triggering of environmental factors can be the absence of a proper "education" of the immune system in early life by commensal microbiota and parasites with whom humans have co-evolved for hundreds of thousands of years. Interestingly, this has led to the trial of novel therapies using parasites that improve immunological tolerance and transplantation of commensal gut bacteria. More fundamentally, the rise in immune disorders observed in

G. Sorci (✉) · E. Guivier · C. Lippens · B. Faivre
Biogéosciences, CNRS UMR 6282, Université Bourgogne Franche-Comté,
6 Bd Gabriel, 21000 Dijon, France
e-mail: gabriele.sorci@u-bourgogne.fr

E. Guivier
e-mail: em.guivier@gmail.com

C. Lippens
e-mail: cedric.lippens@u-bourgogne.fr

B. Faivre
e-mail: bruno.faivre@u-bourgogne.fr

© Springer International Publishing Switzerland 2016
A. Alvergne et al. (eds.), *Evolutionary Thinking in Medicine*,
Advances in the Evolutionary Analysis of Human Behaviour,
DOI 10.1007/978-3-319-29716-3_15

wealthy countries shows that the theories and concepts of evolutionary biology can help understanding the dynamics of many human diseases and guide us to prevent and control them.

15.1 Introduction

Improved hygiene and sanitation, and the discovery of effective drugs and vaccines have profoundly changed human exposure to microorganisms and parasitic agents, even though this has been mostly true for wealthy countries. While improved medical conditions have undoubtedly contributed to the spectacular rise in human longevity that we have witnessed in the last decades, it is striking to note that many debilitating diseases of which aetiology results from a dysregulated immune functioning have increased in frequency during the same laps of time [1]. It is therefore tempting to draw a parallel between the reduced exposure to infectious agents and the rise of immune disorders [2, 3]. However, while epidemiological and experimental results support the view that some parasite species do alleviate the risk of immunopathology, others appear to exacerbate it [4]. The heterogeneity in the outcome of the interaction between infectious agents and the host immune response might reside in the evolutionary interests of the parasites [5]. On the one hand, when parasites rely on a downregulated immune response for their own persistence with the host (in other terms for their own Darwinian fitness), we do expect them to exert a protective role against immune disorders. On the other hand, some parasites adopt different strategies for their own survival, for instance hiding from or escaping the immune response, and a protective role is hard to envisage. Sometimes, pathogens can even benefit from an upregulated immune response in terms of access to privileged sites, or accelerated spread within the host; in this case, pathogens should even promote immunopathology. In addition to this, a long-term coevolution between the immune system and parasitic organisms might have promoted the evolution of tolerance strategies where the host better copes with the pathogen by reducing the infection cost rather than eradicating it [6]. Even though tolerance to an infection and immunological tolerance are not synonymous terms, in many cases tolerant hosts pay a reduced cost of infection because they substantially reduce the cost induced by an overzealous immune response [7, 8]. This very broad view illustrates the need to bring in evolutionary thinking into the study of the immunological and epidemiological determinants of immune diseases.

The aim of this chapter is to provide such an integrative view. We will first briefly introduce the immunological mechanisms that promote or prevent immunological damage. We will then discuss the epidemiological and experimental evidence in support or against the view that the rise in incidence of immune disorders results from a change in the environmental conditions the immune system is experiencing nowadays. We will finally put forward a few possible ideas on how evolutionary thinking, beyond our fundamental understanding of the evolution of immune diseases, might feed into decision-making for policy and practice.

15.2   Research Findings

15.2.1   The Immune System: A Double-Edged Sword

Despite its crucial role for protecting organisms from parasites and malignant cells, and for repairing wound and injuries, the expression of immunity may also inflict severe damages when it fails to distinguish between the self and the non-self [9]. An erroneous recognition of self-antigens by T and B lymphocytes can have devastating effects on host homeostasis, depending on the organ that is targeted by the misoriented immune response. A classic example of such mistargeted immune response is type-1 diabetes where T cells and antibodies attack pancreatic β-cells [9]. In addition, immune disorders can also occur when the immune system, while focusing on the right target (the invading pathogen) produces an overzealous, poorly controlled, response. Overexpressed inflammatory responses that occur during a septic shock are one of the most striking examples of how a dysregulated cytokine storm can inflict overwhelming harm to the vital functions of the host.

Given the potential costs of a misdirected or dysregulated immune response, it is not surprising that hosts have evolved regulatory filters that reduce the likelihood of immune damage. These filters involve both central and peripheral mechanisms. For instance, most autoreactive T lymphocytes are deleted in the thymus, or are energized (i.e. silenced) in peripheral organs if they escaped the deletion in the thymus. Similarly, following a wound, injured cells produce danger signals that attract immune cells into the area. These immune cells further produce signalling molecules that constitute a sort of "switch on" signal that polarizes the immune response towards a Th1/Th17 or a Th2 profile, thereby promoting the clearance of the invading pathogen and the healing of the wound. However, to avoid an overzealous response, the production of "switch on" signals is accompanied by the production of "switch off" signals that contribute to the resolution of inflammation and circumvent its negative effects [10]. Regulatory T cells (Treg) are a particularly important class of lymphocytes that secrete anti-inflammatory cytokines such as interleukin (IL)-10 or transforming growth factor (TGF)-β with strong inhibitory effects on pro-inflammatory cytokines.

It is therefore straightforward to see that ensuring host homeostasis requires a finely tuned balance between effectors that promote the immune response and regulatory feedbacks that prevent the immune response to get out of control. However, the necessity to have regulatory mechanisms to avoid immune damage opens up the possibility for a possible exploitation by pathogens and parasites [11].

The evolutionary success of pathogens and parasites resides in their capacity to survive and multiply within a host and transmit to other hosts. Ineluctably, this requires escaping the host immune response [12, 13]. Therefore, while immune regulation is essential to avoid immunopathology, it can be used by parasites as a Trojan Horse to promote their establishment within the host.

Pathogenic microorganisms and metazoan parasites exhibit an astonishing diversity of mechanisms of interference with the host immune system, and not all of them imply a downregulated immune response. For instance, *Plasmodium* and *Trypanosoma* escape immunity by providing a moving target to immune effectors, that is, by expressing highly polymorphic surface proteins, only one gene being expressed at a time. Consequently, different surface proteins are expressed over the course of the infection and the expression of a novel surface protein leaves the immunity behind, providing an escape route to the parasite [14, 15]. Interestingly, the same parasite species may rely on several escape mechanisms since the rodent malaria (*Plasmodium yoelii*) has also been shown to directly downregulate the host immune response by inducing the differentiation of Th0 lymphocytes into Treg with immunosuppressive properties [16].

Helminths are probably the masters of immune regulation among parasitic organisms [17–19]. Contrary to many microparasites that rapidly multiply within the host, macroparasites usually require a long developmental time to mature into the adult stage and adults can reproduce and shed eggs for weeks, months or even years for some human diseases [20]. This suggests that parasitic worms have to cope with the host immune system for very long periods of time.

Infection with helminths usually polarizes the immune system towards a Th2 response [19]. A Th2 response is indeed required to expel the worms in most cases. Effectors of the Th2 response are crucial for the wound healing process, and it has been recently suggested that the Th2 response may represent a mechanism of tolerance to helminths by contributing to the repair of wound induced by migrating worms [21]. Polarization of the immune response towards a Th2 profile has a number of consequences for the host since Th2 cytokines tend to inhibit the production of Th1, pro-inflammatory, effectors. Therefore, by polarizing the immune response towards a Th2 type of response, helminths downregulate the more harmful Th1-type response. In addition to this, and perhaps more importantly, helminths induce the differentiation and expansion of Treg which further downregulate overzealous inflammatory responses (both Th1 and Th2). Several worm species can induce the production of the anti-inflammatory TGF-β or directly produce a TGF-β mimic [17–19]. They also stimulate alternatively activated macrophages [19]. Finally, helminth secretion and excretion products interfere with dendritic cells during early infective stage by inducing their apoptosis and tolerogenic properties (elicitation and maintenance of immunological tolerance; [22]). These mechanisms generate an immunosuppressive environment beneficial for the helminths, especially during their most vulnerable stage, but also for the host because the harmful Th1-type response is downregulated.

Evidence that induction of Treg is an essential component of parasite fitness has been provided in several experiments. For instance, experimental infection of mice with the nematode *Litomosoides sigmodontis*, a rodent model of human filariasis, induces the differentiation and proliferation of Treg. Once infection has been established, experimental neutralization of Treg results in parasite mortality and reduced parasite burden by comparison with control mice [23]. This study clearly shows that interference with the immune system improves parasite fitness.

Paradoxically, an upregulation of the immune response may in some cases benefit the parasite. This is the case when leucocytes are themselves targeted by the pathogen to spread within the host. For example, Ebola virus, which infects monocytes and macrophages, stimulates the attraction and recruitment of even more macrophages, which provides further potential targets for viral replication and dissemination [24].

Benefits of an upregulated inflammatory response can also arise when pathogens use the host immune response as a weapon against competitors. *Salmonella enterica serovar Typhimurium* triggers an inflammatory response that confers a selective advantage during competition with commensal bacteria. Avirulent strains of the bacterium lack the capacity to trigger inflammation and fail to establish unless the inflammatory response is experimentally induced [25].

The above consideration shows that individual species of parasites may trigger or suppress immune disorders, depending on their own evolutionary interest; and the picture may become very complex once the communities of parasites and commensals exploiting a host are considered.

15.2.2 Epidemiology of Immune Diseases

While it is now well established that many immune diseases have increased in frequency during the last 60 years in Europe and North America, the underlying causes are still debated [26, 27]. Epidemiological studies have highlighted a number of environmental factors that are either positively or negatively associated with the likelihood of being affected by immune disorders.

One of the most popular hypotheses put forward a few decades ago postulated that the improved hygiene and sanitation conditions of households in Western countries have changed the risk of exposure to infectious diseases, thereby producing a dysregulation of the immune system [28]. This hypothesis, first coined "the hygiene hypothesis", has gone through a series of refinements ("old friends" [29] or the "biodiversity" hypothesis [30]). These refined hypotheses postulate that the observed rise in autoimmune and inflammatory disorders is not simply due to improved hygiene but rather to a mismatch between the environment our immune system is currently exposed to and the environment it has co-evolved with for hundreds of thousands of years [29]. This line of argument has received support on the basis of the association between a number of traits associated with lifestyles and the risk of developing allergic and autoimmune diseases. For instance, Strachan [28] reported that household size is negatively correlated with the incidence of hay fever, possibly because in families with several siblings the probability of transmission of microbial diseases (childhood diseases) among sibs increases with the number of children per family. Subsequent work has, however, shown that the benefits of having older siblings (in terms of reduced risk of immune disorders) might be due to the colonization rate of commensal bacteria [31]. Similarly, children living in farms have been shown to have much reduced risk of suffering from allergies and atopy compared to children living in urban settings [32], and these observations have recently been backed up by gene expression studies showing that the expression of regulatory cytokines (such as IL-10 and TGF-β) was increased in children living in farms [33]. The impact of environmental changes on the incidence of atopy and allergic diseases can be astonishingly rapid, as suggested by a recent study. Poland acceded to the European Union in 2004. Becoming a member of the European Union resulted in a number of changes in the agricultural policies adopted by farmers. Sozanska et al. [34] compared the incidence of atopy and asthma between 2003 and 2012 in people living in villages and urban settings. They found that the prevalence of atopy increased from around 8 to 18 % in rural settings, while it remained constant in towns (around 20 %). The increased prevalence of atopy in rural areas correlated with major changes in exposure to animals and animal products (contact with cows decreased from 24 to 4 %, and the percentage of villagers who drank unpasteurized milk decreased from 35 to 9 %). Therefore, a nine-year period was long enough to witness major changes in the risk of immune disorders following environmental exposure to cows and cow products.

At a larger scale, the incidence of immune diseases (i.e. multiple sclerosis and type-1 diabetes) follows a clear north to south cline, with higher disease prevalence in the northern European countries and North America compared to South Europe and South America, respectively [1]. While these geographic trends might be explained by genetic differences between populations, comparisons of disease incidence among migrating and resident populations consistently point towards a major role of environmental conditions, since immigrants tend to rapidly acquire the risk profile of resident populations [35]. For instance, immigrants from Pakistan to the UK tend to acquire the risk of developing type-1 diabetes as local residents, which is 11-fold higher than the incidence of the disease in Pakistan [1].

Although these epidemiological studies strongly suggest a major role of the local environment on the aetiology of immune disorders, they incompletely inform us on the nature of these environmental factors. In support of epidemiological studies, a large body of literature shows that in animal models, exposure to certain microorganisms and parasitic metazoan does indeed reduce the risk of autoimmunity or ameliorates disease symptoms.

15.2.3   Coevolution with "Old Friends"

For hundreds of thousands of years, humans have been exposed to a diversity of microorganisms whose effect ranges from lethal, highly exploitative pathogens, to beneficial symbionts. Faced with this diverse set of foreign organisms, the immune system does not behave in the same way. Actually, in many instances, hosts might benefit from tolerating the infection instead of adopting an aggressive strategy towards the invader. Tolerance is clearly a strategy that confers benefits to both the parasite (it can persist within the host in the presence of a permissive immune response) and the host (it reduces the cost of an overreactive immune response, which in many cases outweighs the direct cost of parasite exploitation [36]).

Gut microbiota and helminths have attracted considerable attention in the last years as potential organisms with tolerogenic effects [26, 27]. Parasitic helminths usually induce long-lasting relatively benign infections, unless the immune system does overreact against them [37, 38]. A classical example of devastating consequences of helminthic infections is illustrated by the filarial nematode *Wuchereria bancrofti*. While the infection is mostly clinically asymptomatic, it may induce lymphedema in hyperreactive hosts and, in its most severe form, elephantiasis. There is no ultimate explanation to this hyper reactivity, but the mechanisms have been partly elucidated. Lymphedema is associated with enhanced Th1/Th17 responses and reduced Treg population [39].

Similarly, our gastrointestinal tract is the natural habitat for an extremely complex community of commensal organisms that include bacteria, viruses, protozoa and helminths. One of the daunting tasks of the immune system is to tease things apart, not only the self from the non-self, but also commensal from potentially pathogenic organisms. Gut bacteria are regularly sensed by the immune system, but they do not generally elicit an improper immune response, unless the immune system overreacts to them. Inflammatory bowel disease is a debilitating syndrome that has increased in prevalence during the last decades and which is believed to result from a dysregulated inflammatory response against gut bacterial stimuli [40]. It is therefore possible that the current "epidemics" of immune disorders does arise as a consequence of the disruption of a long-lasting coevolutionary history between our immune system and these "old friends" [29], the improved hygiene experienced by post-industrial societies being among the proximal causes of this disruption.

There is a wealthy of experimental evidence supporting this view. Disruption of the gut microbiota during infancy due to an overuse of antibiotics has been shown to induce a number of pathologies during adulthood, including a higher risk of several inflammatory diseases [41, 42]. More generally, reduced contact with the natural environment and biodiversity, which characterizes the modern lifestyle in highly urbanized settings, can also disrupt the composition of the microbiota [30, 43]. For instance, the biodiversity around the house of Finnish teenagers suffering from atopy has been shown to be lower compared to healthy individuals, and this goes in parallel with reduced diversity of gamma proteobacteria on the skin [30] (Fig. 15.1).

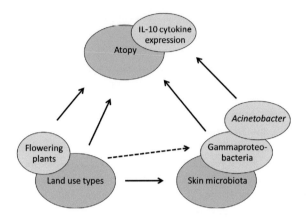

Fig. 15.1 Association pattern between environmental biodiversity, skin microbiota and atopy in Finnish teenagers. Land-use types around the houses of the studied subjects affect the generic diversity of skin microbiota and more marginally the generic diversity of gammaproteo-bacteria (*dashed line*). Skin microbiota correlates with the occurrence of atopy and IL-10 expression. Land-use types and the diversity of flowering plants in the yard of the studied subjects also correlate with the occurrence of atopy. Land use was characterized by the percentage of 15 land cover types (e.g. pastures, forests and permanent crops) estimated in the area of the yard surrounding the houses of the studied subjects. Redrawn with permission from Hanki et al. [30]

15.3 Implications for Policy and Practice

Even though there is a general consensus that understanding why we are susceptible to diseases requires taking into account evolutionary theory, it is less clear whether we could or should bring in evolutionary thinking into policy and practice. Actually, in some cases, our policy already follows such "evolutionary" guidelines. For instance, we use treatments that do not aim at killing pathogens but at reducing symptoms, which parallels the tolerance strategy many hosts adopt when faced with a parasitic attack. Commonly used non-steroidal anti-inflammatory drugs, for example, ibuprofen or aspirin, do boost tolerance. They limit the symptoms of infection, such as inflammation and the associated pain, without directly affecting pathogen burden. Conversely, in other cases, medical practice goes against evolutionary guidelines. For instance, the classical antibiotic treatment against bacterial infection aims to reduce symptoms by eradicating the pathogen. Interestingly, evolutionary thinking also tells us that each of these strategies has its drawbacks (see also Chap. 14). First, tolerating the infection reduces the symptoms but does not stop parasite transmission. When we take an aspirin, we certainly feel better, but we keep spreading the virus, and potentially contribute to the selection of more virulent viral strains. Second, strategies that aim at killing the pathogen strongly select for mechanisms that allow the pathogen to escape them, the evolution of antibiotic resistance being one of the most dramatic examples. In addition, undue or excessive use of antibiotics, especially during infancy (or in utero), has been shown to have profound effects on the community of gut bacteria with potentially negative consequences on the risk of developing several inflammatory diseases in adulthood [41, 42, 44].

Taking into account the coevolutionary history humans have had with parasites and commensals has also promoted the idea that we might use the tolerogenic properties of helminths and bacteria as therapeutic agents (Fig. 15.2). Patients suffering from multiple sclerosis fortuitously infected by intestinal helminths entered a phase of remission compared to non-infected patients and the course of the disease resumed when the patients were treated with anti-helminthic drugs ([45, 46], see also Chap. 17). Similar results have been obtained in many studies using animal models [47, 48]. Mice experimentally induced to develop a form of inflammatory bowel disease (ulcerative colitis) have much attenuated symptoms (reduced diarrhoea and reduced colon inflammation) when infected with the nematode *Heligmosomoides polygyrus* [49, 50].

Following these promising results, several clinical trials have been started with a few helminth species to cure immune diseases, such as inflammatory bowel diseases and multiple sclerosis [51, 52]. However, there are a number of limitations due to the potential undesirable effects of using living organisms. Current efforts are devoted to isolate the effectors that promote the tolerogenic effect of helminths and use them instead of living organisms [53] (Fig. 15.3).

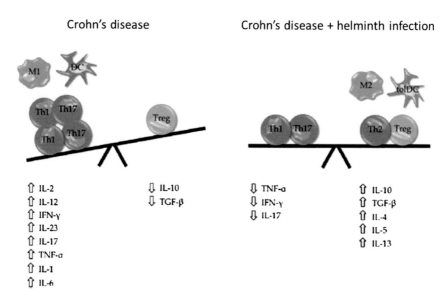

Fig. 15.2 Crohn's disease is a form of inflammatory bowel disease that is characterized by the shift towards a pro-inflammatory status driven by Th1/Th17 cells and cytokines, and M1 macrophages and dendritic cells (DC). Helminth therapy is thought to restore intestinal integrity by expanding Treg, tolerogenic DCs, alternatively activated macrophages (M2) and Th2 cells and cytokines. Redrawn with permission from Heylen et al. [47]

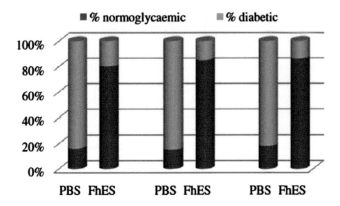

Fig. 15.3 Non-obese diabetic (NOD) mice develop spontaneous hyperglycaemia and type-1 diabetes. However, when treated with the excretory/secretory products of the trematode parasite *Fasciola hepatica* (FhES), only 15 % of the individuals were hyperglycaemic compared to 82 % for control (PBS) mice. Reproduced with permission from Robinson et al. [53]

The tolerogenic properties of helminths also have very interesting potential for the treatment of organ-transplanted patients. Solid organ transplantation is accompanied by long-lasting immunosuppressive drug treatment to prevent rejection. However, immunosuppressive treatment has a number of very undesirable side effects, including the risk of contracting opportunistic diseases or developing cancer. An alternative strategy to immunosuppression might be improved immunological tolerance (silencing the immune response towards the allograft while preserving general immunocompetence). Johnston et al. [54] reviewed the current evidence for a role of helminths on allograft survival and found that several worm species do indeed improve the survival of the allograft in the absence of immunosuppressive drug treatments.

Faecal transplant or bacteriotherapy is another promising avenue to fight a number of immune disorders that might be rooted into a disrupted microbiota [55–57]. Faecal transplants have been successfully tested in randomized controlled trials to treat type-2 diabetes and *Clostridium difficile* intestinal infection, and in case series studies against irritable bowel syndrome, inflammatory bowel disease and multiple sclerosis [56].

Beyond the exciting development of novel therapies for diseases that have proved difficult to cure, evolutionary thinking might provide a useful guide to adopt a lifestyle that might minimize the risk of becoming sick. The spectacular rise in immune disorders is certainly due to the environmental changes that our modern post-industrialized societies are facing. However, it is becoming increasingly clear that improved hygiene per se is not the causal agent but rather the reduced exposure to certain commensals and tolerogenic parasites. It seems that in turn, our immune system has to be educated to learn how to distinguish between harmful and benign tolerogenic microorganisms. It follows that failure to promote early education of the immune system with the proper environmental stimuli can produce the undesirable immune disorders observed nowadays [58, 59].

References

1. Bach J-F (2002) The effect of infections on susceptibility to autoimmune and allergic diseases. N Engl J Med 347:911–920
2. Will-Karp M, Santeliz J, Karp CL (2001) The germless theory of allergic disease: revisiting the hygiene hypothesis. Nat Rev Immunol 1:69–75
3. Yazdanbakhsh M, Kremsner PG, van Ree R (2002) Allergy, parasites, and the hygiene hypothesis. Science 296:490–494
4. Kamradt T, Göggel R, Erb KJ (2005) Induction, exacerbation and inhibition of allergic and autoimmune diseases by infection. Trends Immunol 26:260–267
5. Sorci G, Cornet S, Faivre B (2012) Immune evasion, immunopathology and the regulation of the immune system. Pathogens 2:71–91
6. Schneider DS, Ayres JS (2008) Two ways to survive infection: what resistance and tolerance can teach us about treating infectious diseases. Nat Rev Immunol 8:889–895
7. Ayres JS, Schneider DS (2012) Tolerance of infections. Annu Rev Immunol 30:271–294
8. Vale PF, Fenton A, Brown SP (2014) Limiting damage during infection: lessons from infection tolerance for novel therapeutics. PLoS Biol 12:e1001769
9. Sell S, Max EE (2001) Immunology, immunopathology, and immunity. ASM Press
10. Kim JM, Rasmussen JP, Rudensky AY (2007) Regulatory T cells prevent catastrophic autoimmunity throughout the lifespan of mice. Nat Immunol 8:191–197
11. Belkaid Y (2007) Regulatory T cells and infection: a dangerous necessity. Nat Rev Immunol 7:875–888
12. Finlay BB, McFadden G (2006) Anti-immunology: evasion of the host immune system by bacterial and viral pathogens. Cell 24:767–782
13. Schmid-Hempel P (2008) Parasite immune evasion: a momentous molecular war. Trends Ecol Evol 23:318–326
14. Niang M, Yam XY, Preiser PR (2009) The *Plasmodium falciparum* STEVOR multigene family mediates antigenic variation of the infected erythrocyte. PLoS Pathog 5:e1000307
15. Horn D (2014) Antigenic variation in African trypanosomes. Mol Biochem Parasitol 195:123–129
16. Hisaeda H, Maekawa Y, Iwakawa D, Okada H, Himeno K, Kishihara K, Tsukumo S, Yasutomo K (2004) Escape of malaria parasites from host immunity requires CD4(+)CD25(+) regulatory T cells. Nat Med 10:29–30
17. McSorley HJ, Maizels RM (2012) Helminth infection and host immune regulation. Clin Microbiol Rev 25:585–608
18. McSorley HJ, Hewitson JP, Maizels RM (2013) Immunomodulation by helminth parasites: defining mechanisms and mediators. Int J Parasitol 43:301–310
19. Finlay CM, Walsh KP, Mills KHG (2014) Induction of regulatory cells by helminth parasites: exploitation for the treatment of inflammatory diseases. Immunol Rev 259:206–230
20. Carme B, Laigret J (1979) Longevity of *Wuchereria bancrofti* var pacifica and mosquito infection acquired from a patient with low level parasitemia. Am J Trop Med Hyg 28:53–55
21. Gause WC, Wynn TA, Allen JE (2013) Type 2 immunity and wound healing: evolutionary refinement of adaptive immunity by helminths. Nat Rev Immunol 13:607–614
22. Nono JK, Pletinckx K, Lutz MB, Brehm K (2012) Excretory/secretory-products of *Echinococcus multilocularis* larvae induce apoptosis and tolerogenic properties in dendritic cells in vitro. PLoS Neglected Trop Dis 6:e1516

23. Taylor MD, Harris A, Babayan SA, Bain O, Culshaw A, Allen JA, Maizels RM (2007) CTLA-4 and CD4_CD25_ regulatory T cells inhibit protective immunity to filarial parasites in vivo. J Immunol 179:4626–46341
24. Zampieri CA, Sullivan NJ, Nabel GJ (2007) Immunopathology of highly virulent pathogens: insights from Ebola virus. Nat Immunol 8:1159–1164
25. Stecher B, Robbiani R, Walker AW, Westendorf AM, Barthel M, Kremer M, Chaffron S, Macpherson AJ, Buer J, Parkhill J, Dougan G, von Mering C, Hardt W-D (2007) *Salmonella enterica* Serovar Typhimurium exploits inflammation to compete with the intestinal microbiota. PLoS Biol 5:e244
26. Rook GAW (2008) Review series on helminths, immune modulation and the hygiene hypothesis: the broader implications of the hygiene hypothesis. Immunology 126:3–11
27. Brown EM, Arrieta M-C, Finlay BB (2013) A fresh look at the hygiene hypothesis: how intestinal microbial exposure drives immune effector responses in atopic disease. Semin Immunol 25:378–387
28. Strachan DP (1989) Hay fever, hygiene and household size. Br Med J 299:1259–1260
29. Rook GAW (2010) 99th Dahlem conference on infection, inflammation and chronic inflammatory disorders: Darwinian medicine and the 'hygiene' or 'old friends' hypothesis. Clin Exp Immunol 160:70–79
30. Hanski I, von Hertzen L, Fyhrquist N, Koskinen K, Torppa K, Laatikainen T, Karisola P, Auvinen P, Paulin L, Mäkelä MJ, Vartiainen E, Kosunen TU, Alenius H, Haahtela T (2012) Environmental biodiversity, human microbiota, ad allergy are interrelated. Proc Nat Acad Sci 109:8334–8339
31. Penders J, Gerhold K, Stobberingh EE, Thijs C, Zimmermann K, Lau S, Hamelmann E (2013) Establishment of the intestinal microbiota and its role for atopic dermatitis in early childhood. J Allergy Clin Immunol 132:601–607
32. Von Mutius E (2010) 99th Dahlem conference on infection, inflammation and chronic inflammatory disorders: farm lifestyles and the hygiene hypothesis. Clin Exp Immunol 160:130–135
33. Frei R, Roduit C, Bieli C, Loeliger S, Waser M, Scheynius A, van Hage M, Pershagen G, Doekes G, Riedler J, von Mutius E, Sennhauser F, Akdis CA, Braun-Fahrländer C, Lauener RP, as part of the PARSIFAL study team (2014) Expression of genes related to anti-inflammatory pathways are modified among farmers' children. PLoS One 9:e91097
34. Sozanska B, Blaszczyk M, Pearce N, Cullinan P (2013) Atopy and allergic respiratory disease in rural Poland before and after accession to the European Union. J Allergy Clin Immunol 133:1347–1353
35. Ko Y, Butcher R, Leong RW (2014) Epidemiological studies of migration and environmental risk factors in the inflammatory bowel diseases. World J Gastroenterol 20:1238–1247
36. Graham AL, Allen JE, Read AF (2005) Evolutionary causes and consequences of immunopathology. Annu Rev Ecol Evol Syst 36:373–397
37. Alves Oliveira LF, Moreno EC, Gazzinelli G, Martins-Filho OA, Silveira AMS, Gazzinelli A, Malaquias LCC, LoVerde P, Martins Leite P, Correa-Oliveira R (2006) Cytokine production associated with periportal fibrosis during chronic schistosomiasis mansoni in humans. Infect Immun 74:1215–1221
38. Korten S, Hoerabauf A, Kaifi J, Büttner D (2011) Low levels of transforming growth factor-beta (TGF-beta) and reduced suppression of Th2-mediated inflammation in hyperreactive human onchocerciasis. Parasitology 138:35–45
39. Babu S, Bhat SQ, Pavan Kumar N, Lipira AB, Kumar S, Karthik C, Kumaraswami V, Nutman TB (2009) Filarial lymphedema is characterized by antigen-specific Th1 and Th17 proinflammatory responses and lack of regulatory T cells. PLoS Neglected Trop Dis 3:e420
40. Braus NA, Elliott DE (2009) Advances in the pathogenesis and treatment of IBD. Clin Immunol 132:1–9
41. Stensballe LG, Simonsen J, Jensen SM, Bonnelykke K, Bisgaard H (2013) Use of antibiotics during pregnancy increases the risk of asthma in early childhood. J Pediatr 162:832–838

42. Shaw SY, Blanchard JF, Bernstein CN (2010) Association between the use of antibiotics in the first year of life and pediatric inflammatory bowel disease. Am J Gastroenterol 105:2687–2692
43. Rook GAW (2013) Regulation of the immune system by biodiversity from the natural environment: an ecosystem service essential to health. Proc Nat Acad Sci 110:18360–18367
44. Willing BP, Russell SL, Finlay BB (2011) Shifting the balance: antibiotic effects on host–microbiota mutualism. Nat Rev Microbiol 9:233–243
45. Correale J, Farez M (2007) Association between parasite infection and immune response in multiple sclerosis. Ann Neurol 61:97–108
46. Correale J, Farez M (2011) The impact of parasite infections on the course of multiple sclerosis. J Neuroimmunol 233:6–11
47. Heylen M, Ruyssers NE, Gielis EM, Vanhomwegen E, Pelckmans PA, Moreels TG, De Man JG, De Winter BY (2014) Of worms, mice and man: an overview of experimental and clinical helminth-based therapy for inflammatory bowel disease. Pharmacol Ther 143:153–167
48. Maizels RM, McSorley HJ, Smyth DJ (2014) Helminth in the hygiene hypothesis: sooner or later? Clin Exp Immunol 177:38–46
49. Sutton TL, Zhao A, Madden KB, Elfrey JE, Tuft BA, Sullivan CA, Urban JF Jr, Shea-Donohue T (2008) Anti-inflammatory mechanisms of enteretic *Heligmosomoides polygyrus* infection against trinitrobenzene sulfonic acid-induced colitis in a murine model. Infect Immun 76:4772–4782
50. Donskow-Lysoniewska K, Majewski P, Brodaczewska K, Jozwicka K, Doligalska M (2012) *Heligmosomoides polygyrus* fourth stages induce protection against DSS-induced colitis and change opioid expression in the intestine. Parasite Immunol 34:536–546
51. Weinstock JV, Elliott DE (2013) Translatability of helminth therapy in inflammatory bowel diseases. Int J Parasitol 43:245–251
52. Fleming JO (2013) Helminth therapy and multiple sclerosis. Int J Parasitol 43:259–274
53. Robinson MW, Dalton JP, O'Brien BA, Donnelly S (2013) *Fasciola hepatica*: the therapeutic potential of a worm secretome. Int J Parasitol 43:283–291
54. Johnston CJC, McSorley HJ, Anderton SM, Wigmore SJ, Maizels RM (2014) Helminths and immunological tolerance. Transplantation 97:127–132
55. Borody TJ, Khoruts A (2012) Fecal microbiota transplantation and emerging applications. Nat Rev Gastroenterol Hepatol 9:88–96
56. Smits LP, Bouter KE, de Vos WM, Borody TJ, Nieuwdorp M (2013) Therapeutic potential of fecal microbiota transplantation. Gastroenterology 145:946–953
57. Udayappan SD, Hartstra AV, Dallinga-Thie GM, Nieuwdorp M (2014) Intestinal microbiota and faecal transplantation as treatment modality for insulin resistance and type 2 diabetes mellitus. Clin Exp Immunol 177:24–29
58. Djuardi Y, Wammes LJ, Supali T, Sartono E, Yazdanbakhsh M (2011) Immunological footprint: the development of a child's immune system in environments rich in microorganisms and parasite. Parasitology 138:1508–1518
59. Ege MJ, Mayer M, Normand A-C, Genuneit J, Cookson WOCM, Braun-Fahrländer C, Heederik D, Piarroux R, von Mutius R, GABRIELA Transregio 22 Study Group (2011) Exposure to environmental microorganisms and childhood asthma. N Engl J Med 364:701–709

Chapter 16
Evolutionary Principles and Host Defense

Prof. Neil Greenspan, M.D., Ph.D.

Lay Summary On multiple timescales, evolution plays critical roles in the processes by which humans respond to infectious agents and other foreign substances and the ways that infectious agents try to evade immune mechanisms and exploit the resources of the hosts they infect. The human immune response depends on white blood cells known as lymphocytes, and the success of the responses these cells mount against such well-known causes of infection as HIV-1 depend critically on cellular proliferation, variation in antigen receptor genes, and selection, i.e., differential survival and replication of cells that differ with respect to these immunity-related genes. Similarly, the capacities of bacteria, viruses, and other infectious agents to harm hosts in transforming host resources into pathogen progeny, to transmit to new hosts, and to evade drugs and vaccine responses depend critically on mutation and selection, i.e., pathogen evolution. Some of the evolutionary changes that affect the abilities of infectious agents to successfully infect hosts, replicate, and transmit to additional hosts can occur within the time frame of a single host infection, creating a sort of 'arms race' between the pathogen and the cells of the host immune system.

16.1 Introduction: Evolutionary Timescales and the Immune System

Practitioners of the medically important fields of immunology and microbiology have recognized the relevance of evolution for their disciplines for over a century [1, 2]. As understanding in these fields has progressed to characterizing the roles of cells, genes,

N. Greenspan (✉)
Department of Pathology, Case Western Reserve University,
Cleveland, OH 44106, USA
e-mail: nsg@case.edu

© Springer International Publishing Switzerland 2016   225
A. Alvergne et al. (eds.), *Evolutionary Thinking in Medicine*,
Advances in the Evolutionary Analysis of Human Behaviour,
DOI 10.1007/978-3-319-29716-3_16

and molecules in determining the specificity, magnitude, and quality of immune responses, and the molecular mechanisms responsible for pathogen-mediated tropisms, virulence, transmissibility, and resistance to pharmacologic agents, the relevance of evolution has become even more sharply defined. In this chapter, the goal will be to employ selected examples to illustrate some of the many aspects of host–pathogen relationships for which knowledge of evolutionary mechanisms is relevant and arguably essential.

Evolution shapes the human immune system and the immune response on three timescales (Fig. 16.1). First, over millennia, phylogenetic evolution produced extraordinarily complex and highly networked sets of cellular and molecular responses in vertebrates, to both external and internal stimuli, that we generally refer to as the immune system [3, 4]. This system includes responses that depend on receptors, known as innate immune receptors, for microbial components that are present, with relatively modest structural variations, in many species of pathogen [5, 6]. These structures are often referred to as pathogen-associated or (more accurately) microbial-associated molecular patterns (PAMPs or MAMPs). The innate immune receptors for these microbial molecules are typically referred to as pattern recognition receptors (PRR).

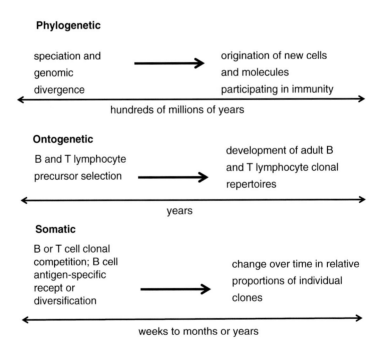

Fig. 16.1 The immune system evolves in clinically relevant ways on three timescales: phylogenetic, ontogenetic, and somatic

Second, the antigen-specific cells of the immune system, B and Tlymphocytes originate primarily from hematopoietic stem cells in the bone marrow, like other cell types found in the blood. They develop into mature forms in primary lymphoid organs (bone marrow for Bcells and thymus for Tcells) through a complex process of diversification of the genes encoding antigen-specific receptors (ASR) and selection. Differential fitness arises in part from differences in antigen-specific receptor (ASR) amino acid sequence and, therefore, antigen specificity [7, 8]. Over a period of weeks to months, these ontogenetic selection events, both positive and negative, determine which B or Tcells contribute to the naïve immune repertoire, which corresponds to a sort of dynamic library of receptor structures that confer on the organism its unique abilities to respond to exogenous molecular stimuli in the form of pathogens and their components, toxins, and other substances originating from non-human species. These processes represent evolution of populations of somatic cells, not populations of organisms. They are relevant to medicine both because they establish the ASR repertoires for B and Tlymphocytes and because aberrations of these developmental pathways can lead to deficiency or malignant diseases.

Third, immune responses involve competition among white blood cells, either B or Tlymphocytes, expressing ASR differing in amino acid sequence. Typically, at a given time, two or more B or Tcells may share the same precise ASR amino acid sequence. Cells displaying ASR with structurally identical antigen-recognition domains are referred to as a clone. Thus, the immune response can be viewed as a process in which B or Tcells displaying ASR with different amino acid sequences (encoded by genes with different nucleotide sequences) exhibit differences in fitness, i.e., a process of clonal competition [9–12]. Particularly in the case of Blymphocytes, the cells that produce antibodies, the clonal competition underlying an immune response can involve both ongoing mutation (through a unique process known as somatic hypermutation) and selection, and represents a truly neo-Darwinian evolutionary process played out (i.e., on the level of somatic cells as opposed to independent organisms), like the processes critical for B and Tlymphocyte repertoire development. This physiological form of evolution occurs in a time frame of days to weeks or months as opposed to the years, centuries, or millennia generally associated with phylogenetic evolution.

The outcome of this somatic cell competition can make the difference between life and death in the setting of an infection by a pathogen for which neutralizing or opsonizing antibodies provide the main mechanism of protective immunity and that produces progeny at a rate comparable to or more typically much greater than the rate at which Blymphocytes proliferate. In theory at least, a tenfold increase in average antibody affinity from an initially modest value can substantially reduce the time necessary to reach the threshold of protective antigen-binding activity by a time interval in which the pathogen burden could increase substantially. Blymphocytes, in mice, require on the order of 7 h to divide [13] while bacterial pathogens can divide as frequently as every half hour and viruses can increase at even more impressive exponential rates. Thus, the selection of somatic cells can influence the fitness of the whole organism and the trajectory of organismal evolution.

16.2 Research Findings and Implications for Policy and Practice

16.2.1 B- and T-Cell Evolution and HIV Vaccine Development

A particularly salient example of the medical relevance of B-cell and immunoglobulin gene evolution is in the context of HIV vaccine development. HIV-1 has an exceptionally high mutation rate [14] and a proclivity for recombination [15]. These two attributes contribute to extremely rapid genomic diversification [14] and evolution, both within-host and between-host [16]. The scale of HIV-1 genomic diversity is exemplified by the claim from Korber et al. that the HIV-1 viral genomes in one infected individual encompass the same approximate extent of nucleotide sequence diversity exhibited by the worldwide population of influenza A viral genomes over the course of a year.

Although a single virus is responsible for the transmission of HIV in most cases [17], diversification of the virus in individual hosts implies that different subjects are likely to be infected by genetically distinguishable viruses. Therefore, a vaccine that elicits antibodies that interact effectively with only a subset of circulating HIV viruses is unlikely to be highly effective on a population basis. These realities have prompted intense interest in identifying and characterizing what are known as potent broadly neutralizing antibodies (pbnAb) that will prevent infection of host cells by the vast majority of extant HIV viruses. A number of prominent investigators are hoping to be able to design HIV-derived immunogens (i.e., viral proteins that can stimulate an immune response) that can elicit pbnAbs.

What has been revealed in studying numerous monoclonal human pbnAb is that they have a very high number of somatic mutations (see Glossary) in the variable domains [18–20], which are the portions of antibodies that are primarily responsible for determining the affinity and specificity of interactions with antigens. Somatic mutations only in the portions of immunoglobulin genes that determine the amino acid sequences of variable domains result from a process that couples highly localized (i.e., affecting only variable and not other antibody domains) genetic variation with intense clonal competition and selection based on access to follicular helper T-cells. These helper cells are CD4+ T lymphocytes that provide signals critical to activating the somatic hypermutation mechanism (see Glossary) and also isotype switching. These processes are most often, although not always, localized to specialized structures in secondary lymphoid tissues known as germinal centers (GC) [21]. Consequently, the antigen elicits an intense evolutionary process that offers one of the several currently plausible pathways to developing a vaccine that can counter the rapid evolution of HIV-1 [22] (Fig. 16.2).

In the case of HIV-1, only a minority of infected individuals develops pbnAbs, and in these individuals, it can take two or more years for pbnAbs to be produced. At that late stage of infection, these antibodies cannot eliminate the virus from all infected cells, some of which harbor the virus in latent forms that do not express

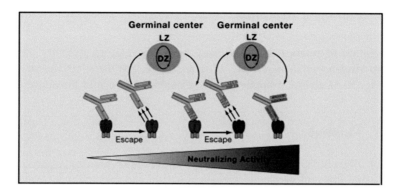

Fig. 16.2 Coevolutionary conflict between HIV and host Blymphocytes and antibodies. Selection of viral envelope variants (indicated by *trimer color* change) by neutralizing antibodies leads to altered viral envelope proteins that in turn select for altered antibody V domains. Selected amino acid substitutions confer better binding along with more potent and broader neutralizing activity. These events occur in the germinal centers (GC) of lymph nodes or other secondary lymphoid tissues. During this process, the Blymphocytes involved proliferate in the *dark zone* of the GC and interact with helper T-cells in the *light zone* (LZ) of the GC. Reprinted from [46] with the permission from RightsLink

sufficient quantities of viral protein for the antibodies or other elements of the immune response to recognize and destroy these cells. In contrast, in the vaccine context, pbnAbs would be present in the circulation prior to initial infection, thereby permitting the antibodies the chance to prevent infection at the start.

Of course, similar if less extensive evolution of Bcells and immunoglobulin genes are associated with antibody responses to most protein antigens. Such evolution is undoubtedly of importance for protective antibody responses elicited by vaccines for many bacterial and viral pathogens.

T-lymphocytes do not exhibit somatic hypermutation (or class switch recombination) but they still engage in intense clonal competition. Therefore, cell-mediated immune responses (which are mediated by T-lymphocytes) can also display changes over time in relative proportions of different clonal lineages; i.e., T-cell populations also evolve in a neo-Darwinian sense in the time frame of an immune response. A particularly interesting example of this process leading to an autoimmune disease, scleroderma, in patients with cancer was described recently [23]. Mutations in a gene that encodes a subunit of RNA polymerase III (an enzyme involved in synthesizing ribosomal and transfer RNA molecules) appear to facilitate tumorigenesis and also contribute to the elicitation of CD4+ T-cell responses that cause the pathology associated with scleroderma. Both the process of tumor formation and the immune response to the new antigen generated by the tumor-promoting mutations in the cancer cells represent examples of somatic cell evolution with clinical consequences.

16.2.2 Pathogen Evolution

The evolution of pathogens affects many medically relevant attributes including pathogen virulence, transmissibility, drug resistance, and the effectiveness of vaccines (Fig. 16.3). Below, I briefly provide examples for each of the above.

16.2.2.1 Virulence

Consider *Toxoplasma gondii*, which is an intracellular protozoan parasite that infects many different vertebrate species asexually and undergoes a sexual cycle after infecting cats [24]. Parasite oocysts are potentially introduced into the human environment in cat feces. *T. gondii* is of interest in clinical medicine because humans can serve as accidental intermediate hosts when they ingest oocysts in, for example, undercooked, contaminated meat or ingest mature parasites in contaminated drinking water. Mother-to-child transmission can also occur.

In most healthy individuals, the infection does not cause illness, but in individuals with immune deficiencies and in fetuses, it can cause substantial morbidity. In the case of congenital infection of a fetus, morbidity, including vision loss, cognitive deficits, and seizures, tends to be more severe with earlier infection. Fetal infection can result in either miscarriage or stillborn birth. Sibley and colleagues [25] have now further clarified the molecular basis for the variation in virulence among different *T. gondii* lineages for mice, an important prey species for cats and therefore an important intermediate host species.

There are three lineages of *T. gondii* in North America and Europe, and these lineages vary substantially in virulence for mice. With respect to typical laboratory mice, Type I is highly virulent, Type II exhibits an intermediate degree of virulence, and Type III is avirulent. In previous work [26, 27], Sibley and his associates used

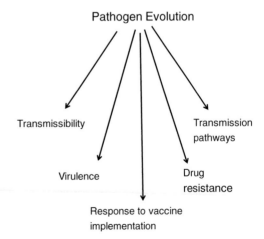

Fig. 16.3 Pathogen evolution influences clinically relevant traits such as transmissibility, virulence, drug resistance, and adaptation in response to the introduction of vaccines. Insights into pathogen evolution can provide a means of determining the origins and pathways of spread for infectious disease outbreaks

genetic crosses among these three lineages to identify key genes that contribute to virulence. Specifically, they identified a gene (*ROP18*) that encodes a protein (ROP18) released from the secretory organelle known as the rhoptry [26, 27]. In the most recent study in this line of investigation, Etheridge et al. reveal that two other genes (*ROP5*, *ROP17*) and their gene products (ROP5, ROP17) contribute critically to virulence in mice.

These parasite proteins promote virulence by inactivating multiple mouse proteins, including those known as immunity-related GTPases (IRGs) that are expressed at increased levels by cells exposed to interferon-gamma (IFNg) and are involved in killing parasites inside mouse cells [28]. However, the detailed molecular basis for *T. gondii* virulence in humans or even other mouse species does not conform to the pattern in the typical inbred mouse strains used in the laboratory. Humans do not appear to have IRG proteins [29]. So, the ROP proteins that are so crucial in laboratory mice are not the key to virulence in humans. Furthermore, the degrees of virulence of *T. gondii* of Types I, II, and III that are so discernible in mice are not important in the human context. While IFNg is important in humans as in mice for cell-autonomous immunity to *T. gondii*, the key effect of IFNg is increased expression of indoleamine 2,3-dioxygenase (IDO), an enzyme that degrades the amino acid tryptophan. Conversely, IDO is not important in cell-autonomous mouse immunity to *T. gondii*.

The preceding illustrates that virulence in *T. gondii* is both a relational property and a dynamic property that depends on the evolution of the parasite and each host species that it infects. Genetic differences among members of each host species and betweenhost species presumably drive the diversification of the parasite into different lineages with different strategies for optimizing virulence and transmissibility.

16.2.2.2 Transmissibility

A central challenge confronting physicians and other healthcare professionals focused on infectious diseases is tracking the transmission of pathogens both inside and outside of hospitals. One key step in the process of controlling outbreaks of infectious diseases in hospitals is determining whether infections in any particular patient are caused by person-to-person transmission. The traditional approach to addressing this challenge for bacterial infections involves taking account of epidemiological data, assessment of antibiotic sensitivities, and identification of alleles at a limited number (<1 %) of bacterial loci using multi-locus sequence typing (MLST).

A 2012 study [30] demonstrates that whole-genome sequencing (WGS) of infectious isolates from patients involved in a presumed outbreak and covering >95 % of pathogen loci can contribute to determining the transmission network. In this particular study, focused on infection by methicillin-resistant *Staphylococcus aureus* (MRSA), both inpatients and outpatients were likely involved.

This study began in 2011 when the authors identified three simultaneous cases of MRSA carriage in the special care baby unit (SCBU) at a university hospital in

England. Infection-control specialists identified thirteen other SCBU patients with one or more positive screen for MRSA. The team was unable to confirm an outbreak stretching over the relevant time period using conventional approaches. Application of WGS to the relevant pathogen isolates permitted the mapping of a plausible network of infection transmission events.

The WGS provided a number of key insights that might have otherwise remained hidden. First, the team identified a new sequence type (ST) of MRSA, ST2371 that was found to be phylogenetically related to a ST of MRSA, ST22, known to be involved in a high percentage of hospital-associated MRSA infections in the United Kingdom. Thus, it is likely that ST2371 was derived from ST22 but differed (i.e., evolved) from ST22 in having acquired genes encoding an exotoxin, Panton-Valentine leucocidin (PVL), which kills host white cells and has been associated with a dangerous form of pneumonia.

Another important contribution derived from analysis of the WGS data was the demonstration that infants in the SCBU transmitted MRSA to their mothers. There were also transmission events from mothers to other mothers in a postnatal hospital unit and from mothers to their partners outside of the hospital.

The result of applying WGS as part of a prospective longitudinal surveillance program in the SCBU was that a new MRSA infection, after a deep clean of the SCBU and an absence of new SCBU MRSA cases over more than two months, was inferred to be part of the outbreak. The authors therefore screened SCBU staff members for MRSA, and one member of the staff, out of 154, was found to be positive for MRSA. WGS confirmed that the staff member carried the outbreak MRSA, ST2371. These results suggested to the infection-control team that this staff member reintroduced the outbreak-associated strain to at least one patient in the SCBU.

16.2.2.3 Drug Resistance

Perhaps the iconic exemplar of the relevance of evolution to biomedicine and even direct clinical care of individual patients is the development of resistance to antibiotics by bacterial pathogens. The first widely used antibiotic, penicillin, was introduced for mass use in 1943 and over the next seventeen years, its use reduced the incidence of infection-related mortality by more than 93 % compared to 1900 [31]. Nevertheless, already in the 1940s it became clear that some bacteria possessed the capacity to degrade and inactivate penicillin. In fact, it is now clear that antibiotics and antibiotic resistance genes have existed for many millions of years [32].

After the introduction of penicillin, as each subsequent antibiotic was introduced in ensuing decades, resistance followed after varying time intervals but with virtual certainty [31]. The close correlation between antibiotic consumption and the rates of antibiotic resistance in a comparison of European countries [33] supports the inference that chemotherapeutic agents intended to kill bacteria will have the consequential effect of selecting resistant variants that will then increase in frequency. This phenomenon exemplifies the broader phenomenon documented in a publication by the American Academy of Microbiology that summarizes a 2012 colloquium on the common mechanisms involved in the development of resistance to antibiotics, antivirals, pesticides, herbicides, and anti-cancer drugs [34].

An example of the intricate molecular mechanisms responsible for the development and spread of resistance is provided by the ability of some influenza A viruses to escape treatment with the antiviral agent, oseltamivir. Beginning in 2007, oseltamivir-resistant influenza A viruses containing N1 neuraminidase (NA), a virion surface glycoprotein involved in releasing maturing virions from the surfaces of infected cells, started to increase in frequency. Mutation of histidine to tyrosine at position 274 (H274Y) of the neuraminidase was associated with this resistance [35]. Yet, when this mutation alone was introduced into viruses and the resulting fitness in mice and ferrets, as assessed both *in vitro*, (in tissue culture), and *in vivo*, was reduced. Bloom et al. [36] investigated this apparent paradox suspecting that other mutations in the neuraminidase might permit the H274Y mutation to occur without greatly diminishing fitness. Analysis of a phylogenetic tree for NAs in seasonal H1N1 viruses from 2006 led to identification of two candidate mutations (V234M and R222Q) that could have counteracted the otherwise negative effects of the H274Y mutation on viral fitness among naturally circulating viruses. Additional experiments verified that both of these mutations restored fitness by preventing loss of either total NA enzymatic activity or replication ability.

Structural studies suggest that the molecular mechanism for H274Y-mediated resistance was based on the ability of the tyrosine to limit the motion of the side chain of the glutamic acid at NA position 276 that is necessary for oseltamivir to bind [37, 38]. Understanding the evolution of resistance in biophysical terms at the molecular level offers the possibility of further efforts in drug development to regain therapeutic efficacy.

16.2.2.4 Vaccine Effectiveness

Vaccination is arguably the single greatest contribution to public health and the reduction of morbidity and mortality made by modern medicine [39, 40]. I would not be surprised if the many investigators in this field regarded the elimination of infection by smallpox virus by a worldwide vaccination campaign the single greatest triumph in the application of immunization. One possible consequence of this spectacular achievement may be to regard the implementation of a successful vaccine as the end of the battle to defeat a particular pathogen, but such a perspective takes insufficient account of the potential evolutionary responses of pathogens to such intervention as long as eradication is not achieved. Because many pathogens have non-human reservoirs and may be much more diverse and rapidly evolving than smallpox virus, it is not plausible, as a public health objective, to eliminate most pathogens.

Consider the challenge posed by *Streptococcus pneumoniae* (pneumococcus). This encapsulated bacterial pathogen exists as at least 90 serotypes, with each serotype expressing a distinctive polysaccharide capsule. Although there are serological cross-reactions among these different capsule structures, antibody-mediated immunity is largely serotype-specific.

As a consequence of this relative serotype specificity in the protective host immune response, the two types of vaccines currently in use clinically to provide immunity to the pneumococcus are both multivalent. The polysaccharide-only vaccine contains 23 different capsular polysaccharides representing the most common serotypes associated with clinical infections. Since this vaccine is relatively ineffective in children less than two years of age, there is also a conjugate vaccine with as many as 13 different capsular polysaccharides independently and covalently attached to a carrier protein.

Studies of the epidemiology of pneumococcal infection after implementation of the conjugate vaccine among young children in various geographic settings have revealed the expected decline in frequency of infection with serotypes represented in the vaccine. Of particular interest for present purposes, infections caused by non-vaccine strains have increased in frequency in some settings [41], a phenomenon called "serotype replacement." Such results suggest that conjugate vaccines can act as reasonably potent agents of selection on the pneumococcal population, thereby shaping the evolutionary trajectory of the pathogen.

The authors of a 2011 [42] study of 240 isolates of a particular multi-drug-resistant strain of pneumococcus from geographic locations on four continents used whole-genome sequencing to address the sources and pace of evolutionary change of this pathogen. Among the interesting findings reported were the following: (1) the majority of genetic variation was due to horizontal gene transfer and recombination between the imported DNA and the host chromosomal DNA as opposed to more classical base substitutions in the chromosomal genes, and (2) loci involved in producing cell surface proteins or capsular polysaccharide were involved more frequently than average loci in such recombination events strongly suggesting selection by the human antibody response. Consistent with the preceding inference, ten of the studied isolates were found to have switched serotypes (determined by the structure of the capsular polysaccharide) to serotypes not represented in the conjugate vaccine available prior to the study.

Such evolution resulting in serotype replacement has implications for vaccine design. In particular, the notion of a universal pneumococcal vaccine based on surface proteins that are relatively conserved in structure among different capsular serotypes may need to include multiple variants of multiple surface proteins encoded in different genomic regions to minimize the chances that one horizontal gene transfer event could eliminate susceptibility to vaccine-elicited antibodies.

16.2.2.5 Chronic Diseases

Beyond the direct relevance of evolution to immunology and microbiology/infectious disease, both immune processes and microbes have been shown in recent years to contribute to the causation of a growing number of medical conditions not previously suspected of involving immune or microbial mechanisms. Examples include type-2 diabetes [43] and obesity [44]. In 2014, a leading authority on infectious diseases, Martin J. Blaser, published a thorough and accessible discussion of one perspective on the increased risks of allergic, autoimmune, and metabolic disease caused by major perturbations in the relationship that evolved over millennia between the many bacterial species that have historically composed the human microbiome and humans [45]. This book documents the growing body of evidence that substantiates major health effects of heavy antibiotic use by drastically altering the selective pressures on bacteria.

16.3 Conclusion

Evolution on multiple timescales is central to immune responses and immunity to pathogens. Reciprocally, evolutionary phenomena are critical to pathogenetic mechanisms of infectious agents, pathogen transmission, and the development of resistance to medical interventions. So for example, the near-certainty that single-drug therapy for viruses and bacteria will elicit drug resistance has critically informed the design of more effective and more resilient combination chemotherapy regimens. In addition, determining the origins of infectious disease outbreaks and making accurate predictions regarding the qualitative and quantitative features of their spread often rely, respectively, on assessing the phylogenetic relationships of disease isolates and understanding the selection pressures to which specific pathogens are subject. Therefore, a grasp of evolutionary concepts and principles is an essential foundation for both biomedical scientists and clinicians interested in specializing in fields of medicine related to immune system function or interactions with infectious agents.

Glossary

Immunity-related GTPases	A family of proteins in humans, mice, and other mammals that are encoded by genes that can become active in response to the cytokine interferon-gamma. These proteins can catalyze the hydrolysis of guanosine triphosphate to guanosine diphosphate and orthophosphate and can be involved in immunity to vacuolar pathogens by triggering the process of autophagy, which also participates in normal cellular recycling of cellular components. Opsonization—the process by which molecules such as antibodies or proteolytically derived components of the serum proteins that are participants in the complement cascade facilitate the ingestion of bacteria, other microbial pathogens, or other particulates by phagocytes, such as neutrophils, monocytes, macrophages, or dendritic cells.

Pathogen	A microbe or macroscopic parasite that can infect host organisms and can cause cellular dysfunction, tissue damage, and fitness reduction in those hosts.
Phylogenetic relationship	A relationship between species, cells, or genes based on relative temporal proximity to shared common ancestors; thus if we consider three species or three genes, A, B, and C, if A and B shared a more recent common ancestor than either A or B shares with C, A and B are more closely related, phylogenetically, to one another than to C.
Somatic cell competition	A process in which survival and proliferation of non-germ cell body cells, such as Blymphocytes, depends on comparative abilities to acquire a limiting resource, such as critical signals from CD4+ T-cells (so-called helper T-cells) in the case of germinal center Bcells.
Somatic hypermutation	A process affecting Blymphocytes following their activation by antigens in which the portions of immunoglobulin encoding genes that determine the structures of the antibody domains responsible for directly binding to antigen are subjected to an increased rate of mutation. This process typically occurs in germinal centers within lymph nodes or other secondary lymphoid tissues.
Tropism	A characteristic of a virus or other pathogen that pertains to which cell types or tissues of which species can support the replication of that pathogen.
Virulence	Frequently regarded as an attribute of a pathogen pertaining to the extent of debilitation that follows infection; in evolutionary terms, virulence is a relational property attributable to a particular host–pathogen pair that measures the extent to which infection of that host with that pathogen reduces host reproductive fitness.

References

1. Tauber AI (1990) Metchnikoff, the modern immunologist. J Leukoc Biol 47(6):561–567
2. Silverstein AM (2003) Darwinism and immunology: from Metchnikoff to Burnet. Nat Immunol. 4(1):3–6 (Review. PubMed PMID: 12496967)
3. Janeway CA Jr (1989) Approaching the asymptote? Evolution and revolution in immunology. Cold Spring Harb Symp Quant Biol. 54(Pt 1):1–13 (Review. PubMed PMID: 2700931)
4. Medzhitov R (2009) Approaching the asymptote: 20 years later. Immunity 30(6):766–775. doi:10.1016/j.immuni.2009.06.004
5. Paul WE (2012) The immune system. Fundamental immunology, 7th edn. In: Paul WE (ed) Wolters Kluwer/Lippincott Williams & Wilkins, Philadelphia, pp 1–21

6. Flajnik MJ, DuPasquier L (2012) The evolution of the immune system. Fundamental immunology, 7th edn. In: Paul WE (ed) Wolters Kluwer/Lippincott Williams & Wilkins, Philadelphia, pp 67–128

7. Hardy RR, Champhekar A (2012) B-Lymphocyte development and biology. Fundamental immunology, 7th edn. In: Paul WE (ed) Wolters Kluwer/Lippincott Williams & Wilkins, Philadelphia, pp 215–245

8. Rothenberg EV (2012) T-Lymphocyte developmental biology. Fundamental immunology, 7th edn. In: Paul WE (ed) Wolters Kluwer/Lippincott Williams & Wilkins, Philadelphia, pp 325–354

9. Burnet FM (2007) A modification of Jerne's theory of antibody production using the concept of clonal selection. Aust J Sci 20:67–69 (Nat Immunol 8:1024–26. Reprinted from 1957)

10. Talmage DW (1957) Allergy and immunology. Annu Rev Med 8:239–256

11. Burnet FM (1959) The clonal selection theory of acquired immunity. Cambridge University Press, Cambridge

12. Talmage DW (1959) Immunological specificity, unique combinations of selected natural globulins provide an alternative to the classical concept. Science 129(3364):1643–1648

13. Park YH, Osmond DG (1987) Phenotype and proliferation of early B lymphocyte precursor cells in mouse bone marrow. J Exp Med 165(2):444–458 (PubMed PMID: 3102670; PubMed Central PMCID: PMC2188517)

14. Korber B, Gaschen B, Yusim K, Thakallapally R, Kesmir C, Detours V (2001) Evolutionary and immunological implications of contemporary HIV-1 variation. Br Med Bull 58:19–42. doi:10.1093/bmb/58.1.19

15. Burke DS (1997) Recombination in HIV: an important viral evolutionary strategy. Emerg Infect Dis 3:253–259. doi:10.3201/eid0303.970301

16. Lemey P, Rambaut A, Pybus OG (2006) HIV evolutionary dynamics within and among hosts. AIDS Rev 8:125–140

17. Keele BF, Giorgi EE, Salazar-Gonzalez JF, Decker JM, Pham KT, Salazar MG (2008) Identification and characterization of transmitted and early founder virus envelopes in primary HIV-1 infection. Proc Natl Acad Sci USA 105:7552–7557. doi:10.1073/pnas.0802203105

18. Scheid JF, Mouquet H, Feldhahn N, Seaman MS, Velinzon K, Pietzsch J, Ott RG, Anthony RM, Zebroski H, Hurley A et al (2009) Broad diversity of neutralizing antibodies isolated from memory B cells in HIV-infected individuals. Nature 458:636–640

19. Xiao X, Chen W, Feng Y, Dimitrov DS (2009) Maturation pathways of cross-reactive HIV-1 neutralizing antibodies. Viruses 1:802–817

20. Xiao X, Chen W, Feng Y, Zhu Z, Prabakaran P, Wang Y, Zhang MY, Longo NS, Dimitrov DS (2009) Germline-like predecessors of broadly neutralizing antibodies lack measurable binding to HIV-1 envelope glycoproteins: implications for evasion of immune responses and design of vaccine immunogens. Biochem Biophys Res Commun 390:404–409

21. Tarlinton D, Good-Jacobson K (2013) Diversity among memory B cells: origin, consequences, and utility. Science 341(6151):1205–1211. doi:10.1126/science.1241146

22. Greenspan NS (2014) Design challenges for HIV-1vaccines based on humoral immunity. Front Immunol 5:335. doi:10.3389/fimmu.2014.00335 (eCollection)

23. Joseph CG, Darrah E, Shah AA, Skora AD, Casciola-Rosen LA, Wigley FM, Boin F, Fava A, Thoburn C, Kinde I, Jiao Y, Papadopoulos N, Kinzler KW, Vogelstein B, Rosen A (2014) Association of the autoimmune disease scleroderma with an immunologic response to cancer. Science 343(6167):152–157. doi:10.1126/science.1246886 (Epub 2013 Dec 5. PubMed PMID: 24310608)

24. http://www.cdc.gov/parasites/toxoplasmosis/, 2013. (last accessed 9/07/14)

25. Etheridge RD, Alaganan A, Tang K, Lou HJ, Turk BE, Sibley LD (2014) The Toxoplasma pseudokinase ROP5 forms complexes with ROP18 and ROP17 kinases that synergize to control acute virulence in mice. Cell Host Microbe 15(5):537–550. doi:10.1016/j.chom.2014.04.002

26. Saeij JP, Boyle JP, Coller S, Taylor S, Sibley LD, Brooke-Powell ET, Ajioka JW, Boothroyd JC (2006) Polymorphic secreted kinases are key virulence factors in toxoplasmosis. Science 314(5806):1780–1783

27. Taylor S, Barragan A, Su C, Fux B, Fentress SJ, Tang K, Beatty WL, Hajj HE, Jerome M, Behnke MS, White M, Wootton JC, Sibley LD (2006) A secreted serine-threonine kinase determines virulence in the eukaryotic pathogen *Toxoplasma gondii*. Science 314 (5806):1776–1780

28. Fentress SJ, Behnke MS, Dunay IR, Mashayekhi M, Rommereim LM, Fox BA, Bzik DJ, Taylor GA, Turk BE, Lichti CF, Townsend RR, Qiu W, Hui R, Beatty WL, Sibley LD (2010) Phosphorylation of immunity-related GTPases by a *Toxoplasma gondii*-secreted kinase promotes macrophage survival and virulence. Cell Host Microbe 8(6):484–495. doi:10.1016/j.chom.2010.11.005

29. Könen-Waisman S, Howard JC (2007) Cell-autonomous immunity to *Toxoplasma gondii* in mouse and man. Microbes Infect 9:1652–1661

30. Harris SR, Cartwright EJ, Török ME, Holden MT, Brown NM, Ogilvy-Stuart AL, Ellington MJ, Quail MA, Bentley SD, Parkhill J, Peacock SJ (2012) Whole-genome sequencing for analysis of an outbreak of meticillin-resistant *Staphylococcus aureus*: a descriptive study. Lancet Infect Dis. pii: S1473–3099(12)70268-2. doi:10.1016/S1473-3099 (12)70268-2 (Epub ahead of print)

31. Xue K (2014) Superbug: an epidemic begins. Harvard Magazine, May–June, pp 40–49

32. Aminov RI (2010) A brief history of the antibiotic era: lessons learned and challenges for the future. Front Microbiol 1:134. doi:10.3389/fmicb.2010.00134 (eCollection. PubMed PMID: 21687759; PubMed Central PMCID: PMC3109405)

33. Goossens H, Ferech M, Vander Stichele R, Elseviers M, ESAC Project Group (2005) Outpatient antibiotic use in Europe and association with resistance: a cross-national database study. Lancet 365(9459):579–587

34. Greene SE, Reid A (2013) Moving targets: fighting the evolution of resistance in infections, pests, and cancer. American Academy of Microbiology, Washington, DC

35. Gubareva LV, Kaiser L, Matrosovich MN, Soo-Hoo Y, Hayden FG (2001) Selection of influenza virus mutants in experimentally infected volunteers treated with oseltamivir. J Infect Dis 183(4):523–531 (Epub 2001 Jan 11)

36. Bloom JD, Gong LI, Baltimore D (2010) Permissive secondary mutations enable the evolution of influenza oseltamivir resistance. Science 328(5983):1272–1275

37. Russell RJ, Haire LF, Stevens DJ, Collins PJ, Lin YP, Blackburn GM, Hay AJ, Gamblin SJ, Skehel JJ (2006) The structure of H5N1 avian influenza neuraminidase suggests new opportunities for drug design. Nature 443(7107):45–49 (Epub 2006 Aug 16)

38. Collins PJ, Haire LF, Lin YP, Liu J, Russell RJ, Walker PA, Skehel JJ, Martin SR, Hay AJ, Gamblin SJ (2008) Crystal structures of oseltamivir-resistant influenza virus neuraminidase mutants. Nature 453(7199):1258–1261 (Epub 2008 May 14)

39. Centers for Disease Control and Prevention (1999) Impact of vaccines universally recommended for children—United States, 1990–1998. MMWR Morb Mortal Wkly Rep 48 (12):243–248

40. Centers for Disease Control and Prevention (1999) Ten great public health achievements—United States, 1900–1999. MMWR Morb Mortal Wkly Rep 48(12):241–243

41. Feikin DR, Kagucia EW, Loo JD, Link-Gelles R, Puhan MA, Cherian T, Levine OS, Whitney CG, O'Brien KL, Moore MR (2013) Serotype replacement study group. Serotype-specific changes in invasive pneumococcal disease after pneumococcal conjugate vaccine introduction: a pooled analysis of multiple surveillance sites. PLoS Med 10 (9):e1001517. doi:10.1371/journal.pmed.1001517 (Epub 2013 Sep 24)

42. Croucher NJ, Harris SR, Fraser C, Quail MA, Burton J, van der Linden M, McGee L, von Gottberg A, Song JH, Ko KS, Pichon B, Baker S, Parry CM, Lambertsen LM, Shahinas D, Pillai DR, Mitchell TJ, Dougan G, Tomasz A, Klugman KP, Parkhill J, Hanage WP, Bentley SD (2011) Rapid pneumococcal evolution in response to clinical interventions. Science 331(6016):430–434

43. Neels JG, Olefsky JM (2006) Inflamed fat: what starts the fire? J Clin Invest 116(1):33–35. doi:10.1172/JCI27280)
44. Turnbaugh PJ, Ley RE, Mahowald MA, Magrini V, Mardis ER, Gordon JI (2006) An obesity-associated gut microbiome with increased capacity for energy harvest. Nature 444 (7122):1027–1031
45. Blaser MJ (2014) Missing microbes: how the overuse of antibiotics is fueling our modern plagues. Henry Holt and Company, New York
46. West AP Jr, Scharf L, Scheid JF, Klein F, Bjorkman PJ, Nussenzweig MC (2014) Structural insights on the role of antibodies in HIV-1 vaccine and therapy. Cell 156(4):633–648

Chapter 17
Helminth Immunoregulation and Multiple Sclerosis Treatment

Jorge Correale, M.D.

Lay Summary Multiple sclerosis (MS) is a disease of the central nervous system (CNS) (brain, spinal cord and optic nerve), characterized by loss of myelin, an insulating cover of fat and proteins surrounding structures of the nervous system. As a consequence of this damage, conduction of nerve impulses is slower and patients present different symptoms, which can lead to irreversible disability. MS is the second cause of disability in young adults after traumatic brain injury. Although its origin remains unknown, there is scientific evidence to support the hypothesis that the individual's own immune system damages myelin, making MS a so-called autoimmune disease. During the last half of the twentieth century, the number of MS cases has increased significantly, probably due to different environmental factors, including decline of infections resulting from better public health practices. Epidemiological studies have demonstrated that patients infected with certain kinds of parasites, particularly those called helminths, present a more benign disease course. Parasites are often long-lived and inhabit hosts with an intact immune system; consequently, it is not surprising that they would acquire modulatory molecules enhancing their survival. Studies of autoimmune diseases in different animal models have corroborated these epidemiological findings. Mice infected with helminths show protection from disease or attenuation of symptoms. These observations have triggered interest in exploring the clinical efficacy of establishing controlled parasite infections in patients with allergic and autoimmune diseases. To date, clinical trials using this approach are based on small sample sizes and oriented to reproduce epidemiological data and confirm observations in animal models. Clearly,

J. Correale (✉)
Dr. Raúl Carrea Institute for Neurological Research, FLENI,
Montañeses 2325, (1428), Buenos Aires, Argentina
e-mail: jcorreale@fleni.org.ar; jorge.correale@gmail.com

© Springer International Publishing Switzerland 2016
A. Alvergne et al. (eds.), *Evolutionary Thinking in Medicine*,
Advances in the Evolutionary Analysis of Human Behaviour,
DOI 10.1007/978-3-319-29716-3_17

241

more prolonged and extensive studies including larger numbers of patients are needed to assess this novel therapeutic strategy. Alternatively, identification of parasite-derived molecules responsible for the modulation of the host immune system would allow the treatment of autoimmune diseases without the risk of potential side effects observed using live parasites. Although positive results have been reported administering parasite products in mouse models of autoimmunity, much remains to be explored before the field can move from experimental animal models to clinical practice.

17.1 Introduction

Multiple sclerosis (MS) is an inflammatory demyelinating disease of the CNS, affecting an estimated 2 million people worldwide and representing the second cause of nervous system disability in young adults after traumatic brain injury. The disease affects mainly young adults between 20 and 40 years of age and is approximately 2–3 times more frequent in females than in males. Its symptoms vary both over time and among patients, and their broad spectrum exerts considerable impact on health-related quality of life experienced by patients and their families compared to other chronic, debilitating diseases. MS course is highly variable, but most classically characterized by a relapsing–remitting (RR) pattern in which acute exacerbations are followed by periods of stability (remissions). However, in up to 50 % of patients, the pattern evolves to a secondary progressive course after 10–15 years of disease, characterized by relentless neurological deterioration over a period of several years, something that may occur from the onset (primary progressive course) in a minority of patients ($\sim$15 %; [1]).

Although the aetiology of MS remains elusive, several lines of evidence support the hypothesis that autoimmunity plays a major role in disease pathogenesis [2]. Autoimmune diseases are currently considered to result from complex interactions between individual genetic susceptibility and external environmental factors [3, 4]. Based on the findings of several genomewide association studies, the genetic component of MS is believed to result from common allelic variants in several genes acting as cooperative networks [5].

One of the most striking illustrations of the importance of the environment in MS pathogenesis is the particular geographical distribution of the disease. Prevalence rates are greater in high-latitude regions and uncommon near the equator [6]. Additionally, population migration studies indicate that individuals moving from areas of low, to areas of high risk, particularly before the age of 15, show similar incidence to host country populations, suggesting the presence of either a protective factor in the region of origin or, alternatively, a harmful factor in the adopted region [7]. Space–time cluster analyses performed both in Norway and in Sardinia have shown clustering between 13 and 20, and 1 and 3 years of age, respectively [8, 9]. The hypothesis analysed in these studies suggests that the higher-than-expected disease prevalence in individuals living close to one another during the same time period may have resulted from exposure to putative environmental risk factor(s) prior to disease onset and considers the probable cause to be infection acquired either in adolescence or in early childhood by individuals not protected through previous infection, depending on population-specific susceptibility. Furthermore, serial cross-sectional comparisons of MS epidemiology from various continents provide compelling evidence in favour of a significant rise in MS incidence and prevalence in recent decades [10]. Given the short duration over which these population changes have occurred, genetic factors alone seem an unlikely cause, indicating that MS risk is likely influenced by the environment. Therefore, MS most likely results from the combination of both genetic and environmental factors. Identifying these environmental factors and elucidating how they increase autoimmune disease risk would help develop new MS treatment strategies. Candidates likely to be responsible for the development of MS, alone or in combination, include most notably sunlight–UV exposure/or vitamin D deficiency, and viral infections, and cigarette smoking. Factors may not only influence disease onset at any time in the life of an individual, but also affect relapse rates in patients presenting relapsing–remitting forms of MS (RRMS) [11–14].

17.2 Research Findings

17.2.1 The Hygiene Hypothesis and the "Old Friends" Hypothesis

An ongoing debate persists as to whether infections prevent or precipitate autoimmune diseases (see also Chap. 15). Several studies implicate infectious environmental factors present during childhood and young adulthood as strong determinants of MS risk. Microbial infections have also been identified as triggers inducing autoimmunity, resulting in clinical disease manifestations in genetically predisposed individuals. Alternatively, infections might accelerate subclinical autoimmune processes [15, 16].

Conversely, however, certain epidemiological and experimental studies support the hygiene hypothesis, which considers infections protect rather than induce or accelerate autoimmune diseases such as MS [17]. In line with this concept, Leibowitz and coworkers suggested, in 1966, that greater MS prevalence correlated with high levels of sanitation in childhood environments [18]. There is increasing evidence that lack of exposure to organisms that were part of mammalian evolutionary history is leading to disordered regulation of the immune system and hence to increases in several chronic inflammatory disorders [19]. Epidemiological data have demonstrated an inverse relation between infections, and allergic and autoimmune diseases in the developed world during the last five decades, even after adjusting for improvements in access to medical attention and diagnostic capabilities [17]. The rise observed in autoimmune disease prevalence is too rapid to be considered secondary to genome alterations, implying some critical environmental change must have taken place. Progressive industrial development has pushed human migration from rural areas to cities, exposing the immune system to new environments, and the decreased incidence of many infectious diseases resulting from better public health practices has likely increased autoimmune disease emergence. Organisms changed or depleted from the modern environment and shown to be relevant to immunoregulation include the following: firstly microbiota, commensal organisms on the skin and in the gut; secondly environmental saprophytes like lactobacilli and many actinomycetes species; and finally helminths [19].

The component of the hygiene hypothesis implicated in faulty induction of immunoregulation is explained as the "old friends" hypothesis [19]. This hypothesis excludes childhood diseases as a requisite factor and focuses on organisms that have coevolved with mammals for a very long time, ones that were always present (lactobacilli, a variety of saprophytic mycobacteria and helminths), were tolerated by the immune system and were absent from the pathogen load in developed nations. Thus, induction of appropriate levels of immunoregulation by "old friends" becomes a physiological necessity, in which genes involved in this immunoregulatory setting are located in certain micro-organisms rather than in the mammalian genome. The theory postulates that reduced exposure to "old friends" would therefore not allow ending appropriately inflammatory episodes, leading to a range of chronic inflammatory disorders.

Allergies and autoimmune diseases have increased both their prevalence and incidence with decreasing helminthic infection. Individuals infected by helminths are less likely to suffer allergic sensitization or allergic disorders [19]. Likewise, epidemiological investigations demonstrating an inverse correlation between the global distribution of MS and that of the parasite *Trichuris trichiura*, a common human pathogen, further strengthen the hygiene hypothesis [20]. MS prevalence appears to fall steeply once a critical threshold of *T. trichiura* prevalence (about 10 %) is exceeded in any given population. Thus, the dichotomous distribution of MS and *T. trichiura* infection suggests that helminth infection protects against MS development. Indeed, regions of the world where poor sanitary conditions generate endemic areas of parasitoses show lower prevalence of allergic and autoimmune diseases. Additionally, evidence for a causal effect of parasites on reducing allergies and autoimmune diseases stems from reports that clearance of infection using antihelminthic treatment increases reactivity to skin tests against different allergens, as well as disease activity in MS patients [21, 22]. This protective effect of helminths might depend on parasite load. Elevated numbers of organisms may trigger regulatory circuits, while lower ones may act as immune adjuvants, enhancing allergic sensitization.

Animal models have confirmed the hypothesis that helminths can dampen allergic manifestations and autoimmune disease by driving immune regulation. Many examples exist in both spontaneous and induced models of human autoimmune diseases, where helminth infection or products thereof influence the course of autoimmune pathology [23–26]. One particular animal model, experimental autoimmune encephalomyelitis (EAE), which mimics essential clinical and pathological characteristics of MS, has been used in several studies investigating the impact of helminth infections, or their products, on disease severity and immunological response. In most EAE models, prior infection with helminths, or exposure to non-viable ova, or to parasite-secreted products reduced both incidence and severity of the disease. These observations would indicate the presence of a systemic anti-inflammatory milieu generated by multiple cell types and molecular mediators, influencing autoimmune response [23, 27]. Heterogeneity of immunological response can be attributed to specific helminth species, helminth-derived products, age at which infection was acquired and infection intensity. Helminths are often considered a homogeneous species, but significant differences exist between organisms. Conversely, there is evidence from mouse models to suggest that helminth infections, under certain conditions, can also exacerbate disease [28]. For these reasons, caution is recommended when interpreting data from animal models.

17.2.2 Induction of Immunoregulatory Circuits by Helminths

The immune system is made up of different cell types able to recognize and eliminate pathogens. Type-1 immune responses protect against intracellular pathogens, type-2 responses are directed mainly against parasites, and type-17 cells are important in the control of extracellular bacteria and fungi. However, these different cell populations can also inflict damage to tissues when acting in uncontrolled manner. T helper (Th)-1 and Th-17 cells release proinflammatory cytokines necessary to attack pathogens. However, inappropriate activation is associated with several autoimmune diseases. Likewise, Th-2 cells drive different antiparasite mechanisms, but when overactivated can lead to allergic disorders. Th cells are under the control of regulatory networks, represented mainly by T cells, which produce different inhibitory molecules.

In endemic areas, many if not most helminth-infected individuals are relatively asymptomatic. Manifest disease occurs often in individuals with reduced immunity, more susceptible to infection or presenting very high worm burden. Maintaining a disease-tolerant or asymptomatic state requires adequate balance between regulatory immune mechanisms present both in the host and in the helminth. Chronic helminth infections cause continuous and profound effects on the immune system function [29, 30]. Finely tuned immune regulatory networks governing susceptibility and resistance to helminths exist, which are both redundant and parallel, in order to exclude parasites while minimizing collateral pathology. Current investigations have shown that peripheral T cells from infected patients are unresponsive to stimulation with parasite antigens, and response to other antigens is also reduced, with the regulatory T cells being one of the most common mechanisms in play [31].

The "old friends" hypothesis suggests helminths are recognized by dendritic cells (DCs), and these in turn mature into regulatory DCs and drive regulatory T cell responses against parasite antigens, leading to the release of regulatory cytokines exerting bystander suppression. Furthermore, regulatory DCs increasingly process self-antigens, further elevating regulatory T cell numbers, specifically triggered by these antigens and downregulating autoimmune response [19].

Induction of alternatively activated macrophages has also been identified as key component of immune regulatory networks functioning during helminth infections [29]. In this case, helminths and their excretory–secretory molecules are endowed with the ability to act through a broad array of cellular mediators to temper host immune responses.

Evidence of these protective mechanisms has been demonstrated in patients suffering from MS [32]. In an observational prospective double-cohort study, we demonstrated that RRMS patients infected with different parasites (*Hymenolepis nana*, *T. trichiura*, *Ascaris lumbricoides*, *Strongyloides stercolaris* and *Enterobius vermicularis*) showed significantly lower number of exacerbations, minimal changes on disability scores as well as significantly lower radiological activity, compared to uninfected MS individuals. Parasite-driven protection leads to the development of interleukin (IL)-10 and transforming growth factor (TGF)-β secreting cells, as well as CD4+CD25+FoxP3+ regulatory T cells, while simultaneously inhibiting T cell proliferation and suppressing interferon (IFN)-γ and IL-12 production [32]. In addition, helminth infection in MS patients induces regulatory B cells capable of dampening the immune response through IL-10 production [33]. Interestingly, when some patients received antihelminthic drug treatment for worsening of parasite-associated symptoms, a major reduction in parasite egg numbers per gram of faeces was observed, as well as significant increase in clinical and radiological MS activity. Flares were accompanied by substantial increase in IFN-γ- and IL-12-producing cell numbers and a decline in IL-10, TGF-β and regulatory T cells, providing evidence of direct autoimmune response suppression as a result of helminth infection [22]. These observations indicate that helminth therapy can induce protection not only through prevention, since helminths can be present before an autoimmune disease develops, but also after the disease is established. Figure 17.1 illustrates the major mechanisms involved in the control of the autoimmune response and allergic processes during helminth infections. Moreover, it is important to note that immunosuppressive effects mediated by parasite infections end once the parasite has been eliminated, suggesting that immune regulation is a transient process requiring constant induction.

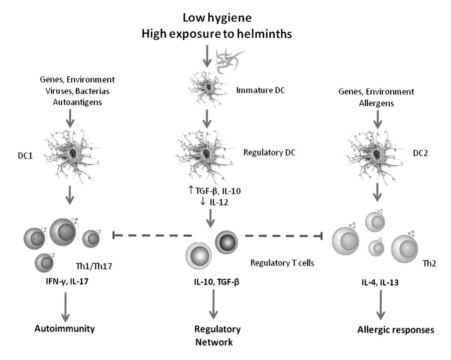

Fig. 17.1 The immune system is educated under different stimuli received by pathogens or the environment. Dendritic cells (DC) can differentiate into different populations according to the type of stimulus they receive and the magnitude of it. Dendritic cells can be stimulated by autoantigens, viruses, bacteria or helminths, inducing the development of Th-1, Th-17, Th-2 or regulatory T cells. Uncontrolled Th-1 or Th-17 activity leads to the development of autoimmunity, whereas an abnormal Th-2 response determines allergic responses. A high parasitic load changes the physiology of the microenvironment, endowing the dendritic cells with the ability to induce regulatory T cells producing IL-10 and TGF-β, generating an anti-inflammatory environment able to inhibit inflammatory responses that affects the development of autoimmunity or allergic reactions. The *dotted lines* represent inhibitory processes

17.3 Implications for Policy and Practice

On the basis of the findings here described, helminth therapy has been used in clinical trials associated with allergic and autoimmune diseases including inflammatory bowel diseases (IBD) and MS [34–36], and several more clinical trials are currently underway in other diseases. It must be pointed out that not all helminth infections can be deemed therapeutically equal, and some might worsen the disease [28, 37]. Therefore, parasite species selection is crucial. In this respect, most current studies use either *Trichura suis* (*T. suis*, pig whipworm) or *Necator americanus* (human hookworm). Using parasites that do not permanently colonize humans (*T. suis*) or organisms that can be administered at low infection intensity and eliminated with antihelminthic drugs (*N. americanus*) decreases the potential for accidental disease transmission to healthy subjects. In this line, *T. suis* ova (TSO) has recently been approved as an investigational medicinal product (IMP) by both the US Food and Drug Administration and the European Medicines Agency, while *N. americanus* has been granted IMP licence by the Medicines and Healthcare Regulatory Authority in the UK.

Nevertheless, possible caveats should be considered when assessing these trials. Data from animal models demonstrating favourable influence on EAE outcome have preferentially been observed during preimmunization and inductive phases [23, 38], suggesting it may be more difficult to suppress ongoing reactions than prevent their development. Moreover, many trials use asymptomatic infection doses, which are significantly lower than those of natural infections and therefore possibly insufficient to suppress pathology [39].

Initially, encouraging results on the effects of helminth infections in IBD [35, 36] led to trials to establish whether *T. suis* had any effect on MS. The first clinical trial of helminth therapy in MS was the helminth-induced immunomodulation therapy (HINT) study [40]. In this small scale, safety-oriented trial, five newly diagnosed RRMS patients received TSO, administered orally every 2 weeks for 3 months. Mean number of gadolinium (Gd)-enhancing MRI lesions fell from 6.6 at baseline, to 2.0 at the end of treatment and rose again to a mean of 5.8 lesions 2 months after TSO treatment was discontinued. Treatment was associated with relative increases in IL-4 and IL-10 levels in serum, as well as elevation of C-reactive protein and antibodies to *T. suis* excretory/secretory products (IgG1 and IgA), indicating robust systemic immune response to *T. suis* colonization. Peripheral CD4+CD5+FoxP3+ cells increased modestly in 2 of the five study subjects, and TSO was well tolerated. Minor gastrointestinal symptoms observed in 3 of 5 subjects were transient. Although MRI study results seem promising, they should be interpreted with caution due to the small sample size and the short follow-up duration. After reviewing HINT study results, regulatory authorities approved a follow-up clinical trial (HINT 2) with 18 relapsing–remitting MS patients, studied for 20 months using a baseline versus treatment design. Final results of this trial are expected to be published in 2015.

In another study (Trichuris suis ova therapy for relapsing multiple sclerosis—a safety study, TRIMS A) conducted at the Danish Multiple Sclerosis Center of Copenhagen University Hospital, 10 RRMS patients were treated with TSO every 2 weeks for 3 months [41]. The primary outcome measure was MRI activity based on the number of new or enlarging T2 lesions, the number of Gd-enhancing lesions and the volume of T2 lesions. Brain MRI testing was performed every 3 weeks. The investigators concluded that TSO seemed to be safe and well tolerated. However, no clinical, MRI or immunological signals indicating benefit were observed.

Investigators at Charite University in Berlin conducted the first exploratory study in secondary progressive MS patients [42]. Four patients were surveyed during 6 months of therapy with TSO, given orally every 2 weeks. The study focused on T cell modulation as well as on innate immune response. Stimulated peripheral blood mononuclear cells showed slight downregulation of Th-1-associated cytokine patterns, with temporal increase of Th-2-associated cytokines such as IL-4.

A double-blind placebo-controlled phase II trial (Trichuris suis ova in recurring–remitting multiple sclerosis and clinically isolated syndrome, TRIOMS) has been initiated by the same investigators [43]. The study will recruit 50 patients with RRMS or clinically isolated syndrome with clinical activity, not undergoing any standard therapy. Patients will receive either TSO every 2 weeks or placebo for a 12-month treatment period and will be followed for an additional 6 months.

Another phase II double-blind placebo-controlled study (worms for immune regulation of multiple sclerosis, WIRMS; NCTO1470521) has begun at the University of Nottingham. The study will enrol 72 RRMS and secondary progressive MS patients with superimposed relapse, who will be treated with dermally administered live larvae of *N. americanus* or placebo. Worms will be allowed to remain within host for a full 9 months. Investigators speculate that this period of residence will establish and maintain immunoregulatory mechanisms of sufficient magnitude to translate into the anti-inflammatory effect and consequently generate therapeutic benefit. The cumulative number of new and active lesions on T2-weighted MRI will be the primary outcome measure. Regulatory network induction (regulatory T cell induction, regulatory B cells, Tr1 cells and natural killer cells) will be the secondary outcome measure.

Clearly, at this time, a number of critical issues need to be addressed in further investigations. Questions remain regarding which helminth is most effective, and at what dose, which is the best route of administration or optimal timing for introducing infection in relation to disease onset, whether helminth-derived molecules have the same efficacy as live parasites and what the optimal treatment schedule should be.

Although effects of helminth therapy on vaccine efficacy have not been evaluated, several studies have shown that helminths can influence vaccine efficacy by modulating host immune response, in particular when Th-1-like and cell-dependent responses are required. *S. mansoni* infection was shown to reduce BCG-induced protective response against *Mycobacterium tuberculosis* in mice [44]. Likewise, helminth infections dramatically reduced malaria DNA vaccine immunogenicity [45]. Moreover, epidemiological studies have demonstrated that *Schistosoma* sp. infections decrease the efficacy of vaccines against tetanus and hepatitis B virus [46, 47]. Overall, effects of helminth therapy on vaccine efficacy need to be further investigated.

TSO safety profiles have been extensively studied in IBD patients, even while on concomitant immunosuppressive drugs, without observing any significant side effects. In the HINT 1 study, some patients presented transient diarrhoea and upper abdominal pain 30–50 days after TSO treatment initiation (lasting 3–5 days), symptoms possibly related to an innate inflammatory immune response in the gut induced after initial *T. suis* larvae colonization [40]. These had not been observed in earlier TSO studies in IBD patients, perhaps because they were occurring in the context of moderate gastrointestinal pathology in the study population [48]. Nor did these symptoms interfere with patient daily life activities. An additional concern in helminth therapy is whether helminth colonization may worsen other pathogenic infections (bacterial, parasitic or viral) especially in immunocompromised hosts.

Enhancement of disease and pathology by coinfection of *T. suis* and *Campylobacter jejuni* or *T. Trichiura* has been described. However, this has never been observed in TSO-treated patients [49, 50].

Another helminth studied in clinical trials thus far is the hookworm *N. americanus*. In previous studies of helminth therapy, the most common hookworm-related side effect was localized maculopapular rash at skin entry site, which began within a day or so of infection and typically lasted 2–5 days. In some patients, rash recurred approximately 2–3 weeks after infection for up to 10 days before disappearing. The most troublesome adverse effects were gastrointestinal symptoms, such as diarrhoea and abdominal pain. The other most commonly reported symptoms were malaise and fatigue, which occurred between week six to seven of treatment and have been associated with systemic eosinophilia, rather than with a direct parasite effect. Dose-ranging studies of therapeutic *N. americanus* infection have shown side effects to be dose dependent. Doses higher than 10 larvae correlated with more frequent and severe adverse events than low-dose inocula. All symptoms disappeared completely after subjects were treated with the anti-helminthic drug, mebendazole [48, 51]. It is strongly recommended at this time that live helminths or parasite ova should not be administered outside strictly monitored controlled clinical trials.

17.4   Future Directions

Helminth infections are often long-lived and inhabit immunocompetent hosts; it is therefore not surprising that these organisms may have acquired modulatory molecules attenuating host responses and enhancing their own survival. Understanding host–parasite interactions and identifying different parasite molecules possessing immunomodulatory effects will help combat allergic and autoimmune diseases, without the costly price of infectious side effect. During recent decades, a number of helminth-derived immunomodulatory molecules have been characterized, in terms of both structure and bioactivity [52, 53]. Although this approach might overcome some of the safety concerns regarding the use of live helminths as therapeutic agents, controversies still exist as to whether live infection is a prerequisite for suppression of inflammatory responses in different disease models of autoimmunity.

ES-62, a glycoprotein from the rodent nematode *Acanthonema vitae*, has been widely investigated as an immunomodulatory molecule. Its administration in mice with collagen-induced arthritis resulted in significant reduction in disease severity and slowing down of progression [54]. Likewise, when rDiAg, a product from the filarial parasite *Dirofilaria immitis*, was administered to non-obese diabetic (NOD) mice, it prevented insulitis and diabetes onset [55]. Furthermore, treatment with soluble products from *T. suis*, *S. mansoni* and *Trichinella spiralis* caused strong reduction in the severity of EAE in mice, significant suppression of pro-inflammatory phenotype in human DCs, as well as subsequent generation of human Th-1 and Th-17 effector cells [56]. Finally, lacto-N-fucopentaose III (LNFP III), a Lewis X-containing glycan found in *S. mansoni* eggs, suppressed EAE through enhancement of IL-10 and Th-2 cytokines [57]. These few examples of helminth-derived immunomodulatory products illustrate the potential of these molecules to serve as drugs, or templates for drug design. Nevertheless, it would be dangerous to ignore the fact that this regulatory environment could hamper essential and necessary responses to other antigens, vaccinations and life-threatening pathogens. As chronic helminth infections establish and accumulate, suppression may become more generalized and render the host susceptible to secondary infections. At this stage, both beneficial and deleterious consequences of helminth infection need to be clearly identified. Individual species may develop very different infection dynamics over time and/or during peak infection intensity. Moreover, the same species may trigger opposite effects under varying conditions. Finally, if susceptibility to autoimmune diseases is genetically influenced, so too must be the propensity for infections to modulate disease immunopathology, making particular infections protective only in certain genotypes, rather than in the population at large. It is evident that future studies in this area will be required to establish whether certain infections, particularly those produced by helminths during critical periods of infancy, truly exert a protective effect against MS. Clearly, much remains to be explored to move the field from observations in animal models to clinical practice; issues relating to in vivo stability and helminth-derived molecule pharmacodynamics, delivery methods, as well as immunogenicity need to be overcome, if new therapeutic modalities are to be developed.

Overall, these observations raise a paradox, namely that deworming populations with helminth-associated morbidity could cause emergence of chronic inflammatory conditions and autoimmune diseases [58]. Therefore, well-designed trials in the context of large-scale deworming programs are warranted, to assess benefits of the intervention, weighed against potential adverse effects such as increased chronic inflammatory disease risk.

Glossary

Actinomycetes	Group of terrestrial or aquatic bacteria
Allergy	Hypersensitivity disorder of the immune system. Symptoms include red eyes, itchiness, runny nose, eczema or asthma attacks
Antibodies	Protein produced by the immune system to recognize or neutralize foreign molecules or micro-organisms
Antigen	Any substance inducing a response of the immune system

Autoimmune disease	Diseases arising as a result of an abnormal immune response by the body, directed against substances and tissues normally present in the body of the host
Clinically isolated syndrome	First episode of a demyelinating disease. It is the step prior to MS development
Commensal organisms	Organisms that obtain benefits from other organisms without producing any damage
Cytokines	Broad and loose category of small proteins that are important in cell signalling. They are released by several cells but particularly by cells of the immune system and affect the behaviour of other cells, or even of the releasing cell itself
Deworming	To give an antihelminthic drug to a person or animal to rid it of intestinal parasites
Demyelinating disease	Disease produced by the loss of myelin
Dendritic cell	Cells that recognize, process and present antigens to T cells, inducing their activation
Experimental autoimmune encephalomyelitis (EAE)	Animal model that mimics essential clinical and pathological characteristics of MS. It can be induced in different species such as rabbit, guinea pigs, rats, mice or monkeys
Gadolinium	Contrast medium used in magnetic resonance imaging studies to highlight abnormalities or disease process. In the brain, a gadolinium-positive lesion means the presence of inflammation
Gut microbiota	Also known as gut flora is a complex group of micro-organism species inhabiting the digestive tract of animals and humans and the largest reservoir of commensal micro-organisms
Helminths	Wormlike organisms living in and feeding on living hosts
Incidence	Measure of the risk of developing some new condition within a given time period
Interferon-γ	A type of cytokine that has most often a pro-inflammatory effect

Interleukins	A group of cytokines produced by white blood cells
Macrophage	Type of white blood cell that engulfs and digests cellular debris, foreign substances, microbes and cancer cells in a process called phagocytosis
Magnetic resonance imaging (MRI)	Also known as nuclear magnetic resonance imaging (NMRI) is a medical imaging technique used in radiology to investigate the anatomy and physiology of the body. MRI scanners use strong magnetic fields and radio waves to form images of the body without exposure to ionizing radiation
Multiple sclerosis	Inflammatory disease in which the insulating covers of nerve cells in the brain, spinal cord and optic nerve are damaged. This damage disrupts the ability of parts of the nervous system to communicate, resulting in a wide range of signs and symptoms
Mycobacteria	Widespread bacteria, typically living in water and food sources. Some, however, including tuberculosis and the leprosy bacteria, appear to be obligate parasites and are not found as free-living members of the genus
Natural killer cell	Type of cytotoxic lymphocytes that induce the death of tumour cells or virally infected cells
Prevalence	Proportion of a population found to have a condition, typically a disease or a risk factor
Saprophytes	Organisms obtaining nutrients from dead organic matter
T helper cell (Th)	Type of T cell that helps the activity of other immune cells by releasing T cell cytokines. These cells help, suppress or regulate immune responses
Transforming growth factor-β	Is a cytokine that controls proliferation, cellular differentiation and other functions in most cells

References

1. Lublin FD, Reingold SC (1996) Defining the clinical course of multiple sclerosis: results of an international survey. National Multiple Sclerosis Society (USA) Advisory Committee on Clinical Trials of New Agents in Multiple Sclerosis. Neurology 46:907–911
2. McFarland H, Martin R (2007) Multiple sclerosis: a complicated picture of autoimmunity. Nat Immunol 8:913–919
3. Marrie RA (2004) Environmental risk factors in multiple sclerosis. Lancet Neurol 3:709–718
4. Oksenberg JR, Baranzini SE, Sawcer S, Hauser SL (2008) The genetics of multiple sclerosis: SNP to pathways to pathogenesis. Nat Rev Gen 9:516–526
5. Gourraud PA, Harbo HF, Hauser SL, Baranzini SE (2012) The genetics of multiple sclerosis: an up-to-date review. Immunol Rev 248:87–103
6. Rosati G (2011) The prevalence of multiple sclerosis in the world: an update. Neurol Sci 22:117–139
7. Alter M, Kahana E, Lowenson R (1978) Migration and risk of multiple sclerosis. Neurology 28:1089–1093
8. Riise T, Groonning M, Klauber MR, Barret-Connor E, Nyland H, Albreksten G (1991) Clustering of residence of multiple sclerosis patients at age 13 to 20 in Hordaland, Norway. Am J Epidemiol 133:932–939
9. Pugliatti M, Riise T, Sotgiu MA et al (2006) Evidence of early childhood as the susceptibility period in multiple sclerosis: space-time cluster analysis in Sardinian population. Am J Epidemiol 164:326–333
10. Melcon MO, Correale J, Melcon CM (2014) Is it time for a new global classification of multiple sclerosis? J Neurol Sci 344:171–181
11. Ascherio A, Munger KL (2007) Environmental risk factors for multiple sclerosis. Part II: Noninfectious factors. Ann Neurol 61:504–513
12. Ascherio A (2013) Environmental factors in multiple sclerosis. Expert Rev Neurother 13(12 Suppl):3–9
13. DeLuca GC, Kimball SM, Kolasinski J, Ramagopalan SV, Ebers GC (2013) Review: the role of vitamin D in nervous system health and disease. Neuropathol Appl Neurobiol 39:458–484
14. Koch MW, Metz LM, Agrawal SM, Yong VW (2013) Environmental factors and their regulation of immunity in multiple sclerosis. J Neurol Sci 324:10–16
15. Christen U, von Herrath MG (2005) Infections and autoimmunity-good or bad? J Immunol 174:7481–7486
16. Correale J, Fiol M, Gilmore W (2006) The risk of relapses in multiple sclerosis during systemic infections. Neurology 67:652–659
17. Bach JF (2002) The effect of infections on susceptibility to autoimmune and allergic diseases. N Engl J Med 347:911–920
18. Leibowitz U, Atanovsky A, Medalie JM, Smith HA, Halpern L, Alter M (1966) Epidemiological study of multiple sclerosis in Israel. II. Multiple sclerosis and the level of sanitation. J Neurol Neurosurg Psychiatry 29:60–68
19. Rook GAW (2009) The changing microbial environment, Darwinian medicine and the hygiene hypothesis. In: Rook GAW (ed) The hygiene hypothesis and Darwinian medicine. Birkhäuser, Basel, pp 1–27
20. Fleming JO, Cook TD (2006) Multiple sclerosis and the hygiene hypothesis. Neurology 67:2085–2086
21. van den Bigeelar AH, Rodrigues LC, van Ree R et al (2004) Long-term treatment of intestinal helminthes increases mite skin-test reactivity in Gabonese schoolchildren. J Infect Dis 189:892–900
22. Correale J, Farez M (2011) The impact of parasite infections on the course of multiple sclerosis. J Neuroimmunol 233:6–11

23. La Flamme AC, Ruddenklau K, Bäckström BT (2003) Schistosomiasis decreases central nervous system inflammation and alters the progression of experimental autoimmune encephalomyelitis. Infect Immun 71:4996–5004
24. Zaccone P, Fehervari Z, Jones FM et al (2003) *Schistosoma mansoni* modulate the activity of the innate immune response and prevent onset of type 1 diabetes. Eur J Immunol 33:1439–1449
25. Osada Y, Shimizu S, Kumagai T, Yamada S, Kanazawa T (2009) Schistosoma mansoni infection reduces severity of collagen-induced arthritis via down-regulation of por-inflammatory mediators. Int J Parasitol 39:457–464
26. Elliot DE, Li J, Blum A, Metwali A, Qadir K, Urban JF, Weinstock JV (2003) Exposure to schistosome eggs portects mice from TNBS-induced colitis. Am J Physiol Gastrointest Liver Physiol 284:G385–G391
27. Hasseldam H, Hansen CS, Johansen FF (2013) Immunomodulatory effects of helminths and protozoa in multiple sclerosis and experimental autoimmune encephalomyelitis. Parasite Immunol 35:103–108
28. Hunter MM, Wang A, Hirota CL, McKay DM (2007) Helminth infection enhances disease in a murine TH2 model of colitis. Gastroenterology 132:1320–1330
29. McSorley HJ, Maizels RM (2012) Helminth Infections and host immune regulation. Clin Microbiol Rev 25:585–608
30. Maizels RM, Hewitson JP, Smith KA (2012) Susceptibility and immunity to helminth parasites. Curr Opin Immunol 24:459–466
31. Maizels RM, Yazdanbakhsh M (2008) T-cell regulation in helminth parasite infections: implications for inflammatory diseases. Chem Immunol Allergy 94:112–123
32. Correale J, Farez M (2007) Association between parasite infection and immune responses in multiple sclerosis. Ann Neurol 61:97–108
33. Correale J, Farez M, Razzitte G (2008) Helminth infections associated with multiple sclerosis induce regulatory B cells. Ann Neurol 64:187–199
34. Bager P, Arnved J, Rønborg S et al (2010) *Trichuris suis* ova therapy for allergic rhinitis: a randomized, double-blind, placebo-controlled clinical trial. J Allergy Clin Immunol 125: 123–130
35. Summers RW, Elliot DE, Urban JF Jr, Thompson RA, Weinstrock JV (2005) *Trichuris suis* therapy for active ulcerative colitis: a randomized controlled trial. Gastroenterology 128: 825–832
36. Summers RW, Elliot DE, Urban JF Jr, Thompson R, Weinstrock JV (2005) *Trichuris suis* therapy in Crohn's disease. Gut 54:87–90
37. Graepel R, Leung G, Wang A et al (2013) Murine autoimmune arthritis is exaggerated by infection with the rat tapeworm, Hymenolepis diminuta. Int J Parasitol 43:593–601
38. Fleming JO (2013) Helminth therapy in multiple sclerosis. Int J Parasitol 43:259–274
39. Tilp C, Kapur V, Loging W, Erb KJ (2013) Prerequisites for the pharmaceutical industry to develop and commercialise helminthes and helminth-derived product therapy. Int J Parasitol 43:319–325
40. Fleming JO, Isaak A, Lee JE, Luzzio CC et al (2011) Probiotic helminth administration in relapsing-remitting multiple sclerosis: a phase 1 study. Mult Scler 17:743–754
41. Voldsgaard A, Bager P, Kapel C et al (2012) Trichuris suis ova therapy for relapsing multiple sclerosis—a safety study. Neurology 78:S30.005
42. Benzel F, Erdur H, Kholer S, Frentsch M et al (2012) Immune monitoring of trichuris suis egg therapy in multiple sclerosis patients. J Helminthol 86:339–347
43. Rosche B, Wernecke KD, Ohlaraun S, Dörr JM, Paul F (2013) Trichuris suis ova in relapsing-remitting multiple sclerosis and clinically isolated syndrome (TRIONS): study protocol for a randomized controlled trial. Trials 14:112
44. Elias D, Akuffo H, Pawlowski A, Haile M, Schön T, Britton S (2005) *Schistosoma mansoni* infection reduces the protective efficacy of BCG vaccination against virulent *Mycobacterium tuberculosis*. Vaccine 23:1326–1334

45. Cruz-Chan JV, Rosado-Vallado M, Dumonteil E (2010) Malaria vaccine efficacy: overcoming the helminth hurdle. Exp Rev Vaccines 9:707–711
46. Sabin EA, Araujo MI, Carvalho EM, Pearce EJ (1996) Impairment of tetanus toxoid-specific Th1-like immune responses in human infected with *Schistosoma mansoni*. J Infect Dis 173:269–272
47. Bassily S, Hyams N, El-Ghorab M, Mansour MM, El-Masry NA, Dunn MA (1987) Immunogenicity of hepatitis B vaccine in patients infected with *Schistosoma mansoni*. Am J Trop Med Hyg 36:549–553
48. Mortimer K, Brown A, Feary J et al (2006) Dose ranging study for trials of therapeutic infection with Necator americanus in humans. Am J Trop Med Hyg 75:914–920
49. Shin J, Gardiner GW, Deitel W, Kandel G (2004) Does whipworm increase the pathogenicity of *Campylobacter jejuni*? A clinical correlate of an experimental observation. Can J Gastroenterol 18:175–177
50. Mansfield LS, Gauthier DT, Abner SR, Jones KM, Wilder SR, Urban JF (2003) Enhancement of disease and pathology by synergy of *Trichuris suis* and *Campylobacter jejuni* in the colon of immunologically naive swine. Am J Trop Med Hyg 63:70–80
51. Wolff MJ, Braodhurst MJ, Loke P (2012) Helminthic therapy: improving mucosal barrier function. Trends Parasitol 28:187–194
52. Johnston MJG, MacDonald JA, McKay DM (2009) Parasitic helminthes: apharmacopeia of anti-inflammatory molecules. Parasitology 136:125–147
53. Danilowiz-Luebert E, O'Regan NL, Steinfelder S, Hartmann S (2011) Modulation of specific and allergy-related immune responses by helminths. J Biomed Biotechnol 2011:821578
54. Pineda MA, McGrath MA, Smith PC et al (2012) The parasitic helminth product ES-62 suppresses pathogenesis in collagen-induced arthritis by targeting the interleukin-17-producing cellular network at multiple sites. Arthritis Rheum 64:3168–3178
55. Imai S, Tezuka H, Fujita K (2001) A factor inducing IgE from a filarial parasite prevents insulin-dependent diabetes mellitus in nonobese diabetic mice. Biochem Biophys Res Commun 286:1051–1058
56. Kuijk LM, Klaver EJ, Kooij G (2012) Soluble helminth products suppress clinical signs in murine experimental autoimmune encephalomyelitis and differentially modulate human dendritic cell activation. Mol Immunol 51:210–218
57. Zhu B, Trikudanathan S, Zozulya AL et al (2012) Immune modulation by Lacto-N-fucopentose III in experimental autoimmune encephalomyelitis. Clin Immunol 142:351–361
58. Wammes LJ, Mpairwe H, Elliot AM, Yazdanbakhsh M (2014) Helminth therapy or elimination: epidemiological, immunological, and clinical considerations. Lancet Infect Dis 14:70771–70776

Part VII
Geriatrics

Chapter 18
Inflammaging and Its Role in Ageing and Age-Related Diseases

Prof. Claudio Franceschi, Ph.D., Zelda Alice Franceschi, Ph.D., Paolo Garagnani, Ph.D. and Cristina Giuliani, Ph.D.

Abbreviations

IS	Immune system
NES	Neuro-endocrine system
POMC	Pro-opiomelanocortin
BA	Biogenic amines
NOS	Nitric oxide synthase
ACTH	Adrenocorticotrophic hormone
CRH	Corticotrophin-releasing hormone

C. Franceschi (✉) · P. Garagnani
Department of Experimental, Diagnostic and Specialty Medicine (DIMES),
University of Bologna, Bologna, Italy
e-mail: claudio.franceschi@unibo.it

P. Garagnani
e-mail: paolo.garagnani2@unibo.it

C. Franceschi · P. Garagnani
Interdepartmental Center "Luigi Galvani" (CIG), University of Bologna, Bologna, Italy

C. Franceschi
IRCCS Institute of Neurological Sciences, and CNR-ISOF, Bologna, Italy

Z.A. Franceschi
Department of History, Cultures and Civilizations (DISCI),
University of Bologna, Bologna, Italy
e-mail: zelda.franceschi@unibo.it

C. Giuliani
Department of Biological, Geological and Environmental Sciences (BiGeA),
Laboratory of Molecular Anthropology and Centre for Genome Biology,
University of Bologna, Bologna, Italy
e-mail: cristina.giuliani2@unibo.it

© Springer International Publishing Switzerland 2016
A. Alvergne et al. (eds.), *Evolutionary Thinking in Medicine*,
Advances in the Evolutionary Analysis of Human Behaviour,
DOI 10.1007/978-3-319-29716-3_18

Lay Summary The growing number of elderly and the increase in age-related diseases are pressing issues for medicine and public health. Inflammaging (i.e. a chronic, low-grade inflammatory status that occurs during ageing) represents a common mechanism to the vast majority of age-related disorders. Accordingly, inflammaging and inflammation are strategic targets for prevention and treatment of these conditions. Inflammation and stress response are the result of a complex network of interactions between genes and environment. They are ancestrally interconnected and can be considered a most ancient and evolutionary-conserved maintenance/repair mechanism owing to its crucial capability to cope with and neutralize damaging agents. Evolutionary adaptation is thus treated as a "plural model", according to Ingold [1]. In this chapter, we illustrate the evolution of inflammation/stress response, the role of inflammation during ageing, including what we propose to call "immunological biography", which includes all the immune adaptive mechanisms that occur lifelong. We contextualize inflammaging within a human eco-anthropological perspective to better understand the role that changes occurred in the human environment in the last 200 years played in the demographic explosion of the elderly population. This comprehensive view on inflammation and inflammaging can have a far-reaching beneficial impact in the medical field and, in particular, could represent a strong, evolutionary-based conceptual framework to identify the most effective strategies (e.g. dietary interventions) to slow ageing and avoid/postpone age-related pathologies.

18.1   Introduction

Population ageing and age-related diseases are pressing issues for modern healthcare systems. Accumulating evidence suggests that ageing and the major age-related diseases are characterized by a chronic low-grade inflammation, referred to as inflammaging [2] to indicate its central role in the ageing process. High levels of inflammatory markers such as IL-6 and others are the powerful predictors of mortality and morbidity [3]. The major sources of inflammatory stimuli with age and inflammaging and are as follows: (1) endogenous self-derived debris that accumulate during ageing as a consequences of an increased production of and a reduced clearance of dead cells and damaged organelles, (2) *senescent cells* and cells which harbour a DNA damage and are capable of a *"DNA damage response"* that secrete a variety of pro-inflammatory cytokines that alter the microenvironment or tissues and organs, (3) persistent infections that accelerate immunosenescence, (4) products of the gut microbiota that undergo profound changes during ageing and (5) increasing activation of the coagulation system [4]. Thus, we have proposed the new word *garbaging* to indicate the exogenous and endogenous inflammatory stimuli that, as a whole, increase progressively with age and trigger inflammaging.

In the medical field, inflammation is often considered a pathological process that contributes to diseases, but from a biological point of view, inflammation has to be considered a fundamental process crucial for survival and an adaptive/protective response to damaging agents/stressors. Indeed, as first suggested by Metchnikoff, inflammation is an evolutionary-conserved "positive" phenomenon, enabling the body to react to and neutralize foreign damaging agents. It can be predicted that the capability to mount a strong inflammatory process can contribute to fitness and survival at younger ages (development and reproduction) and that genetic variants that codes for this trait have been positively selected for.

In this chapter, we will first describe the evolution of stress response to understand the evolutionary mechanisms that lead to inflammation. The stress response is at the heart of the matter to understand the physiological phenomena of immunosenescence and inflammaging within an evolutionary context. Then, we will described the pool of molecules involved in inflammatory pathway that are common to different systems (NES and IS). This pool of molecules, that is well conserved during evolution, reduces the semantic boundaries between inflammation, stress response and immune response. The bow tie architecture revealed how the organisms/individuals could mediate between the wide range of signals originating from the environment and the limited pool of effector molecules. The antagonist pleiotropy theory helps us in understanding how certain genetic characteristics and biological responses could exert a beneficial effect at young age and a deleterious effect later in life (and vice versa), and we surmise that this scenario can explain the double and apparently contradictory (beneficial and deleterious) role of inflammation lifelong. The beneficial effects of inflammation early in life and in adulthood become a detrimental/damaging process late in life, in a period where positive selection fades. However, inflammation and inflammaging are likely not the only cause of an unhealthy ageing, and the "two-hits" theory was proposed to explain such a complex outcome. Then, we will describe the link between inflammation and environment, introducing the concept of "immunological biography". We and others [5] suggest that the inflammatory response in humans has been optimized through evolution to cope with an environment that has been largely modified in the last ten thousand years and particularly in the last two hundred years and that in most cases does not exist anymore. Here, the variety of cultural habits of human beings will be discussed, to connect inflammation and inflammaging with cultural and anthropological settings, with particular attention to the effects of transitions.

In conclusion, it is fundamental to posit this intricate scenario in a broader context both at the macro- and micro-evolutionary level. Using the well-defined framework of evolutionary biology could be crucial to explain in depth the phenomena observed by clinicians, such as inflammaging and to identify public health and individual strategies to reduce the progressive age-related increase of inflammatory process that constitute an inherent characteristic of major chronic age-related. We suggest nutrition and dietary interventions as important tools for intervention to slow down ageing rate of several domains (cognitive, cardiovascular, immunological and among others).

18.2 Research Findings

18.2.1 The Evolution of Stress Responses

The stress response is at the heart of the physiological phenomena of immunosenescence and inflammaging in an evolutionary context. A brief outline is needed to understand the meaning and the complexity of stress response. In the past, changes in the immune system (IS) have often been proposed to be the direct consequences of the activation of the neuro-endocrine system (NES) after a stress signal, and the IS has often been suggested to be a target of the stress response itself.

However, about 15 years ago, new hypotheses and data on the influence of the IS on the NES emerged, leading to something of a conceptual revolution in the field. The idea that the IS was not only a target of the stress response but also an active component of this response began spreading. It is now accepted that a stress response can be induced (1) by cognitive stimuli through the five senses and (2) by non-cognitive stimuli that may impact the IS. An example of the latter is illustrated by antigens, which might be considered as stressors from an evolutionary point of view [6]. Several studies showed that antigens, like stressors, are capable of inducing an increase in the blood concentration of ACTH and corticosterone, and a concomitant increase of the overall electrical activity of neurons. Furthermore, antigens and stressors are able to activate a complex network of common responses, which include chemotaxis, phagocytosis, the release of biogenic amines (BA) and others. This is the main reason why antigens can be considered as one particular type of a broader category of stressors. The interaction between the IS and the NES is also supported by various data, including the possibility of neuropeptides and hormones to bind to receptors on immune cells (lymphocytes). It follows that the IS may be considered as a "*sensory organ*" which alerts the organism of those danger signals coming from the inside (i.e. antigens) and cannot be perceived by the classical five senses, as already hypothesized by Blalock [7].

The stress response is an adaptive mechanism that has played a key role for the survival of species, as demonstrated by the fact that it is a highly conserved mechanism from a phylogenetic point of view. Across our evolutionary history, while the structural and hierarchical organization of the immune–neuro-endocrine system has changed, the pool of effector molecules remained largely conserved [8]. From invertebrates to humans, the cellular response to a number of stressors appears to be highly maintained and involves the up-regulation of a variety of evolutionary-conserved mediators, such as oxygen-free radicals, nitric oxide (NO), pro-inflammatory cytokines (IL-1, IL-6, TNFα), propiomelanocortin-derived peptides (ACTH, β-endorphin, α-MSH), steroids (cortisol), BA (noradrenaline, adrenaline, dopamine) and neuropeptides (CRH) [2]. These observations formed the basis for the hypothesis of "*a common origin of the immune and neuro-endocrine systems*" [6, 8]. Macrophages, i.e. the cells with phagocytic activity first described by Metchnikoff and present from invertebrates to man, play a primary role in defence mechanisms and are able to release all the above-mentioned molecules. Thus, we argued that this cell could be considered the eyewitness of the common evolutionary origin of the immune and neuro-endocrine systems. For this reason, we argued that the macrophage is the best candidate to play a central role in inflammaging, a condition linked to the chronic activation of the macrophage with age ("*macrophaging*") [2]. We also argued that owing to the capability of the macrophage to secrete such a variety of immune and neuro-endocrine molecules, this cell could be considered a sort of "*mobile immune brain*". The link between macrophages and inflammation is particularly striking in the fat tissue. The visceral fat that increases with age is a source of inflammation as this tissue tends to become infiltrated of macrophages that, in turn, produce large amounts of pro-inflammatory cytokines.

The increasing of complexity observed through evolution, it is likely to be linked to a host–pathogen co-evolution. In this complex system, an anatomical subdivision between the immune and the neuro-endocrine systems as well as a new organization at the level of organs and systems (thymus, hypothalamus, pituitary and adrenal gland) and new uses of the above-mentioned "old" conserved molecules emerged during evolution. In conclusion, a progressive increase in complexity is the main difference between vertebrates and invertebrates, while the immune–neuro-endocrine responses continue to use the same pool of basic molecular mediators such as POMC-products, NOS, CRH and cytokines, which are still fundamental for the maintenance of body homeostasis (Fig. 18.1). Thus, the NES and the IS are deeply

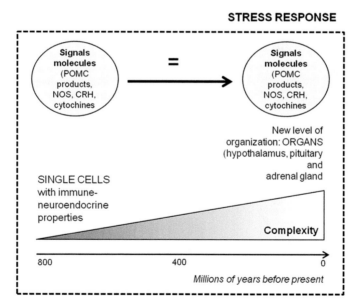

Fig. 18.1 The ancestral response to stress is exerted by a multifunctional cell called "the mobile immune brain" to indicate its immune–neuro-endocrine functions. During the evolution of vertebrates, moving towards a higher level of complexity, the same immune–neuro-endocrine properties have been organized in new interconnected organs (thymus, pituitary adrenal gland, etc.)

interconnected by a bidirectional communication system [7, 9]. Within such a conceptual evolutionary framework, the distinction between "hormone", "neurotransmitter" and "cytokine" is more blurred and less defined than previously assumed.

The pool of molecules shared by the IS and the NES—cytokines, CRH, ACTH, BA, NO and glucocorticoids—plays a major role in inflammation as well as in natural immunity, supporting the view that natural immunity, inflammation and the stress response can be considered as highly linked and connected processes. It is clear that a limited pool of molecules has to react to many input signals. The organism needs to be able to dynamically respond to an enormous amount of ecologically defined environmental stimuli that could potentially constitute a danger for the organism itself (input signals) [10]. Concomitantly, a high variety of fine-tuned responses are required in order to ensure survival (output signals). These complex input and output signals are mediated by the above-mentioned limited pool of molecules (which may be conceived as a kind of "compressed" information network). It has been suggested that these processes could be represented by the bow tie architecture (Fig. 18.2), as originally proposed by Csete and Doyle [11]. The advantage of this architecture is to save on the energy costs associated with the stimulation of the neuro-endocrine and/or the IS, thereby reducing and minimizing the pool of mediators involved. This process is closely dependent on the environment, the context in which the body is living and therefore this type of phenomena is often referred to as ecoimmunology [12].

18.2.2   Inflammaging, Immunosenescence and the Antagonistic Pleiotropy Theory

From an evolutionary perspective, *immunosenescence* could be considered as the progressive impairment of the IS (at the level of cells, organs and system) which

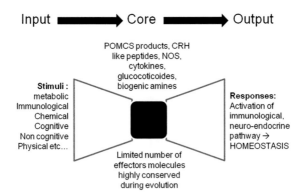

Fig. 18.2 Bow tie architecture. The highly conserved pool of molecules involved in natural immunity, inflammation and the stress response constitutes the core of this architecture. The bow tie architecture ensures that the high number of stimuli is compressed, but at the same time allows a high variety of responses in order to maintain the homeostasis of the organism. Figures adapted from [12]

occurs with age. It is important to note that the ancestral innate immunity is highly preserved, while the recent adaptive immunity (lymphocyte-centred) deeply deteriorates during ageing [13]. T cell reactivity indicates that the body is able to react to antigenic load even at high and extreme age, but at the same time, the response tends to run out with ageing. The evidence is that during ageing an accumulation of clones of memory cells that fill the "immunological space" occurs, while naive T cells are much less represented, contributing to the reduced response to new antigens/stressors, as observed during vaccination towards a variety of bacteria and viruses [13]. If antigens are compared to stressors, immunosenescence could be considered to be the result of the continuous attrition caused by the exposure to biological (bacteria, viruses, parasites and xenobiotic) and non-biological (life events, emotional and socio-economical stress) stressors lifelong.

Age-related diseases and longevity could be conceptualized as the result of the adaptation of the body to continuous stimuli and stressors occurring over time. Age remodelling is a universal physiological process that occur in all individuals and represents the basis of the *"remodelling theory of ageing"* [14, 15] proposed in 1995 to conceptualize the results emerged from studies on human immunosenescence. This hypothesis suggests that immunosenescence is the result of the continuous adaptation of the body to the deteriorative changes occurring over time. The continuous stimulation by antigens/stressors leads to the decline/exhaustion of the adaptive IS and concomitantly to the increased pro-inflammatory status (innate immunity) that we indicated as inflammaging. Healthy old subjects and centenarians could be considered as those individuals with the best ability to adapt to damaging agents and immunological and non-immunological stressors [15] by setting up a variety of anti-inflammatory responses capable of neutralizing, at least in part the detrimental effects of inflammaging [16].

From an evolutionary and ecological perspective, each individual is characterized by an ability to cope with environmental insults, and this ability may move the effect of stress towards *hormesis*. "Hormesis refers to the beneficial effects of a treatment that at a higher intensity is harmful. In one form of hormesis, sublethal exposure to stressors induces a response that results in stress resistance. The principle of stress-response hormesis is increasingly finding application in studies of ageing, where hormetic increases in life span have been seen in several animal models" [17]. Thus, in order to understand the subtle shift from a harmful outcome to a hormetic beneficial effect of a given stimulus, a characterization of the environment where an individual is embedded is mandatory. Environment is expressed by local practices (i.e. "concrete matrixes" for a set of specific skills) and can be understood as a silent and incorporated knowledge within a specific context [18]. The term "dwelling" is used in cultural anthropology in general and by Ingold in particular to illustrate how human beings are always "embedded" in their experience of being with a specific body within a specific environment.

This ability to adapt to the surrounding environment through a pro-inflammatory status and then to deal with the associated stress is presumably a complex trait with a genetic and socio-cultural component. It is also presumed that genotypes that can generate a strong inflammatory response have been positively selected [19]. From

Fig. 18.3 The two-hits hypothesis of inflammaging. A strong inflammatory background and a genetic make-up not apt to counteract external or internal stressors are the two conditions that lead to unhealthy ageing

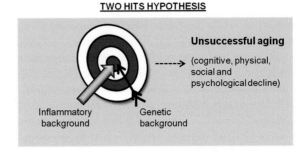

an evolutionary perceptive, the framework just described can be understood by invoking the antagonistic pleiotropy theory, first suggested by Williams in 1957 [20–23]. This theory suggests that many of the genes that have deleterious effects later in life may be favoured by natural selection because those same genes are associated with beneficial effects at a young age. Ageing is thus thought to have evolved as the result of optimizing fitness early in life. Accordingly, it can be predicted that inflammatory genes have been positively selected because they contributed to the survival of individuals, at least until they attain reproductive age [19]. A two-hits theory was proposed to better explain this phenomenon (Fig. 18.3): the first hit is constituted by the inflammatory background, and a second hit is necessary for the onset of age-related diseases. The second hit can be identified in the absence of robust gene variants and/or the presence of frail gene variants.

18.2.3   Immunological Biography and Socio-Cultural Life Histories: A Case Study

The study of what the body recognizes as "self" or as a stressor is very important. The attempt to understand the immune response in an evolutionary perspective leads to a reappraisal of the concept of the *immunological self*. Indeed, an abstract and well-established dichotomy between self and non-self is simplistic, and to reduce the concept of the immunological self to a fixed entity within a static scenario is misleading. It is better to adopt an integrated dynamic approach (the immunological biography), where the immunological self can undergo continuous changes in a personalized space- and time-dependent manner [24]. All the processes involved in the immune response are and have been shaped by, at least, two dimensions: (1) the spatial and ecological dimension constituted by the relations between human beings and their environment, according to the geography and history of the populations to which they belong (nutrition, climate, lifestyle, etc.) and (2) the temporal dimension since each individual is the result of a lifelong process of adaptation to the biological and non-biological (cultural/anthropological) determinants of his/her environment(s) (Fig. 18.4).

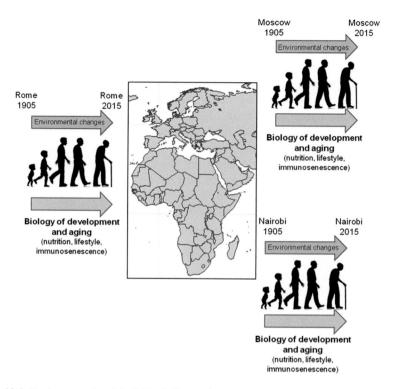

Fig. 18.4 Environmental and individual changes in space and time shape the immunological biography. Inflammation and the stress response are complex processes that must be analysed as a whole: an individual undergoes a lot of changes during development, adulthood and ageing. These changes could include a physiological dimension (nutrition, lifestyle) as well as a pathological one (infection, diseases). Moreover, individuals live in environments that have changed according to past migrations, adaptation and biodemography, and are changing during an individual's lifetime

Indeed, it is true that each individual constantly changes over time according to the different ages and phases of life (nutrition, school, work, housing, marriage(s), diseases, stress stimuli), concomitantly with changes in the surrounding environment. The environment undergoes profound remodelling during an individual's lifetime, but also in a wider timescale, such as the major changes that have occurred during the period spanning from the Palaeolithic to the Neolithic or the changes that have taken place over the last 200 years. Knowing these significant environmental modifications allows us to understand the conditions in which humans have evolved and adapted. Ultimately, this approach enables us to understand how the cultural and anthropological context may have shaped the human genetic and immunological make-up. If we consider the life of an individual, for example a living centenarian, it is likely that during his/her life, a very profound reshaping of the social and cultural scenario occurred.

In more general terms, the industrialization process, which started about 250 years ago in Europe has led to rapid and extreme changes in the social, biotic and abiotic environments. Epidemiological studies generally identify two main transitions during the history of modern populations. The first concerns the transition from Palaeolithic (200,000 years ago) to Neolithic times (10,000 years ago), from a subsistence mode based mainly on hunting and gathering to a diversification of subsistence modes, including pastoralism, horticultural and agriculture [25, 26]. In populations relying on large-scale agriculture, sedentarism and greater contact with animals have increased the risk of infection via the oro-faecal route. With the organization in small cities (about 3000 years ago), the incidence of communicable diseases (such as influenza and mumps) has significantly increased to become endemic. The second major epidemiological transition took place at the beginning of 1900. This transition indicates a shift from the rural small cities towards a more organized idea of industrialized cities, and hygiene behaviour itself changed dramatically in the West (clean and chlorinated water, soaps, detergents, antibiotics, etc). These changes have greatly reduced the rate of death from infectious diseases, placing the chronic and degenerative diseases at the first place among the leading causes of death in the developed world [27, 28].

The picture is further complicated by population genetics, that is, the result of biodemographic dynamics and adaptations characterizing the different history of human populations. The immunological biographies, as fluid, dynamic and malleable entities, need to be rethought and reconceptualized in relation with socio-cultural and anthropological aspects, and an ethnographic example is proposed below, regarding a population we have direct experience of and we have thoroughly studied from a biological, cultural and anthropological point of view [29, 30].

The population here examined is that of the Wichí in the Argentine Chaco. Traditionally hunters, fishers and gatherers, homogeneous in their social and political subdivision and organized in bands, nowadays they are 50,419 (2010 Census) in the Chaco region, and about 3500 in the Misión Nueva Pompeya village, where a series of ethnographical surveys have been carried out since 2004 [30, 31]. Up to the mid-nineteenth century, the Wichí of the Argentine Chaco engaged in their traditional activities which they carried out following the rhythm of the seasons and consequently practicing a regular semi-nomadic lifestyle. From the mid-nineteenth century, though, salaried jobs on the sugar cane and later cotton plantations, as well as the creation of missions (both Catholic and Anglican) on their territories, slowly but inevitably led them to become sedentary; this sedentary lifestyle was then reinforced by ever increasing schooling. The indigenous populations were to be educated through schools and religion to abandon their ancestral customs (whether religious, economic or political). Indigenous lands have been confiscated by the state, and today, the biggest multinational companies are cultivating soy on these lands (especially in the Salta area) and what is not cultivated has been cleared, falling prey to the animals farmed by the Criollo breeders who have settled in this area together with the Wichí. We are now facing a native population that, although trying to save the integrity of its knowledge (of local flora and fauna, and in general their mythical corpus) lives in an environment that is degraded [32], has changed lifestyle and consequently its nutritional habits [33, 34]. The Wichí no longer live on the proceeds of hunting and gathering, rather they feed themselves thanks to state-issued pensions, with which they are able to buy most of the food they consume (bread, pasta, flour, corn meal, meat); the balanced, seasonally driven, protein-rich diet has given way to a high-carbohydrate one. Immune responses in this sense are certainly moulded by the environment, by the climate, by local and global policies. Issues relating to identity are also dictated by incidental political and economical convenience. Thus, the ethnic blending between the Wichí and the Criollo often allows them to define as indigenous or Creole according to the needs of the moment [30]. The relationship with one's environment, with one's body (a "rotund" body certainly signals economic wellbeing), with traditional knowledge, the taste of new food and the memory of ancient foods, these are all elements that must be taken into account to understand the answers in cultural and biological terms. What we have very succinctly summarized here has happened very rapidly. The transition undergone by the Wichí has not been neither slow nor gradual, without any of the features of the above-mentioned historical transitions (Palaeolithic/Neolithic). The social and cultural history of the elderly Wichí who have lived the semi-nomadic lifestyle, who experienced the salaried work on the plantations and who today receive state pensions is that of grandparents who today see their children or their grandchildren malnourished and obese: misfits, both from an evolutionary and a cultural and environmental point of view.

This and other examples of social structure (another well-known example is the caste system in India) is of great importance to study inflammation and the stress response because the social structure could easily be linked to the so-called social stress that occurs as a consequence of social confrontation. The importance of social stress for immunology has been clearly demonstrated in studies on fish. Experiments showed that two fish randomly selected from a group sharing the same environment when put into a small aquarium start to combat for establishing hierarchy. After a fighting, one fish emerges as dominant (alpha) and the second as subordinate (beta). The most important observation is that many immunological alterations, due to the stress response, were present only in the beta fish and not in the alpha fish, despite the fact that both fishes experienced the fight. Moreover, only the alpha fish was able to successfully counteract the infection of pathogenic bacteria, while the beta fish succumbed. Thus, every type of intervention to slow down inflammaging and age-related diseases should be planned into a population-context perspective. Strategies that consider the population and the individual dimension can best maximize health benefits, preventing healthcare resources from being wasted.

18.3 Implications for Policy and Practice

Inflammation is a process that is beneficial, and also stress, in a broader sense, can be considered a type of positive/adaptive reaction. What is relevant is the threshold above which an individual is no longer able to cope with inflammatory stimuli and stressors and inflammaging and its deleterious effects start. This threshold is strictly linked to the concept of immunological biography, depending on social and cultural histories of communities and individuals, and marks the divide between healthy and unhealthy ageing (inflammaging and anti-inflammaging).

Inflammaging must also be contextualized from a population point of view, because the evolutionary history of each population may have shaped the ability to cope with specific stressors (e.g. infectious agents and food) typical of the context and of the specific environment. The study of inflammaging and immunosenescence and the understanding of rearrangements that occur lifelong are often neglected and should be carefully considered to appreciate not only the individual capability to adapt to damaging agents and stressors but also to quantify the whole immunological/stressful load, which is likely critical to explain most of the chronic pathologies which start usually decades before their clinical onset. Therefore, the population genetic structure and evolutionary processes such as selection, drift and migrations should be taken into account in the clinical setting. This is especially true in our century, where globalization and migration are increasing, and people of different ancestry need to cope with environments different from those they were exposed to in early/adult life, and despite a genetic background that might not be the best to cope with the new context. Thus, the concepts of healthy ageing and inflammaging should be analysed as a dynamic and integrated process with the final aim of promoting public health.

Thus, we can argue that a profound link exists between garbaging, inflammaging and the demographic revolution in which humans experienced a more than a doubling of human life expectancy in about 100 years and four generation [35]. During this period, people experienced a decreased rate of inflammaging, due to a decreased production/exposure to external danger signals, such as:

- less microorganisms and "hygienized" environment;
- sanitation;
- better nutrients and better gut microbiota;
- improvement in life conditions of: (1) pregnant women and (2) newborns and children in their first years of life

and to internal/self-generated danger signals such as:

- less cellular and molecular damages as a consequence of (1) less physically and emotionally stressful work, (2) more, comfortable housing, etc.

Thus, the identification of the major sources of inflammaging paves the way for a variety of possible interventions aimed at eliminating and neutralizing the "unnecessary" "excessive" inflammatory stimuli, thus slowing down the rate of inflammaging. This approach has the potential to postpone or even avoid the onset of age-related phenotypes and pathologies altogether instead of one by one. Possible strategies include the following: (1) elimination of senescent cells (2) healthy nutrition (Mediterranean diet) and physical exercise and (3) gut microbiome modulation (Fig. 18.5).

Fig. 18.5 Possible strategies to modulate inflammation and to slow down the rate of inflammaging. This approach has the potential to postpone or avoid the onset of age-related pathologies

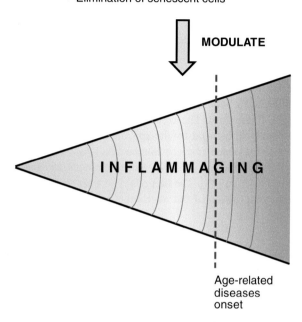

- Healthy nutrition (Mediterranean diet)
- Physical exercise
- Gut microbiome modulation
- Elimination of senescent cells

MODULATE

INFLAMMAGING

Age-related diseases onset

Nutritional interventions are expected to give promising results targeted towards some impairments of the aged IS. Such interventions could have a relevant impact on age-related diseases such as diabetes, brain disorders and heart diseases as dietary interventions hit the process of inflammation that is central for many systems. An example of nutritional intervention capable of neutralizing genetic risk factors regards the role of the Mediterranean diet in preventing a specific age-associated disease. A recent study observed that individuals having the TT genotype of the TCF7L2 gene (rs790316), a top ranking gene for type 2 diabetes risk in all the genome-wide association studies (GWAS) reported to date, have also a higher risk of cardio- and cerebrovascular diseases. However, it has been reported that TCF7L2 TT subjects that consumed and where strictly adherent to a Mediterranean diet neutralized completely the risk of stroke, in comparison with TCF7L2 TT who followed a controlled diet [36], suggesting that a healthy diet can counteract a strong genetic risk for an important age-related disease. Evidence that nuts by themselves or as part of a cholesterol-lowering diet may significantly reduce markers of inflammation has also been reported [37–39]. Fibre consumption was inversely associated with inflammatory markers (such as C-reactive protein and interleukin-6) [40] and molecular mechanisms to explain this phenomenon have been proposed [41]. Curcumin is also a widely studied phytochemical which has a variety of anti-inflammatory activities [42].

Dietary interventions have also the potential to modulate the gut microbiota composition, and targeting the gut microbiota biodiversity could constitute a new therapy for the modulation of inflammation during ageing. Indeed, we showed that a decrease of gut microbiota diversity and of "good" bacteria capable of producing short chain fatty acids (SCFA) and an increase of "bad" bacteria (pathobionts) is a characteristic of the gut microbiota of the elderly, highly correlated with the increase of pro-inflammatory cytokines in the blood [43, 44]. The IS in fact has evolved not only in response to external pathogens but also in response to the gut microbiota, which is also highly connected with many organs including the brain (gut–brain axis). The co-evolution of the gut microbiota with their host has been strongly influenced by the specific diet of the different populations during human evolution since modern humans had to face different environmental challenges (such as food availability, climate changes and pathogen loads) after they moved out of Africa [45]. The diet could be a target for new therapy that aims to modulate the composition of the microbiota to reduce inflammaging.

Glossary

Antagonistic pleiotropy theory of ageing	This hypothesis was first proposed by Williams in 1957 who suggested that a certain gene variant can be beneficial in early life (fitness), while it can become detrimental later in life
Biodemographic dynamics	The analysis of demography within an evolutionary biology context, with particular attention on events that impact on human population structure, such as colonization, migrations and expansion
Ecoimmunology	Discipline that integrates the analysis of the immune system function within animal biology and that considers the interaction between an organism and their ecological environment during evolution
Garbaging	The exogenous and endogenous inflammatory stimuli that, as a whole, increase progressively with age and trigger inflammaging
Heterochronic parabiosis	Parabiosis is an experimental model where two animals, here mice, are joined together surgically to create shared blood circulation. Heterochronic parabiosis indicates that the two animals joined together are of different ages (one old and one young)

Hormesis	Literally from Greek it means "to stimulate", and it indicates the ability of respond positively to low amounts of substances (or stressors) that would otherwise be highly toxic at higher concentrations
IL-6	Interleukin 6 (IL-6) is a cytokine with various biological functions, among which a role in the acute phase response. It is secreted by T cells and macrophages to stimulate the immune response and acts as both a pro-inflammatory and an anti-inflammatory cytokine. It is also a myokine discharged into the bloodstream after muscle contraction and acts to increase the breakdown of fats and to improve insulin resistance
Immunological space	is a metaphor to conceptualize the IS as a whole from a spatial (volume) point of view. During immunosenescence, a progressive accumulation of clones of memory cells tends to fill the "immunological space", reducing the number of other immune cells, such as naïve T cells, and the possibility to respond to new antigens
Inflammaging	Human ageing is characterized by a chronic, low-grade inflammation (high levels of inflammatory cytokines and other inflammatory markers such as C-reactive protein), and this phenomenon has been termed "inflammaging". Inflammaging is a highly significant risk factor for both morbidity and mortality in the elderly, as the vast majority of age-related diseases share an inflammatory pathogenesis

References

1. Cartledge B (1998) Mind, brain, and the environment. Oxford University Press, Oxford/New York
2. Franceschi C, Bonafè M, Valensin S et al (2000) Inflamm-aging. An evolutionary perspective on immunosenescence. Ann NY Acad Sci 908:244–254
3. Morrisette-Thomas V, Cohen AA, Fülöp T et al (2014) Inflamm-aging does not simply reflect increases in pro-inflammatory markers. Mech Ageing Dev 139:49–57. doi:10.1016/j.mad.2014.06.005
4. Franceschi C, Campisi J (2014) Chronic inflammation (inflammaging) and its potential contribution to age-associated diseases. J Gerontol A Biol Sci Med Sci 69:S4–S9. doi:10.1093/gerona/glu057

5. Okin D, Medzhitov R (2012) Evolution of inflammatory diseases. Curr Biol CB 22: R733–R740. doi:10.1016/j.cub.2012.07.029

6. Ottaviani E, Franceschi C (1996) The neuroimmunology of stress from invertebrates to man. Prog Neurobiol 48:421–440

7. Blalock JE (1989) A molecular basis for bidirectional communication between the immune and neuroendocrine systems. Physiol Rev 69:1–32

8. Ottaviani E, Malagoli D, Franceschi C (2007) Common evolutionary origin of the immune and neuroendocrine systems: from morphological and functional evidence to in silico approaches. Trends Immunol 28:497–502. doi:10.1016/j.it.2007.08.007

9. Holzer P, Farzi A (2014) Neuropeptides and the microbiota-gut-brain axis. Adv Exp Med Biol 817:195–219. doi:10.1007/978-1-4939-0897-4_9

10. Kitano H (2007) Towards a theory of biological robustness. Mol Syst Biol 3:137. doi:10.1038/msb4100179

11. Csete M, Doyle J (2004) Bow ties, metabolism and disease. Trends Biotechnol 22:446–450. doi:10.1016/j.tibtech.2004.07.007

12. Ottaviani E, Malagoli D, Capri M, Franceschi C (2008) Ecoimmunology: is there any room for the neuroendocrine system? BioEssays 30:868–874. doi:10.1002/bies.20801

13. Franceschi C, Bonafè M, Valensin S (2000) Human immunosenescence: the prevailing of innate immunity, the failing of clonotypic immunity, and the filling of immunological space. Vaccine 18:1717–1720

14. Franceschi C, Monti D, Barbieri D et al (1995) Immunosenescence in humans: deterioration or remodelling? Int Rev Immunol 12:57–74. doi:10.3109/08830189509056702

15. Franceschi C, Valensin S, Bonafè M et al (2000) The network and the remodeling theories of aging: historical background and new perspectives. Exp Gerontol 35:879–896

16. Franceschi C, Capri M, Monti D et al (2007) Inflammaging and anti-inflammaging: a systemic perspective on aging and longevity emerged from studies in humans. Mech Ageing Dev 128:92–105. doi:10.1016/j.mad.2006.11.016

17. Gems D, Partridge L (2008) Stress-response hormesis and aging: "that which does not kill us makes us stronger". Cell Metab 7:200–203. doi:10.1016/j.cmet.2008.01.001

18. Ingold T (2000) The perception of the environment: essays on livelihood, dwelling and skill. Routledge, London/New York

19. Raj T, Kuchroo M, Replogle JM et al (2013) Common risk alleles for inflammatory diseases are targets of recent positive selection. Am J Hum Genet 92:517–529. doi:10.1016/j.ajhg.2013.03.001

20. Rose M, Charlesworth B (1980) A test of evolutionary theories of senescence. Nature 287:141–142

21. Hughes KA, Alipaz JA, Drnevich JM, Reynolds RM (2002) A test of evolutionary theories of aging. Proc Natl Acad Sci USA 99:14286–14291. doi:10.1073/pnas.222326199

22. Parsons PA (2007) Antagonistic pleiotropy and the stress theory of aging. Biogerontology 8:613–617. doi:10.1007/s10522-007-9101-y

23. Capri M, Salvioli S, Monti D et al (2008) Human longevity within an evolutionary perspective: the peculiar paradigm of a post-reproductive genetics. Exp Gerontol 43:53–60. doi:10.1016/j.exger.2007.06.004

24. Grignolio A, Mishto M, Faria AMC et al (2014) Towards a liquid self: how time, geography, and life experiences reshape the biological identity. Front Immunol 5:153. doi:10.3389/fimmu.2014.00153

25. Ulijaszek SJ (1995) Human energetics in biological anthropology. Cambridge University Press, Cambridge/New York

26. Ulijaszek SJ, Mann N, Elton S (2012) Evolving human nutrition: implications for public health. Cambridge University Press, New York

27. Trevathan W, Smith EO, McKenna JJ (2008) Evolutionary medicine and health: new perspectives. Oxford University Press, New York

28. Banwell C, Ulijaszek S, Dixon J (2013) When culture impacts health—global lessons for effective health research. Elsevier, Amsterdam

29. Moretti E, Castro I, Franceschi C, Basso B (2010) Chagas disease: serological and electrocardiographic studies in Wichì and Creole communities of Misión Nueva Pompeya, Chaco, Argentina. Mem Inst Oswaldo Cruz 105:621–627
30. Sevini F, Yao DY, Lomartire L et al (2013) Analysis of population substructure in two sympatric populations of Gran Chaco. PLoS ONE, Argentina. doi:10.1371/journal.pone.0064054
31. Franceschi ZA, Dasso MC (2010) Etno-grafías: la escritura como testimonio entre los wichí. Corregidor, Buenos Aires
32. Morello JH, Rodriguez AF (2009) El Chaco sin bosques: La Pampa o el desierto del futuro, 1a ed. Orientación Gráfica Editora, Buenos Aires
33. Torres GF, Santoni ME, Romero LN (2007) Los Wichi del Chaco Salteño: ayer y hoy: alimentación y nutrición. Crisol, Salta
34. Franceschi ZA, Peveri V (2014) Raccontare di gusto - Arti del cibo e della memoria in America latina e Africa, ETS
35. Burger O, Baudisch A, Vaupel JW (2012) Human mortality improvement in evolutionary context. Proc Natl Acad Sci USA 109:18210–18214. doi:10.1073/pnas.1215627109
36. Corella D, Carrasco P, Sorlí JV et al (2013) Mediterranean diet reduces the adverse effect of the TCF7L2-rs7903146 polymorphism on cardiovascular risk factors and stroke incidence: a randomized controlled trial in a high-cardiovascular-risk population. Diab Care 36:3803–3811. doi:10.2337/dc13-0955
37. Jenkins DA, Kendall CC, Marchie A et al (2003) EFfects of a dietary portfolio of cholesterol-lowering foods vs lovastatin on serum lipids and c-reactive protein. JAMA 290:502–510. doi:10.1001/jama.290.4.502
38. Ros E (2009) Nuts and novel biomarkers of cardiovascular disease. Am J Clin Nutr 89:1649S–1656S. doi:10.3945/ajcn.2009.26736R
39. Kendall CWC, Josse AR, Esfahani A, Jenkins DJA (2010) Nuts, metabolic syndrome and diabetes. Br J Nutr 104:465–473. doi:10.1017/S0007114510001546
40. Wannamethee SG, Whincup PH, Thomas MC, Sattar N (2009) Associations between dietary fiber and inflammation, hepatic function, and risk of type 2 diabetes in older men: potential mechanisms for the benefits of fiber on diabetes risk. Diab Care 32:1823–1825. doi:10.2337/dc09-0477
41. Ghanim H, Chaudhuri A, Dandona P et al (2010) Associations between dietary fiber and inflammation, hepatic function, and risk of type 2 Diabetes in older men: potential mechanisms for the benefits of fiber on diabetes risk response to Wannamethee et al. Diab Care 33:e43–e43. doi:10.2337/dc09-2127
42. Salvioli S, Sikora E, Cooper EL, Franceschi C (2007) Curcumin in cell death processes: a challenge for CAM of age-related pathologies. Evid-Based Complement Altern Med ECAM 4:181–190. doi:10.1093/ecam/nem043
43. Biagi E, Nylund L, Candela M et al (2010) Through ageing, and beyond: gut microbiota and inflammatory status in seniors and centenarians. PLoS ONE 5:e10667. doi:10.1371/journal.pone.0010667
44. Collino S, Montoliu I, Martin F-PJ et al (2013) Metabolic signatures of extreme longevity in northern Italian centenarians reveal a complex remodeling of lipids, amino acids, and gut microbiota metabolism. PLoS ONE 8:e56564. doi:10.1371/journal.pone.0056564
45. Quercia S, Candela M, Giuliani C et al (2014) From lifetime to evolution: timescales of human gut microbiota adaptation. Front Microbiol 5:587. doi:10.3389/fmicb.2014.00587

Chapter 19
Dementias of the Alzheimer Type: Views Through the Lens of Evolutionary Biology Suggest Amyloid-Driven Brain Aging Is Balanced Against Host Defense

Prof. Caleb E. Finch, Ph.D. and George M. Martin, M.D.

Lay Summary The major risk factor for Alzheimer's disease (AD), like so many other chronic diseases, is advancing age. Here, we ask whether certain key pathological features of the disorder also appear in other species of primates and how they are related to their marked variations in life spans. Concentrating on two such alterations, one dealing with a "sticky" type of protein called amyloid beta and a protein of importance in the transport of cargo along nerves called tau, we find that there is indeed such evidence and that the times of their appearance support the notion that they are responding to some basic processes of aging, or even some yet unknown process. These changes, however, do not reach the advanced pathology seen in our species and which define AD. Why is that? First, environmental factors are important influences upon disorders of aging: Captive, caged monkeys and apes show significant amounts of amyloid beta in their brains. Second, genetic differences are of major significance as to why only some old people develop AD. For the common sporadic, late-onset forms of the disease, by far the most important risk factor involves genetic variations in apolipoprotein E (apoE), which delivers lipids to our neurons. The gene comes in three "flavors" (alleles) known as *E2*, *E3*, and *E4*. While neither necessary nor sufficient to cause the disease, people with the *E4* allele are far more likely to develop AD as they age. Surprisingly, our primate cousins and our ancient human precursors appear to have an allele that is closer to this "bad" *E4* allele. Why has such allele not been eliminated by natural selection? We suggest that its lower efficiency in delivering lipids to certain infectious agents may be the reason

C.E. Finch (✉)
Department of Neurobiology, Davis School of Gerontology, The Dornsife College,
The University of Southern California, Los Angeles, CA, USA
e-mail: cefinch@usc.edu

G.M. Martin
Department of Pathology, University of Washington, Seattle, WA, USA
e-mail: gmmartin@uw.edu

© Springer International Publishing Switzerland 2016
A. Alvergne et al. (eds.), *Evolutionary Thinking in Medicine*,
Advances in the Evolutionary Analysis of Human Behaviour,
DOI 10.1007/978-3-319-29716-3_19

APOE4 evolved and persisted in certain populations of humans subjected to great hazards of infectious diseases, including malaria. A great deal more research is needed to clarify the relationships between infectious agents and AD. In any case, it is important to consider an infectious etiology for AD.

19.1 Introduction: Species Variations in Neurodegenerative Disorders of Aging

Aging is a lifelong process that plays out differently in each tissue. The ovary and the brain both have irreplaceable cell populations: The ovarian oocytes are fully formed before birth and are lost irreversibly at exponential rates with exhaustion at menopause by midlife [1]. In contrast, brain neurons are rarely replaced in adults and are largely stable into midlife, when they show increasing risk for damage during pathological processes of brain aging [2]. Although humans may be unique among mammals in developing the extremes of neurodegeneration found in later clinical stages of AD [1], many primates also develop varying degrees of AD-like changes. We focus here on the senile plaques containing amyloid β-peptide (Aβ) and neurofibrillary tangles (NFT) with hyperphosphorylated tau, which are diagnostic of AD when they reach certain levels in particular brain structures.

Some levels of Aβ deposits and NFTs also arise at later ages in primate clades (Fig. 19.1). This similarity in outcomes of aging implies that mild AD-like processes are evolutionarily ancient and may have arisen in early mammals before 60 million years ago. The adult ages of mild AD-like changes range several-fold and approximate the life span of each species. Primates show extreme variations in brain aging over a fourfold range of life spans from pro-simians to great apes. Chimpanzees and other great apes at advanced ages have modest accumulations of brain Aβ aggregates that would not qualify as the senile plaques of AD [3, 4]. Moreover, among the great apes, NFT are rare [5]. While baboons show more extensive neurofibrillary changes [6], aging macaque monkeys have modest synapse loss [7], and little evidence of neuron loss [8]. However, definitive proof awaits higher resolution analysis of neuron and synapse density by optical fractionator techniques. Of great note, human neuronal loss is modest in individuals who have no significant cognitive impairment at advanced ages, but among AD individuals, neuronal loss becomes severe early, during preclinical stages. The present evidence supports Rapoport's hypothesis [9] that AD is unique to humans. We can exclude one possibility that the Aβ peptide has undergone evolutionary change in primates. Remarkably, the Aβ sequence is identical across most vertebrates from zebra fish to human. However, the β-amyloid precursor protein (APP) gene has species differences relevant to Aβ production [10, 11].

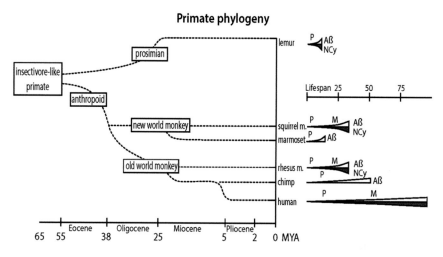

Fig. 19.1 Brain aging of primates organized by phylogeny. Aβ, deposits of the fibrillar Aβ aggregates in brain parenchyma; M, menopause; P, puberty; NCy, neurocytoskeletal disorgani-zation, identified as tau hyperphosphorylation only in species marked as *t*+. Mouse lemur, *Microcebus marinus*, *t*+ [95–98]; squirrel monkey, *Saimiri sciureus* [99–102]. Common marmoset, *Callithrix jacchus* [103–107]. Rhesus monkey, *Maccaca mulata*, *t*+ [108]. Chimpanzee, *Pan troglodytes* [3, 5, 109]

Conclusions about possible AD-like changes in other species must be tempered by two caveats. First, all studies are from captive animals under conditions of housing and diet that differ widely from their natural habitat [12: 1]. Most captive apes and monkeys, for example, are obese (see also Chap. 21), and small cage spaces increased brain amyloid several-fold in monkeys [13]. While some field observations suggest that older chimpanzees retain competence in foraging and in complex social behaviors [14, 15], their numbers are too few for generalizations. None from natural habitats have been killed for examination of their brains (un-thinkable in this era of endangered species). Second, the advanced ages reached by captives would represent an extreme minority in nature, where background mor-tality is high from predators, injuries, and infections.

That said, the evidence shows that brain aging in primates and other mammals varies widely in some proportion with differing species life spans and emerges later in life span when most reproduction has been achieved by young adults, which are always the majority age group. Thus, brain aging, as observed in captivity, arises at later ages that are less subject to natural selection. The wide range of species differences in primate brain aging, as well as life span differences, must represent genetic differences between species that were at some level shaped by natural selection. Many other phylogenetic comparisons such as those shown in Fig. 19.1 underscore the plasticity of aging in ovaries and other tissues [16, 17].

19.2 Research Findings: Evolutionary Hypotheses for the Origins of AD-like Dementias

19.2.1 Antagonistic Pleiotropy

We consider an evolutionary hypothesis to explain how *APOE4*, as the major genetic risk factor for AD, could persist globally despite its adverse associations. The possibility that this "bad gene" has benefits at an early stage of life but also confers disease risk later in life is described in the antagonistic pleiotropy hypothesis of aging.

The antagonistic pleiotropy hypothesis for the evolution of life spans was proposed by George Williams five decades ago to involve hypothetical gene variants showing advantage early in life but harm at older ages [18]. Extending the classical concept of genetic pleiotropy as a multisystem effect of a gene variant, Williams hypothesized that some alleles selected for benefits to early survival may have later adverse consequences that were not selected against because most of the reproduction is accomplished by young adults [18, 19]. Although not discussed by Williams, we emphasize that in natural populations the major causes of mortality are infections, as observed in wild chimpanzees and in human foragers living with limited access to modern medicine [20]. With the great reduction of infections in the twentieth century that allowed the doubling of life spans, humans are experiencing greatly reduced selective pressure for resistance to infections. Moreover, with greatly improved infant survival allowing reduction of family size, fecundability is also under diminishing natural selection. *APOE* alleles can be considered as having potential roles in both host resistance and reproduction.

The three *APOE* alleles may be the largest "public" allele system involved in both AD and longevity: Depending on the population, a single copy of *APOE4* increases the risk of AD by about twofold [21, 22] and shortens life expectancy by about 5 years [23]. Women *APOE4* carriers have twofold–fourfold greater vulnerability to AD than men [24, 25]. The minor allele *APOE2*, however, is AD-protective [26, 27] and is also associated with exceptional longevity [23].

More than 10 genes are now associated with late-onset AD [28]. Among them, the *APOE* allele system is the best understood in terms of its evolutionary history and trade-offs, which show evidence of antagonistic pleiotropy. Another example may be the Aβ peptide, which, as noted above, occurs in most vertebrates from fish to great apes to humans. An adaptive role in host defense is suggested by its antimicrobial activity against common human pathogenic infections [29].

19.2.2 Selective Advantage in Host Defense

The evolutionary history of the *APOE* alleles points to *APOE4* as the human ancestral allele. Here, we argue that the major genetic risk factor for the late-onset sporadic forms of AD, the *APOE4* allele, was selected across evolutionary history because of its selective advantage in host defense. The APOE protein is secreted by the liver. By its binding to the LDL receptor, APOE plays a major role in the blood transport and clearance of cholesterol and triglycerides. While APOE3 binds preferentially to the phospholipid-rich HDL, APOE4 binds the triglyceride-rich VLDL [30]. It is also important in the brain, where it is secreted by astrocytes to supply lipids to neurons. While a role for resistance to *Trypanosoma brucei* was initially suggested as a cogent example [31], a more significant selective advantage may involve resistance of *E4* carriers to the malaria plasmodia, which require host lipids for replication. Several lines of evidence support this hypothesis.

First, we consider the evolutionary history of *APOE*. Chimpanzees, our closest great ape relatives, share with human APOE4 the two arginine (R) residues 112 and 158 (Table 19.1). However, unlike humans, chimpanzees and other primates have not shown APOE isoforms. By microsatellite dating, the human *APOE3* allele with 112-cysteine emerged about 225,000 years ago (range 180,000–580,000 years ago) [32]. This range spans the emergence of anatomically modern *H. sapiens* and precedes our immigration from Africa [20]. Fossil DNA sequences of two Denisovans further show that *APOE4* existed even earlier in our genus, consistent with the earlier limit above [33]. *APOE2* appears to have arisen after *APOE3* [32]. In modern populations, *APOE3* is predominant (60–90 %), while APO4 prevalence ranks second (5–40 %) and *APOE2* is generally third (1–10 %) [34].

Second, while chimpanzees and other great apes share R112 and R158 with humans, they differ at residue 61, which is R in human (R61R) but threonine (R61T) in great apes (Table 19.1). The Mahley research group showed the critical role of R61 in lipid binding with transgenic mice, which also resemble great apes at these residues: By targeted APOE replacement of R61T, the mouse APOE protein acquired human APOE4-like lipid binding [35]. To test these ideas further, the Finch Lab made a mouse with targeted replacement of the chimpanzee *APOE* gene. Preliminary data show that chimpanzee *APOE* resembles human *APOE4* more than *APOE3* in supporting neurite outgrowth [36]. These experiments confirmed the Mahley group findings that human *APOE3* is more neurotrophic than *APOE4* by more efficient lipid delivery [37, 38]. Furthermore, we must consider other differences between human and chimpanzee *APOE*. Of the 8 residues that show evidence of positive selection in human *APOE*, half are in the lipid-binding C-terminus [39]. While *APOE4* is likely to be the human ancestral isoform, the unknown effect of

Table 19.1 Apolipoprotein E residues for contemporary human, Denisovan, and Chimpanzee alleles

	Amino acid 61	112	158
Human E2	R	C	C
Human E3	R	C	R
Human E4	R	R	R
Denisovan	R	R	R
Chimpanzee	T	R	R

C Cysteine; *R* Arginine; *T* Threonine; [33]

these other amino acid differences in lipid binding could also have influenced lipophilic steps in host defense.

Thirdly, the *APOE* alleles show regional geographic gradients. Within Europe, for example, *APOE4* shows a fourfold longitudinal cline, from 5 to 10 % below 30° latitude north [40] to 20–30 % at higher latitudes [34]. Longitudinal gradients of APOE4 are found in China [41] and India [42]. Latitudinal gradients are also seen, e.g., 41 % *APOE4* in Aka Pygmies of western Congo versus 33 % for Zairians [43].

Fourth, the *APOE4* allele is associated with resistance to certain infectious agents. A case is building for a role of *APOE* alleles in malarial resistance, because Plasmodium parasites are dependent (auxotrophic) on cholesterol as a nutrient during the blood and liver stages of replication [44]. A population from Gabon highly infected with *Plasmodium falciparum* showed evidence for an epistatic gene interaction of *APOE4* with sickle-cell hemoglobin (HbAS, malarial resistant vs. HbAA): The HbAS carriers who were also *APOE4* had a 40 % lower infection index than HbAA carriers, while *APOE3* carriers did not differ [45, 46]. In vitro, plasma from *APOE4* carriers, but not from homozygotes for the E3 allele, inhibited the growth of the parasite *P. falciparum* [47]. There is some evidence that *APOE4* carriers also have higher blood levels of IL-13 [48], a cytokine with antiparasitic activity. Moreover, *APOE4* carriers have greater induction of TNFα in response to bacterial endotoxins, suggesting greater innate immune activation [49]. Another example may be the resistance of *APOE4* carriers to progressive fibrosis in chronic hepatitis C [50, 51]. Contrarily, *APOE4* may increase susceptibility to HIV [52, 53] and herpes simplex virus [53], as discussed below.

Besides their potential roles in host defense, *APOE* alleles may also influence brain development. An MRI observational study of normal children showed that the entorhinal cortex was consistently thinner in *APOE4* versus *APOE3* carriers [54]. A thinner temporal cortex was also found in neonatal *APOE4* carriers [55]. Because of neurodegeneration in this brain region during the early stages of AD, the thinner cortex implies a smaller neuronal reserve in *APOE4* carriers. In mice transgenic for human *APOE* alleles, the E4 mice have synaptic deficits relative to E3 mice [56]. Frustratingly, we lack cell-level details for *APOE* alleles at stages of human brain development. A trade-off in *APOE4* carriers between resistance to infections and brain development is suggested for Brazilian slum children who commonly suffer diarrhea: The *APOE4* carriers had better cognitive responses to micronutrient supplementation than the *APOE3* carriers [57, 58]. Thus, under conditions of infection that differ from privileged healthier populations, the APOE4 protein could be advantageous for brain development. This association is also shown in the resistance of malnourished mice to cryptosporidial infections, a common cause of diarrhea, in which E4 mice grew best [59].

Lastly, we note a very recent report that blood progesterone levels are higher in *APOE4* carriers during the luteal phase of the ovulatory cycle [60]. The women in this study from Poland were healthy and not carrying a burden of infections. These authors proposed that the potential increase of fecundability in *APOE4* carriers because of elevated progesterone during the luteal phase represents the beneficial component in its antagonistic pleiotropy.

19.2.3 Trade-Offs from Variegated Gene Expressions as a Form of Bet-Hedging

Evolutionary biologists use the term "bet-hedging" in several different ways [61]. A simple way to think about it is the old adage "don't put all your eggs in one basket"! Our use of the term here fits best with the notion of a "probabilistic diversification of the phenotypes expressed by a single genotype" [61]. We would include stochastic variations in gene expression (transcriptional, translational or post-translational in origin) among populations of otherwise identical cell types. For example, in the *C. elegans* nematode, in which all individuals are genetically identical, individual worm-to-worm variations in expression of the heat-shock gene hsp-16.2 predicted survival over a fourfold range [62].

Martin [63] used the term "epigenetic gambling" to describe such phenomena, implying a strong conceptual bias in favor of a transcriptional origin for such variegated gene expression. Furthermore, while such bet-hedging may have evolved as an adaptive trait as it ensures phenotypes that could survive in the face of unpredictable environmental challenges, the phenomenon of "epigenetic drift" [64, 65] when extended to the post-reproductive period of the life course would partially escape the force of natural selection, thus contributing to diverse geriatric pathologies, including AD and other later-onset dementias [66] (Fig. 19.2). Such a scenario could represent the action of antagonistic pleiotropic genes. A similar role of epigenetic drift in the pathogenesis of AD was independently proposed, together with supporting data [67]. Gerontologists have become aware that stochastic events during development are important determinants of individual outcomes of aging [20, 68], but this topic remains underdeveloped in theory and experiment.

19.2.4 Mutations and Polymorphisms Which Are not Phenotypically Expressed Until the Post-reproductive Period

Peter Medawar's influential book An Unsolved Problem in Biology [69] clearly delineated this major mechanism of biological aging. However, it was J.B.S. Haldane who first developed the fundamental concept based upon Huntington's disease [70]. Haldane was puzzled by the unusual high frequency of this severe autosomal dominant disorder (about 1/18,000 in the British population). He concluded that the disease represented a late-acting mutation that had escaped the force of natural selection. Medawar elaborated upon this idea. Specifically, he hypothesized that for deleterious mutations and polymorphisms to be maintained, suppressor alleles at various loci are likely to have evolved such that deleterious mutations could partially escape the force of natural selection if the associated adverse phenotypes emerged later in life. According to Medawar, many such mutations associated with

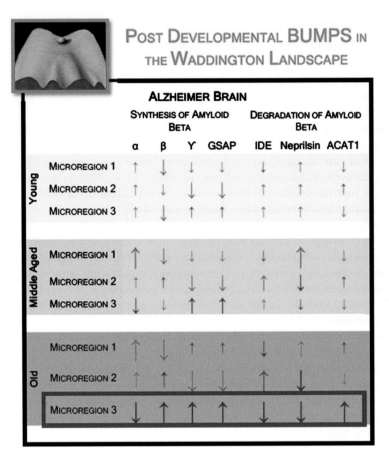

Fig. 19.2 This figure illustrates what might be described as the genesis of "A Perfect Storm" of age-related drifts in gene expression, sufficient to lead to the quasi-stochastic multifocal distributions of neuritic plaques in the hippocampus of patients with LOAD. We imagine a series of "bumps" in the Waddington landscape, which eventually become sufficient to produce localized fibrillary deposits of amyloid beta that create neuritic plaques. The left set of columns illustrate variable degrees of drift (either increased or decreased gene expression) for four loci involved in the *synthesis* of Aβ peptides, while the right set of columns illustrate variable degrees of drift for four loci involved in the *degradation* of Aβ. GSAP is a gamma secretase activating protein [110]. ACAT1 (Acyl-CoA:cholesterol acyltransferase) exemplifies how the down-regulation of an enzyme can enhance Aβ clearance via autophagy-mediated lysosomal proteolysis in microglia [111]. The diagram is an oversimplification, as other loci, including APOE, are involved. An earlier draft of this figure was published in [112]

late adverse phenotypes could collectively contribute to biological aging. This theory predicts that the patterns of mutations and phenotypes are likely to be idiosyncratic among populations, consistent with the views of geriatricians and pathologists that no two human beings age in precisely the same ways.

19.2.5   Evolutionary Biology and the Common Sporadic, Late-Onset Forms of Alzheimer's Disease (LOAD)

To varying degrees, all phenotypes are the products of diverse interactions between alleles at multiple loci (Gene × Gene) and between genes and changing environments (Gene × Environment). For complex phenotypes like AD, with its diverse spectrum of neuropathological lesions and highly variable times of onset, rates of progression, and patterns of clinical presentations, one would anticipate large numbers of genetic contributions and environmental interactions. This is particularly valid for LOAD, given the Medawar proposition of selection for suppressor loci that push the times of expression later in the life course. Indeed, large-scale genome-wide association studies (GWAS) and other emerging genomic approaches, although far from having exhausted its potential for the discovery of both rare and common variants, have already identified more than 20 loci, each with variations associated with LOAD risk. These loci can be functionally grouped as lipid metabolism, inflammation/immune response, endocytosis/intracellular trafficking, tau metabolism/microtubular structure function, synaptic plasticity, and metabolism of the β-APP [71–75].

A speculative but heuristic attempt to integrate these various discoveries in terms of evolutionary biology and antagonistic pleiotropic gene action invokes a range of host responses to challenges by infectious agents. We have outlined above our hypothesis according to which the *APOE4* allele has a protective role against pathogens, including agents that rely upon host lipids for their replication. In addition, a protective role conferred by both the immune system and the inflammatory response would seem obvious, and similarly, it could be argued that genes associated with endocytosis and intracellular trafficking have a protective role against certain intracellular pathogens. Given recent evidence for "infectious proteins" in AD [76, 77], microtubular-associated proteins such as tau could be incorporated in such a theory. The fact that different molecular forms or "strains" of tau aggregates exhibit different efficiencies of spread within the brain and that they can be transmitted via retrograde axonal transport from peripheral nerves has attracted a review by a specialty journal in virology [78].

Lastly, we note intriguing findings that the Aβ peptide, implied as a neurotoxic factor in AD, also has antimicrobial activities in vitro against Gram-negative and Gram-positive bacteria and yeast [29] and for replication of the influenza A virus [79]. As noted earlier, the Aβ42 peptide sequence is remarkably conserved in vertebrates. Because *APOE* isoforms have differential binding to APP [80], novel human *APOE* isoforms could have been selected for host defense against infections, as may have been the case for *APOE4*. We anticipate progress in the biology of *APOE* with the powerful new genomic techniques to identify novel infectious agents with the potential to cause LOAD. Indeed, a role for the herpes simplex virus in AD was suggested decades ago [81, 82] and an association with the E4 allele was shown twenty years ago [83]. Although some viruses appear to require host fatty acid synthesis, there is as yet no evidence that Herpes simplex is a lipophilic agent,

An enhanced susceptibility to AD in *APOE4* carriers might be related to a deficient delivery of lipids following neuronal damage to the host [37, 38, 84]. Another infectious candidate detected in AD brains is *Chlamydia pneumonia*, a spirochetal bacteria, which a meta-analysis associated with fourfold higher risk of AD [85]. The role of infections remains controversial in AD, as it is for atherosclerosis, where infectious agents have also been associated with arterial lesions [86].

19.3 Implications for Policy and Practice: Brain Aging Is Highly Plastic

Our discussion of evolution in brain aging and AD-like processes shows a huge range of plasticity, i.e., a dissociation from any strict clock of aging. While there is no systematic evidence to link the rates of AD-like processes to species-specific life spans, it is clear for humans that aging remains by far the major risk factor for AD. Continued research on fundamental processes of aging warrants high priority.

We anticipate that even for early-onset familial AD, there is untapped plasticity. As noted earlier, monkeys restricted to small cages exhibit higher brain amyloid burdens and less synaptic protein [13]. The restricted social interactions and limited mental challenge may be a model for the strong inverse association of AD risk with education levels [87–89]. Corresponding evidence for the impact of cognitive activity on brain structure and chemistry is emerging, based on small clinical studies from brain imaging with PiB binding. In healthy elderly, the brain Aβ load varied inversely with life-time cognitive activity by >50 %, specifically in cortical regions with multimodal nodes (lateral–medial prefrontal and parietal cortex; lateral temporal cortex) [90]. Other aspects of brain plasticity are given by Stern's analysis of cognitive reserve [91, 92].

Thus, we propose that the human environment will prove to influence most if not all risk factors for AD. Those risk factors are likely to include exposures to infectious diseases. While we have argued that *APOE4* is likely to have evolved because of the protection, it confers against certain pathogens, the evidence of an association of *APOE4* with a herpes zoster virus provides a rationale to search for pathogenetic roles for this and other viral agents using the emerging power of next-generation genome sequencing. We anticipate that new drugs such as the apoE peptide mimetics [93] will be used in concert with environmental interventions to delay the ages of onset and slow down the rates of progress to LOAD.

Acknowledgements CEF thanks Troy Locker-Palmer for excellent graphics in Fig. 19.1. CEF is grateful for support from the National Institutes of Health (P01-AG026572) to Roberta D Brinton, PI, Project 2 (CEF), and encouragement by the Center for Academic Research and Training in Anthropogeny (CARTA).

Glossary

Amyloids and Aβ peptides	Amyloid refers to a large group of proteins or peptides that are rich in β-helical sheets of amino acids and are prone to form insoluble aggregates within tissues. The aggregates that are among the classical diagnostic features of Alzheimer's disease are known as Amyloid beta, or Aβ. There are several varieties, all of which are derived from a much larger precursor protein (APP, or the β-APP). The two Aβ peptides that have been most investigated are Aβ 1–40 (with forty amino acids) and Aβ 1–42 (forty-two amino acids). The name amyloid is derived from a chemical test for starch (an iodine test) used by the famous pathologist Rudolph Virchow in the mid-nineteenth century to determine the nature of waxy material he found in livers of some autopsied subjects. Virchow was prescient in using the suffix "–oid" (like starch), as this and all other types of amyloids turned out to be proteins. Amyloids, however, co-precipitate with hyaluronan sulfate proteoglycans, hence the positive iodine test!
Aging	Aging can be defined as a collection of gradual and insidious declines in multiple cellular physiological functions, which reduce responses to stress, increase the risk of chronic disease, and accelerate the probability of death during later adult ages
Amino acid residues	Proteins consist of strings of amino acids. When amino acids are linked, molecules of water are lost and the resulting amino acids are referred to as amino acid residues
APOE	This is the accepted abbreviation for a gene that codes for the apolipoprotein E protein. By convention, the abbreviations for genes are both capitalized and italicized; the related protein is also capitalized, but not italicized. Human populations have three different forms (alleles) of this gene, each differing slightly by amino acid sequence. The most common allele is E4, where E stands for epsilon. The least common allele, E2, is associated with lesser risk for Alzheimer's disease. E4 is the major risk factor for late-onset Alzheimer's disease
APP	Abbreviation for β-amyloid precursor protein (see Aβ peptides, above)

Astrocytes	The name of this major brain cell represents its typical star-like shapes. They provide neurons with lipids carried by apolipoprotein E
Entorhinal cortex	This is the part of the medial temporal cortex of the brain that connects the hippocampus to other areas of the cerebral cortex and is therefore an essential hub in the networks involved in learning and memory
Epigenetic drift	Epigenetics literally means "on top of" the genes. It involves chemical changes to the basic nucleotide base pairs of the DNA and of its associated proteins known as histones. In so doing, these chemical alterations change the expression of genes during cell differentiation and in certain pathological conditions. These chemical marks and their associate alterations in gene expression gradually change during aging, a process known as epigenetic drift
Epistatic gene interaction	Genes do not work in a vacuum. They are dependent upon interactions with other genes, variations at which can modulate the phenotype of the organism. For a fuller account of the origins and evolution of this concept, its terminology and its classifications, consult [94]
Lipid-binding terminus	Proteins and fatty substances (lipids) can be bound together in the same molecule. Characteristic sequences of amino acids have evolved to provide specificity for such interactions. These sequences can occur at different regions of the protein. For the case of apolipoprotein E (APOE), this binding occurs near the carboxyl end (C-terminus)
Neurite	A neurite is a projection from a neuron. These can be axons, the long neurites along which impulses are conducted from the cell body to other cells, or they can be dendrites, the short, branched extensions that transmit signals across the synapse
PiB	Abbreviation for Pittsburgh (Pi) Compound B, a radioactive compound related to a dye that has long been used to stain deposits of amyloid for the microscopic detection of amyloids. When used with positron emission tomography (PET scans), PiB detects deposits of amyloid in the brains of living patients and thus can help with the diagnosis of Alzheimer's disease in very early stages. Thus, PiB can also document the effects of therapies designed to reverse or slow the rate of progression of the disease

Plasticity	In evolutionary biology, plasticity usually refers to the process of one genotype leading to various phenotypes depending on the environment. A broader use has developed among biogerontologists to represent the different timing of aging processes within phylogenetic clades, as shown in Fig. 19.1
Public allele system	This term represents a polymorphic gene (one with several variants, each of which has frequencies greater that $\sim 1\ \%$) that is widely found in different human populations. Its effects upon a given phenotype are generally predictable, as in the classic example of sickle-cell hemoglobin (see trade-offs)
Selective advantage	This term applies to alleles or groups of alleles or to certain phenotypes whose gene actions lead to a greater probability of survival in a given environment
TNFα	This gene encodes a multifunctional proinflammatory cytokine that belongs to the tumor necrosis factor (TNF) superfamily
Trade-offs	This term is used here to refer to gene actions that can exhibit differential effects on phenotypes depending upon the environment or the stage of the life cycle. A classic example is sickle-cell hemoglobin, in which heterozygotes have resistance to malaria, whereas homozygotes suffer painful tissue damage

References

1. Finch CE, Austad CE (2015) Is Alzheimer disease uniquely human? Commentary Neurobiol Aging 36:553–555
2. Finch CE (2007) The biology of human longevity: nutrition, inflammation, and aging in the evolution of lifespans. Academic Press, San Diego
3. Gearing M, Tigges J, Mori H, Mirra SS (1996) A beta40 is a major form of beta-amyloid in nonhuman primates. Neurobiol Aging 17:903–908
4. Perez SE, Raghanti MA, Hof PR, Kramer L, Ikonomovic MD, Lacor PN, Erwin JM, Sherwood CC, Mufson EJ (2013) Alzheimer's disease pathology in the neocortex and hippocampus of the western lowland gorilla (*Gorilla gorilla* gorilla). J Comparative Neurol 521:4318–4338
5. Rosen RF, Farberg AS, Gearing M, Dooyema J, Long PM, Anderson DC, Davis-Turak J, Coppola G, Geschwind DH, Pare JF et al (2008) Tauopathy with paired helical filaments in an aged chimpanzee. J Comp Neurol 509:259–270
6. Schultz C, Hubbard GD, Tredici KD, Braak E, Braak H (2001) Tau pathology in neurons and glial cells of aged baboons. Adv Exp Med Biol 487:59–69

7. Morrison JH, Baxter MG (2012) The ageing cortical synapse: hallmarks and implications for cognitive decline. Nat Rev Neurosci 13:240–250

8. Gazzaley AH, Thakker MM, Hof PR, Morrison JH (1997) Preserved number of entorhinal cortex layer II neurons in aged macaque monkeys. Neurobiol Aging 18:549–553

9. Rapoport SI (1989) Hypothesis: Alzheimer's disease is a phylogenetic disease. Med Hypotheses 29:147–150

10. Huang B, Marois Y, Roy R, Julien M, Guidoin R (1992) Cellular reaction to the Vascugraft polyesterurethane vascular prosthesis: in vivo studies in rats. Biomaterials 13:209–216

11. Jacobsen KT, Iverfeldt K (2009) Amyloid precursor protein and its homologues: a family of proteolysis-dependent receptors. Cell Mol Life Sci: CMLS 66:2299–2318

12. Finch CE, Austad SN (2012) Primate aging in the mammalian scheme: the puzzle of extreme variation in brain aging. Age (Dordrecht, Netherlands) 34:1075–1091

13. Merrill DA, Masliah E, Roberts JA, McKay H, Kordower JH, Mufson EJ, Tuszynski MH (2011) Association of early experience with neurodegeneration in aged primates. Neurobiol Aging 32:151–156

14. Goodall J (1986) The Chimpanzees of combe. Harvard University Press, Massachusetts

15. Finch CE, Stanford CB (2004) Meat-adaptive genes and the evolution of slower aging in humans. Q Rev Biol 79:3–50

16. Finch CE (1990) Longevity, senescence, and the genome. University of Chicago Press, Chicago

17. Finch CE, Holmes DJ (2010) Ovarian aging in developmental and evolutionary context. Ann NY Acad Sci 1204:82–94

18. Williams GC (1957) Pleiotropy, Natural selection, and the evolution of senescence. Evolution 11:398–411

19. Kirkwood TB, Rose MR (1991) Evolution of senescence: late survival sacrificed for reproduction. Philos Trans R Soc Lond B Biol Sci 332:15–24

20. Finch CE (2010) Evolution in health and medicine Sackler colloquium: evolution of the human lifespan and diseases of aging: roles of infection, inflammation, and nutrition. Proc Natl Acad Sci USA 107(Suppl 1):1718–1724

21. Poirier J, Miron J, Picard C, Gormley P, Theroux L, Breitner J, Dea D (2014) Apolipoprotein E and lipid homeostasis in the etiology and treatment of sporadic Alzheimer's disease. Neurobiol Aging 35(Suppl 2):S3–10

22. Roses AD, Lutz MW, Saunders AM, Goldgaber D, Saul R, Sundseth SS, Akkari PA, Roses SM, Gottschalk WK, Whitfield KE et al (2014) African-American TOMM40'523-APOE haplotypes are admixture of West African and Caucasian alleles. Alzheimer's Dement J Alzheimer's Assoc 10(592–601):e592

23. Schachter F, Faure-Delanef L, Guenot F, Rouger H, Froguel P, Lesueur-Ginot L, Cohen D (1994) Genetic associations with human longevity at the APOE and ACE loci. Nat Genet 6:29–32

24. Payami H, Zareparsi S, Montee KR, Sexton GJ, Kaye JA, Bird TD, Yu CE, Wijsman EM, Heston LL, Litt M et al (1996) Gender difference in apolipoprotein E-associated risk for familial Alzheimer disease: a possible clue to the higher incidence of Alzheimer disease in women. Am J Hum Genet 58:803–811

25. Altmann A, Tian L, Henderson VW, Greicius MD (2014) Sex modifies the APOE-related risk of developing Alzheimer disease. Ann Neurol 75:563–573

26. Suri S, Heise V, Trachtenberg AJ, Mackay CE (2013) The forgotten APOE allele: a review of the evidence and suggested mechanisms for the protective effect of APOE varepsilon2. Neurosci Biobehav Rev 37:2878–2886

27. Huang Y, Mahley RW (2014) Apolipoprotein E: structure and function in lipid metabolism, neurobiology, and Alzheimer's diseases. Neurobiol Dis 72(A):3–12

28. Naj AC, Jun G, Reitz C, Kunkle BW, Perry W, Park YS, Beecham GW, Rajbhandary RA, Hamilton-Nelson KL, Wang LS et al (2014) Effects of multiple genetic loci on age at onset in late-onset Alzheimer disease: a genome-wide association study. J Am Med Association, Neurol 71:1394–1404

29. Soscia SJ, Kirby JE, Washicosky KJ, Tucker SM, Ingelsson M, Hyman B, Burton MA, Goldstein LE, Duong S, Tanzi RE et al (2010) The Alzheimer's disease-associated amyloid beta-protein is an antimicrobial peptide. PLoS One 5:e9505

30. Mahley RW, Weisgraber KH, Huang Y (2009) Apolipoprotein E: structure determines function, from atherosclerosis to Alzheimer's disease to AIDS. J Lipid Res 50(Suppl):S183–188

31. Martin GM (1999) APOE alleles and lipophylic pathogens. Neurobiol Aging 20:441–443

32. Fullerton SM, Clark AG, Weiss KM, Nickerson DA, Taylor SL, Stengard JH, Salomaa V, Vartiainen E, Perola M, Boerwinkle E et al (2000) Apolipoprotein E variation at the sequence haplotype level: implications for the origin and maintenance of a major human polymorphism. Am J Hum Genet 67:881–900

33. McIntosh AM, Bennett C, Dickson D, Anestis SF, Watts DP, Webster TH, Fontenot MB, Bradley BJ (2012) The apolipoprotein E (APOE) gene appears functionally monomorphic in chimpanzees (Pan troglodytes). PLoS One 7:e47760

34. Singh PP, Singh M, Mastana SS (2006) APOE distribution in world populations with new data from India and the UK. Ann Hum Biol 33:279–308

35. Raffai RL, Dong LM, Farese RV Jr, Weisgraber KH (2001) Introduction of human apolipoprotein E4 "domain interaction" into mouse apolipoprotein E. Proc Natl Acad Sci USA 98:11587–11591

36. Cacciottollo M, Morgan TE, Finch CE (2013) Chimpanzee apolipoprotein E comparison with human apoE3 and -E4 for effect on neuronal outgrowth. In: 4th Annual Symposium: "ApoE, ApoE Receptors, and Neurodegeneration", Georgetown University, Washington, DC, 3–5 June 2013

37. Bellosta S, Nathan BP, Orth M, Dong LM, Mahley RW, Pitas RE (1995) Stable expression and secretion of apolipoproteins E3 and E4 in mouse neuroblastoma cells produces differential effects on neurite outgrowth. J Biol Chem 270:27063–27071

38. Pitas RE, Ji ZS, Weisgraber KH, Mahley RW (1998) Role of apolipoprotein E in modulating neurite outgrowth: potential effect of intracellular apolipoprotein E. Biochem Soc Trans 26:257–262

39. Vamathevan JJ, Hasan S, Emes RD, Amrine-Madsen H, Rajagopalan D, Topp SD, Kumar V, Word M, Simmons MD, Foord SM et al (2008) The role of positive selection in determining the molecular cause of species differences in disease. BMC Evol Biol 8:273

40. Mastana SS, Calderon R, Pena J, Reddy PH, Papiha SS (1998) Anthropology of the apoplipoprotein E (apo E) gene: low frequency of apo E4 allele in Basques and in tribal (Baiga) populations of India. Ann Hum Biol 25:137–143

41. Hu P, Qin YH, Jing CX, Lu L, Hu B, Du PF (2011) Does the geographical gradient of ApoE4 allele exist in China? A systemic comparison among multiple Chinese populations. Mol Biol Rep 38:489–494

42. Singh P, Gerdes U, Mastana SS (2001) Apolipoprotein E polymorphism in India: high APOE*E3 allele frequency in Ramgarhia of Punjab. Anthropologischer Anzeiger; Bericht uber die biologisch-anthropologischen Literatur 59:27–34

43. Zekraoui L, Lagarde JP, Raisonnier A, Gerard N, Aouizerate A, Lucotte G (1997) High frequency of the apolipoprotein E*4 allele in African pygmies and most of the African populations in sub-Saharan Africa. Hum Biol 69:575–581

44. Wozniak MA, Riley EM, Itzhaki RF (2004) Apolipoprotein E polymorphisms and risk of malaria. J Med Genet 41:145–146

45. Rougeron V, Woods CM, Tiedje KE, Bodeau-Livinec F, Migot-Nabias F, Deloron P, Luty AJ, Fowkes FJ, Day KP (2013) Epistatic Interactions between apolipoprotein E and hemoglobin S Genes in regulation of malaria parasitemia. PLoS One 8:e76924

46. Aucan C, Walley AJ, Hill AV (2004) Common apolipoprotein E polymorphisms and risk of clinical malaria in the Gambia. J Med Genet 41:21–24

47. Fujioka H, Phelix CF, Friedland RP, Zhu X, Perry EA, Castellani RJ, Perry G (2013) Apolipoprotein E4 prevents growth of malaria at the intraerythrocyte stage: implications for

differences in racial susceptibility to Alzheimer's disease. J Health Care Poor Underserved 24 (Suppl. 4):70–78

48. Soares HD, Potter WZ, Pickering E, Kuhn M, Immermann FW, Shera DM, Ferm M,Dean RA, Simon AJ, Swenson F, Siuciak JA, Kaplow J, Thambisetty M, Zagouras P, Koroshetz WJ, Wan HI, Trojanowski JQ, Shaw LM, Biomarkers Consortium Alzheimer's Disease Plasma Proteomics Project (2012) Plasma biomarkers associated with the apolipoprotein E genotype and Alzheimer disease. Arch Neurol 69:1310–1317

49. Gale SC, Gao L, Mikacenic C, Coyle SM, Rafaels N, Murray Dudenkov T, Madenspacher JH, Draper DW, Ge W, Aloor JJ, Azzam KM, Lai L, Blackshear PJ, Calvano SE, Barnes KC, Lowry SF, Corbett S, Wurfel MM, Fessler MB (2014) APOε4 is associated with enhanced in vivo innate immune responses in human subjects. J Allergy Clin Immunol 134:127–134

50. Fabris C, Vandelli C, Toniutto P, Minisini R, Colletta C, Falleti E, Smirne C, Pirisi M (2011) Apolipoprotein E genotypes modulate fibrosis progression in patients with chronic hepatitis C and persistently normal transaminases. J Gastroenterol Hepatol 26:328–333

51. Zhao Y, Ren Y, Zhang X, Zhao P, Tao W, Zhong J, Li Q, Zhang XL (2014) Ficolin-2 inhibits hepatitis C virus infection, whereas apolipoprotein E3 mediates viral immune escape. J Immunol 193:783–796

52. Burt TD, Agan BK, Marconi VC, He W, Kulkarni H, Mold JE, Cavrois M, Huang Y, Mahley RW, Dolan MJ et al (2008) Apolipoprotein (apo) E4 enhances HIV-1 cell entry in vitro, and the APOE epsilon4/epsilon4 genotype accelerates HIV disease progression. Proc Natl Acad Sci USA 105:8718–8723

53. Kuhlmann I, Minihane AM, Huebbe P, Nebel A, Rimbach G (2010) Apolipoprotein E genotype and hepatitis C, HIV and herpes simplex disease risk: a literature review. Lipids Health Dis 9:8

54. Shaw P, Lerch JP, Pruessner JC, Taylor KN, Rose AB, Greenstein D, Clasen L, Evans A, Rapoport JL, Giedd JN (2007) Cortical morphology in children and adolescents with different apolipoprotein E gene polymorphisms: an observational study. Lancet Neurol 6:494–500

55. Knickmeyer RC, Wang J, Zhu H, Geng X, Woolson S, Hamer RM, Konneker T, Lin W, Styner M, Gilmore JH (2014) Common variants in psychiatric risk genes predict brain structure at birth. Cereb Cortex 24:1230–1246

56. Wang C, Wilson WA, Moore SD, Mace BE, Maeda N, Schmechel DE, Sullivan PM (2005) Human apoE4-targeted replacement mice display synaptic deficits in the absence of neuropathology. Neurobiol Dis 18:390–398

57. Oria RB, Patrick PD, Oria MO, Lorntz B, Thompson MR, Azevedo OG, Lobo RN, Pinkerton RF, Guerrant RL, Lima AA (2010) ApoE polymorphisms and diarrheal outcomes in Brazilian shanty town children. Braz J Med Biol Res 43:249–256

58. Mitter SS, Oria RB, Kvalsund MP, Pamplona P, Joventino ES, Mota RM, Goncalves DC, Patrick PD, Guerrant RL, Lima AA (2012) Apolipoprotein E4 influences growth and cognitive responses to micronutrient supplementation in shantytown children from northeast Brazil. Clinics (Sao Paulo, Brazil) 67:11–18

59. Azevedo OG, Bolick DT, Roche JK, Pinkerton RF, Lima AA, Vitek MP, Warren CA, Oriá RB, Guerrant RL (2014) Apolipoprotein E plays a key role against cryptosporidial infection in transgenic undernourished mice. PLoS One 9(2):e89562

60. Jasienska G, Ellison PT, Galbarczyk A, Jasienski M, Kalemba-Drozdz M, Kapiszewska M, Nenko I, Thune I, Ziomkiewicz A (2015) Apolipoprotein E (ApoE) polymorphism is related to differences in potential fertility in women: a case of antagonistic pleiotropy? Proc R Soc Lond B Biol Sci 282(1803):20142395

61. Seger JB, Jane H (1987) What is bet-hedging? In: Partridge PHHL (ed) Oxford surveys in evolutionary biology. Oxford University Press, Oxford, pp 182–211

62. Rea SL, Wu D, Cypser JR, Vaupel JW, Johnson TE (2005) A stress-sensitive reporter predicts longevity in isogenic populations of *Caenorhabditis elegans*. Nat Genet 37:894–898

63. Martin (2009) Epigenetic gambling and epigenetic drift as an antagonistic pleiotropic mechanism of aging. Aging Cell 8:761–764

64. Fraga MF, Ballestar E, Paz MF, Ropero S, Setien F, Ballestar ML, Heine-Suner D, Cigudosa JC, Urioste M, Benitez J et al (2005) Epigenetic differences arise during the lifetime of monozygotic twins. Proc Natl Acad Sci USA 102:10604–10609

65. Bahar R, Hartmann CH, Rodriguez KA, Denny AD, Busuttil RA, Dolle ME, Calder RB, Chisholm GB, Pollock BH, Klein CA et al (2006) Incsed cell-to-cell variation in gene expression in ageing mouse heart. Nature 441:1011–1014

66. Martin (2012) Stochastic modulations of the pace and patterns of ageing: impacts on quasi-stochastic distributions of multiple geriatric pathologies. Mech Ageing Dev 133:107–111

67. Wang SC, Oelze B, Schumacher A (2008) Age-specific epigenetic drift in late-onset Alzheimer's disease. PLoS One 3:e2698

68. Kirkwood TB, Finch CE (2002) Ageing: the old worm turns more slowly. Nature 419:794–795

69. Medawar PB (1952) An unsolved problem in biology: an inaugural lecture delivered at University College. H.K. Lewis and Company, London

70. Haldane JBS (1941) New paths in genetics. George Allen & Unwin Limited, London

71. Chouraki V, Seshadri S (2014) Genetics of Alzheimer's disease. Adv Genet 87:245–294

72. Karch CM, Cruchaga C, Goate AM (2014) Alzheimer's disease genetics: from the bench to the clinic. Neuron 83:11–26

73. Karch CM, Goate AM (2015) Alzheimer's disease risk genes and mechanisms of disease pathogenesis. Biol Psychiatry 77:43–51

74. Lord J, Lu AJ, Cruchaga C (2014) Identification of rare variants in Alzheimer's disease. Front Genet 5:369

75. Reitz C (2014) Genetic loci associated with Alzheimer's disease. Future Neurol 9:119–122

76. Irwin DJ, Abrams JY, Schonberger LB, Leschek EW, Mills JL, Lee VM, Trojanowski JQ (2013) Evaluation of potential infectivity of Alzheimer and Parkinson disease proteins in recipients of cadaver-derived human growth hormone. JAMA Neurol 70:462–468

77. Watts JC, Condello C, Stohr J, Oehler A, Lee J, DeArmond SJ, Lannfelt L, Ingelsson M, Giles K, Prusiner SB (2014) Serial propagation of distinct strains of Abeta prions from Alzheimer's disease patients. Proc Natl Acad Sci USA 111:10323–10328

78. Morales R, Callegari K, Soto C (2015) Prion-like features of misfolded Abeta and tau aggregates. Virus Res

79. White MR, Kandel R, Tripathi S, Condon D, Qi L, Taubenberger J, Hartshorn KL (2014) Alzheimer's associated beta-amyloid protein inhibits influenza A virus and modulates viral interactions with phagocytes. PLoS One 9:e101364

80. Theendakara V, Patent A, Peters Libeu CA, Philpot B, Flores S, Descamps O, Poksay KS, Zhang Q, Cailing G, Hart M et al (2013) Neuroprotective sirtuin ratio reversed by ApoE4. Proc Natl Acad Sci USA 110:18303–18308

81. Middleton PJ, Petric M, Kozak M, Rewcastle NB, McLachlan DR (1980) Herpes-simplex viral genome and senile and presenile dementias of Alzheimer and Pick. Lancet 1:1038

82. Ball MJ (1982) Limbic predilection in Alzheimer dementia: is activated herpesvirus involved? Can J Neurol Sci 9:303–306

83. Lin WR, Shang D, Wilcock GK, Itzhaki RF (1995) Alzheimer's disease, herpes simplex virus type 1, cold sores and apolipoprotein E4. Biochem Soc Trans 23:594s

84. Sanchez EL, Lagunoff M (2015) Viral activation of cellular metabolism. Virology 479-480C:609–618

85. Maheshwari P, Eslick GD (2015) Bacterial infection and Alzheimer's disease: a meta-analysis. J Alzheimer's Dis: JAD 43:957–966

86. Lathe R, Sapronova A, Kotelevtsev Y (2014) Atherosclerosis and Alzheimer diseases with a common cause? Inflammation, oxysterols, vasculature. BMC Geriatrics 14:36

87. Zhang MY, Katzman R, Salmon D, Jin H, Cai GJ, Wang ZY, Qu GY, Grant I, Yu E, Levy P et al (1990) The prevalence of dementia and Alzheimer's disease in Shanghai, China: impact of age, gender, and education. Ann Neurol 27:428–437
88. Letenneur L, Launer LJ, Andersen K, Dewey ME, Ott A, Copeland JR, Dartigues JF, Kragh-Sorensen P, Baldereschi M, Brayne C et al (2000) Education and the risk for Alzheimer's disease: sex makes a difference. EURODEM pooled analyses. EURODEM Incidence Research Group. Am J Epidemiol 151:1064–1071
89. Jefferson AL, Gibbons LE, Rentz DM, Carvalho JO, Manly J, Bennett DA, Jones RN (2011) A life course model of cognitive activities, socioeconomic status, education, reading ability, and cognition. J Am Geriatric Soc 59:1403–1411
90. Landau SM, Marks SM, Mormino EC, Rabinovici GD, Oh H, O'Neil JP, Wilson RS, Jagust WJ (2012) Association of lifetime cognitive engagement and low beta-amyloid deposition. Arch Neurol 69:623–629
91. Stern Y (2012) Cognitive reserve in ageing and Alzheimer's disease. Lancet Neurol 11:1006–1012
92. Steffener J, Barulli D, Habeck C, O'Shea D, Razlighi Q, Stern Y (2014) The role of education and verbal abilities in altering the effect of age-related gray matter differences on cognition. PLoS One 9:e91196
93. Vitek MP, Christensen DJ, Wilcock D, Davis J, Van Nostrand WE, Li FQ, Colton CA (2012) APOE-mimetic peptides reduce behavioral deficits, plaques and tangles in Alzheimer's disease transgenics. Neuro-Degenerative Dis 10:122–126
94. Gao H, Granka JM, Feldman MW (2010) On the classification of epistatic interactions. Genetics 184(3):827–837
95. Calenda A, Jallageas V, Silhol S, Bellis M, Bons N (1995) Identification of a unique apolipoprotein E allele in *Microcebus murinus*; ApoE brain distribution and co-localization with beta-amyloid and tau proteins. Neurobiol Dis 2:169–176
96. Bons N, Rieger F, Prudhomme D, Fisher A, Krause KH (2006) *Microcebus murinus*: a useful primate model for human cerebral aging and Alzheimer's disease? Genes Brain Behav 5:120–130
97. Kraska A, Dorieux O, Picq JL, Petit F, Bourrin E, Chenu E, Volk A, Perret M, Hantraye P, Mestre-Frances N et al (2011) Age-associated cerebral atrophy in mouse lemur primates. Neurobiol Aging 32:894–906
98. Bertrand A, Pasquier A, Petiet A, Wiggins C, Kraska A, Joseph-Mathurin N, Aujard F, Mestre-Frances N, Dhenain M (2013) Micro-MRI study of cerebral aging: ex vivo detection of hippocampal subfield reorganization, microhemorrhages and amyloid plaques in mouse lemur primates. PLoS One 8:e56593
99. Selkoe DJ, Bell DS, Podlisny MB, Price DL, Cork LC (1987) Conservation of brain amyloid proteins in aged mammals and humans with Alzheimer's disease. Science 235:873–877
100. Elfenbein HA, Rosen RF, Stephens SL, Switzer RC, Smith Y, Pare J, Mehta PD, Warzok R, Walker LC (2007) Cerebral beta-amyloid angiopathy in aged squirrel monkeys. Histol Histopathol 22:155–167
101. Chambers JK, Kuribayashi H, Ikeda S, Une Y (2010) Distribution of neprilysin and deposit patterns of Abeta subtypes in the brains of aged squirrel monkeys (*Saimiri sciureus*). Amyloid: Int J Exp Clin Investig 17:75–82
102. Ndung'u M, Hartig W, Wegner F, Mwenda JM, Low RW, Akinyemi RO, Kalaria RN (2012) Cerebral amyloid beta(42) deposits and microvascular pathology in ageing baboons. Neuropathol Appl Neurobiol 38:487–499
103. Mestre-Frances N, Keller E, Calenda A, Barelli H, Checler F, Bons N (2000) Immunohistochemical analysis of cerebral cortical and vascular lesions in the primate *Microcebus murinus* reveal distinct amyloid beta1-42 and beta1-40 immunoreactivity profiles. Neurobiol Dis 7:1–8
104. Geula C, Nagykery N, Wu CK (2002) Amyloid-beta deposits in the cerebral cortex of the aged common marmoset (*Callithrix jacchus*): incidence and chemical composition. Acta Neuropathol 103:48–58

105. Wu CK, Nagykery N, Hersh LB, Scinto LF, Geula C (2003) Selective age-related loss of calbindin-D28k from basal forebrain cholinergic neurons in the common marmoset (*Callithrix jacchus*). Neuroscience 120:249–259
106. Leuner B, Kozorovitskiy Y, Gross CG, Gould E (2007) Diminished adult neurogenesis in the marmoset brain precedes old age. Proc Natl Acad Sci USA 104:17169–17173
107. Knuesel I, Nyffeler M, Mormede C, Muhia M, Meyer U, Pietropaolo S, Yee BK, Pryce CR, LaFerla FM, Marighetto A et al (2009) Age-related accumulation of Reelin in amyloid-like deposits. Neurobiol Aging 30:697–716
108. Carlyle BC, Nairn AC, Wang M, Yang Y, Jin LE, Simen AA, Ramos BP, Bordner KA, Craft GE, Davies P et al (2014) cAMP-PKA phosphorylation of tau confers risk for degeneration in aging association cortex. Proc Natl Acad Sci USA 111:5036–5041
109. Gearing M, Rebeck GW, Hyman BT, Tigges J, Mirra SS (1994) Neuropathology and apolipoprotein E profile of aged chimpanzees: implications for Alzheimer disease. Proc Natl Acad Sci USA 91:9382–9386
110. Hussain I, Fabregue J, Anderes L, Ousson S, Borlat F, Eligert V, Berger S, Dimitrov M, Alattia JR, Fraering PC et al (2013) The role of gamma-secretase activating protein (GSAP) and imatinib in the regulation of gamma-secretase activity and amyloid-beta generation. J Biol Chem 288:2521–2531
111. Shibuya Y, Chang CC, Huang LH, Bryleva EY, Chang TY (2014) Inhibiting ACAT1/SOAT1 in microglia stimulates autophagy-mediated lysosomal proteolysis and increases Abeta1-42 clearance. J Neurosci 34:14484–14501
112. Martin (2014) Nature, nurture and chance: their roles in interspecific and intraspecific modulations of aging. In: Sprott R (ed) Annual review of gerontology and geriatrics: genetics, vol 34. Springer, New York

Part VIII
Psychology and Psychiatry

Chapter 20
The Evolutionary Etiologies of Autism Spectrum and Psychotic Affective Spectrum Disorders

Prof. Bernard J. Crespi, Ph.D.

Lay Summary The mental and behavioural traits that have evolved in humans, such as complex sociality, make us vulnerable to corresponding mental disorders, such as disorders that involve too little, or too much, social thinking. Autism can be considered as a disorder where complex sociality does not develop, while schizophrenia, bipolar disorder, and depression can be considered as the opposite: pathologically overdeveloped social thought and behaviour, as seen, for example, in paranoia and hearing voices. Evolutionary biology is fundamentally important in understanding, defining, and treating mental disorders because it helps us to determine what the brain has evolved to do, which informs us about the different ways that brain functions can become dysregulated in disease.

20.1 Introduction

20.1.1 The Standard Medical Model and the Reification of Psychiatric Disorders

The standard medical model for understanding and treating disease focuses on determining its proximate physiological and developmental causes, in terms of how functional systems have become dysregulated [1]. High blood glucose levels, for example, may be due to type 1 diabetes, which results from specific, well-characterized physiological and molecular biological causes and, as a result, can be

B.J. Crespi (✉)
Department of Biological Sciences, Simon Fraser University, Burnaby British Columbia V5A 1S6, Canada
e-mail: crespi@sfu.ca

© Springer International Publishing Switzerland 2016
A. Alvergne et al. (eds.), *Evolutionary Thinking in Medicine*,
Advances in the Evolutionary Analysis of Human Behaviour,
DOI 10.1007/978-3-319-29716-3_20

299

unambiguously diagnosed. Understanding the normal functioning of blood glucose regulation, or any other physiological system, thus represents a key precondition for determining aetiology and effective treatments.

How can the standard medical model be applied to psychiatric disorders? The medical model assumes that illness can be objectively and unequivocally quantified. By contrast, the causes and patterns of brain functions that underlie psychiatric disorders are only dimly understood. Psychiatric disorders are, instead, abstract, heuristic, descriptive constructs that are more or less useful for guiding research, diagnoses, and treatments. The clearest evidence for such artificiality is the Diagnostic and Statistical Manual of Mental Disorders (DSM) criteria for diagnosing psychiatric disorders, which comprise detailed lists of symptoms, some set of which are considered necessary and sufficient to infer the presence of disease.

Despite these considerations, it is commonplace for psychiatric conditions to be reified—that is—considered as real, for research, medical, and societal purposes [2]. Such pragmatic reification can be considered as innocuous, but it is not: it constrains and biases how researchers think about mental disorders and their associated research agendas and leads to misconceptions of psychiatric disorders as objectively defined, purely pathological 'diseases' that people 'have' comparable in some fundamental way to diseases like diabetes, cancer, or atherosclerosis that can be objectively and physiologically quantified in terms of their causes and effects.

Under current paradigms, determining the 'causes' of mental disorders often becomes conflated with characterizing mental pathologies or deficits, at levels from genes, to neurodevelopment and function, to cognitive functions, and to deleterious environments. By contrast, according to the standard medical model, mental disorders should instead be conceptualized and analysed, in terms of what functional mental systems have become dysregulated and what forms such dysregulations take. In this regard, for example, to better understand autism, we must also better understand the development of neurotypical social cognition, and to understand bipolar disorder and depression, we must also understand the adaptive functions of normal, contextual variation in mood.

20.1.2   The Evolution of Mental Adaptations

Adaptive functions of the human mind and brain, like those of glucose regulation, have, of course, evolved. Most generally, this meaning of 'adaptive' means that such systems show, and have for many, many past generations shown, genetically based variation among individuals that has influenced survival and reproduction. Such variation has thus been subject to natural selection, which leads, across generations, to increases in, or maintenance of, the adaptive 'fit' or 'match' between organismal phenotypes and aspects of their environments. For example, the beaks of Darwin's finches are 'fit' in their sizes and shapes for different food sources. Similarly, specific regions of the human neocortex adaptively function to recognize individual faces (the fusiform gyrus), or to infer the thoughts and intentions of other humans (the medial prefrontal cortex). Specific mental adaptations, like the insulin pathway, are real and quantifiable and the subject of intense interest in disciplines such as cognitive neuroscience.

Natural selection of human physiology and morphology is expected, under basic evolutionary considerations, to have led to the maximization of functional robustness, homoeostatic ability, and efficiency, as well as optimal flexibility under variable circumstances, all in the service of survival and reproduction. But what, then, is natural selection—the driver of adaptation—expected to maximize with regard to human cognition, emotion, and behaviour? We usually think of mental disorders as centrally involving unhappiness of the subject as well as their social circle, which motivates the seeking of help from the medical community. However, natural selection is by no means expected to maximize happiness, simply because increased happiness is by no means a primary means or route to increased survival and reproduction [3]. Instead, natural selection is predicted, by basic theory, to maximize condition-dependent human striving for the goals that have led, across many past generations in relevant environments, to high survival and reproduction, relative to other humans.

In the context of striving, human emotional systems have evolved to motivate and modulate goal-seeking, dynamically across different circumstances. Such motivation is mediated by the human 'liking' and 'wanting' reward systems, as well as by unhappiness or dissatisfaction with current situations. Human cognitive systems, by contrast, represent sets of evolved mechanisms for information processing, causal thinking, and decision-making that subserve identification of appropriate goals and tactics for reaching them. Both emotional and cognitive systems develop across infancy, childhood, and adolescence, whereby genes, environments, and gene-by-environment interactions mediate neurodevelopment. To understand human psychiatric disorders from an evolutionary perspective, it thus becomes necessary to connect these psychological trajectories and adaptations with their corresponding maladaptations (lacks of fit of phenotypes to the environment), expressed as developmental, emotional, and cognitive dysfunctions that revolve around human striving and cognition. What adaptations, then, are dysregulated in major human mental disorders and how?

Evolutionary biology is useful in medicine for two main reasons: (1) it teaches us how to think about human medically relevant phenotypes and diagnoses, in novel, productive ways, and (2) it indicates specific new data to collect and new approaches for therapies. In this chapter, I focus on the evolutionary biology of psychiatric disorders centrally involving social cognition, affect, and development. I first describe the primary types of causes of mental disorders, from evolutionary medical thinking. Next, I describe autism spectrum disorders and psychotic affective spectrum disorders, in the context of these causes, with reference to recent findings in genetics, neuroscience, and psychology, and in the context of which human-evolved adaptations have been subject to what forms of alteration in each case. Third, I describe and evaluate hypotheses for the relationships of these disorders with one another—relationships that define evolved axes of human development, affect, and cognition that structure variation in adaptive and maladaptive human mental functioning. Finally, I make specific suggestions for research and clinical therapies that follow directly from these considerations.

20.2 Research Findings

20.2.1 Evolutionary Causes of Mental Disorders

The evolutionary causes of psychiatric disorders represent the 'ultimate' sources of these conditions, which indicate why, given their evolutionary history, humans exhibit particular forms of mental disorders with particular symptoms and severities. Each of the six main causes described below centres on explanations for deviations from mental adaptation and health, in the context of how maladaptations can arise, and be maintained, in populations.

20.2.1.1 Deleterious Alleles

Mutations generate novel alleles that usually cause reduced genetic function, because the perturbations randomly alter a system that would otherwise develop reasonably well. Highly penetrant mutations, with large effects, are especially likely to be highly deleterious, and considerable evidence attests to important roles for de novo, deleterious mutations, such as copy number variants or changes to highly conserve amino acid residues, in the causes of mental illness (e.g. [4]). Highly deleterious alleles that are associated with relatively severe mental illnesses include monogenic causes of autism or schizophrenia that evolve under mutation–selection balance: rare mutations arise and are selected against because their bearers exhibit greatly reduced reproduction.

Rare, deleterious alleles such as copy number variants have been estimated to account for a small percentage of cases of major mental illness [5]. Most inferred 'risk alleles' for mental disorders, such as those identified with genomewide association studies are, however, relatively common (at frequencies above 1 % or 5 %) and have small effects on risk through one dimension of their multifaceted impacts on neurodevelopment, neuronal function, and other systems. The degree to which such alleles can be considered as deleterious to health overall—given all of their effects—remains an open question; for example, neurodegenerative disease risk trades off with cancer risk, such that higher risks in one domain of disease may commonly entail lower risks in another [6]. Presumably, if psychiatric risk alleles were purely deleterious, they would indeed not be common in populations. Risk alleles may also exhibit positive effects, on health and reproduction, when expressed in genetic relatives of individuals with mental illness [7]; these findings indicate that 'risk' alleles do not simply confer increased risk of disease, but may, depending on the context, confer benefits as well. Such considerations can help to explain the high heritabilities of psychiatric conditions including autism, bipolar disorder, and schizophrenia, on the order of 50–80 % (e.g. [8]).

20.2.1.2 Mismatched Environments

Populations and individuals are always adapted to past environments, and if environments change more rapidly than they can be tracked by selection and genetic response to selection, then populations will be maladapted. Human environments have changed radically over the past few hundred years, which is expected to lead to higher risk of psychiatric disorders to the extent that the novel environments include risk factors such as increased social stress and isolation, or toxins such as lead and mercury that degrade neurodevelopment. For example, some of the highest rates of schizophrenia are found among visible-minority (e.g. different skin colour) immigrants, who appear to be subject to relatively severe psychosocial stresses due to their novel, challenging environments [9].

20.2.1.3 Extremes of Adaptations

Some psychiatric conditions, such as generalized anxiety disorder, or some manifestations of obsessive–compulsive disorder such as excessive hygienic behaviour, clearly represent extremes of normally adaptive behaviour: anxiety functions to modulate arousal and attention under challenging conditions [10], and hygiene reduces risks of infection [11]. This conceptual framework has been generalized to connect normal personality variation along a spectrum to personality disorders and to severe psychiatric disorders, by demonstrating which aspects of personality are amplified, reduced, or otherwise distorted to generate mental dysfunction [12]. This approach has successfully described continua in personality traits from normal to maladaptive extremes, although the adaptive significance, in terms of fitness-related benefits and costs of personality variation among normal individuals, remains largely unstudied. Maladaptive extremes can also be considered more directly in the context of human evolutionary history, in that the evolution of human-specific traits, such as large brain size and language, has generated potential and scope for loss of these specific traits, as in microcephaly and specific language impairment, as well as potential and scope for dysfunctional overdevelopment, as in macrocephaly and the disordered and exaggerated components of speech in schizophrenia [13, 14].

20.2.1.4 Trade-Offs

Trade-offs have been well characterized for developmental and physiological phenotypes, whereby, for example, increased resource allocation in one domain takes away from another. For neurological and psychological phenotypes, however, conceptual paradigms based on trade-offs have yet to be developed, despite evidence for trade-offs of verbal–social with visual–spatial skills [15], empathic with systemizing (rule-based) interests and abilities [16], neural flexibility with stability [17], as well as trade-offs between neural activation of the internally, self-directed default mode network, and the outwardly focused task-positive network [18]. Cognitive and emotional trade-offs are important because they structure the brain's functional architecture and generate coincidences of relative strengths with relative deficits; for example, Kravariti et al. [19] found that having closer relatives with schizophrenia was strongly associated with better verbal skills relative to visual–spatial skills. Trade-offs are stronger under resource-related constraints, which may commonly follow from dysfunctional neurodevelopment, and their extremes are expected to characterize some psychiatric conditions. Autism, for example, has been strongly associated with a combination of high systemizing and low empathizing, whereas some combination of dysfunctionally high empathizing and low systemizing appears to characterize some psychotic affective conditions [20], especially borderline personality disorder and depression [21].

20.2.1.5 Conflicts

Genetically based conflicts, whereby two parties exhibit different optima for some genetically based phenotype, generate risk of maladaptation because one party may more or less 'lose' the conflict, resources are wasted on conflictual interactions, and conflict mechanisms generate novel targets for dysregulation and disease [22]. The forms of evolutionary genetic conflict most salient to psychiatric conditions include parent–offspring conflict (e.g. [23]), genomic imprinting conflict [24, 25], and sexual conflict [26]. Dysregulated genomic imprinting, for example, underlies the expression of Prader–Willi syndrome, one of the strongest genetic causes of psychosis [27], and this syndrome represents only an extreme case of such psychiatric effects [13]. Similarly, a recent epidemiological study of over two million individuals demonstrated that unaffected sisters (but not brothers) of individuals with schizophrenia and bipolar disorder exhibit higher fertility than controls, a pattern that is uniquely predicted by a hypothesis of 'sexually antagonistic' alleles that impose costs on males but benefit females [7].

20.2.1.6 Defences Mistaken as Symptoms

This last 'cause' of disease is only apparent: some psychiatric symptoms represent conditionally adaptive defences for alleviating problematic conditions, rather than deleterious manifestations of disease. Thus, in the same way that fever represents a conditionally adaptive bodily response to infection, with health benefits that usually outweigh its costs, some psychiatric symptoms can be interpreted as conferring benefits, relative to their absence or reduction. Examples of such phenomena include the following: (a) repetitive behaviour in autism, which serves to dampen excessively high levels of autonomic and sensory arousal [28], (b) dissociation, as a psychological mechanism to reduce deleterious effects of trauma [29], (c) delusion formation in psychosis, as a means to mentally cope with the exaggerated and disordered perceptions of salience (causal meaning) [30], and (d) mild depression (low mood), as a conditionally adaptive response to circumstances that favour disengagement from failing or unreachable goals—which escalates to full depression if useless goal-seeking persists [31]. The danger of conceptualizing defences, like fever, as purely deleterious symptoms is that treating them is expected to make the situation specifically worse unless the underlying cause of the disorder (and defence) is addressed, such as the sensory hypersensitivity in autism, the trauma in dissociation, or the challenging life events and personal motivational structure that underlie liability to low mood and depression.

These six causes of aetiology and symptoms of psychiatric conditions converge in their emphases on determining what evolved genetic, developmental, neural, cognitive and emotional systems are altered, and how they are altered, in psychiatric conditions. These causes also provide our framework for determining how nominal, DSM-designated psychiatric conditions are related to one another in their causes, as independent and separate, partially overlapping, or diametric to one another in the same general way as the development or activity of any biological system or pathway can be altered in two opposite directions.

20.2.2   Autism Spectrum Conditions

Autism is defined, and commonly reified, as a combination of deficits in social reciprocity and communication with high levels of restricted interests and repetitive behaviour (Fig. 20.1). The degree to which this combination represents a cohesive syndrome, with causally shared rather than independent symptoms and causal factors, remains unclear [32]. Beyond these two commonalities, autism presents diverse features, with overall intellectual abilities varying from very low to above average, cognitive enhancements (above neurotypical) in sensory and visual–spatial abilities in a substantial fraction of individuals and a sex ratio that is highly male-biased overall but much less so among more severely affected individuals [33].

The most straightforward connection between the major features of autism, and human evolution, is that our evolutionary history has been characterized by elaboration of the 'social brain': the distributed, integrated set of neural systems that subserve the acquisition, processing, and use of social information. It is these social brain phenotypes that are specifically underdeveloped in autism. As such, autism can be conceptualized as the expression of maladaptive extremes of social brain underdevelopment, which, in principle, may be caused in a proximate way by reduction or loss of any of the myriad systems that are necessary or sufficient for human social brain development. Autism thus exhibits many single-gene, syndromic causes due to deleterious mutation, but it is also commonly underlain by combined effects from the hundreds or thousands of genes bearing alleles that affect

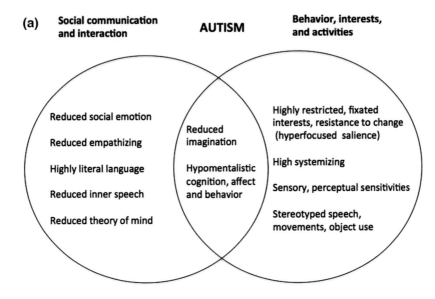

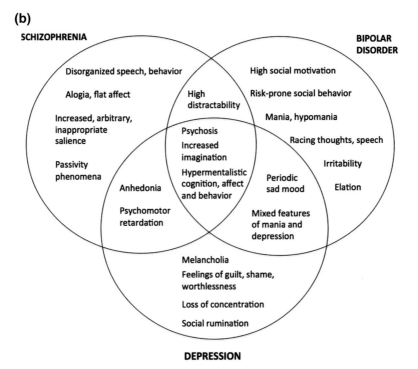

Fig. 20.1 Phenotypes that describe **a** the autism spectrum and **b** the psychotic affective spectrum, based on DSM-V diagnoses, evolutionary considerations, and the hypothesized relationships between the two sets of disorders

social brain development [34]. As such, there can be no primary, proximate physiologically based cause of autism (as there may be, for example, for type 1 diabetes), and the search for causes becomes a differential characterization, subdivision, and prioritization of the diverse genetic, epigenetic, and environmental influences that converge on underdevelopment of the social brain.

As social cognition is underdeveloped under all psychologically based theories for autism, it can also be conceptualized, and studied, in terms of developmental heterochrony, whereby child cognitive development is not completed in autism, and childhood characteristics, including reduced social cognition, are retained into adulthood [35, 36]. In this context, other human-elaborated traits including highly developed, regulated social striving and goal-seeking, guided by perceived reward-associated or cost-associated (aversive) salience (inferred, causative meaning) of social stimuli, remain underdeveloped as well on the autism spectrum. External stimuli may thus have salience predominantly in terms of perceived sensations, or specific, highly restricted non-social interests, especially foci of highly selective attention [37]. Frith [38] indeed sees a weak drive to discern meaning in the world as epitomizing the weak central coherence theory of autism, which has been supported by a wide range of evidence.

A central, unresolved question in the study of autism is whether a single, central, psychological, or cognitive-level factor can explain the apparently inexplicable combination of reduced sociality with restricted interests and repetitive behaviour. In the context of social brain underdevelopment, increased restricted interests and repetitive behaviours, and sensory, visual–spatial, and mechanistic cognition enhancements in autism, can be explained by several hypotheses.

First, increases in asocial phenotypes may pre-empt the development of social phenotypes, such as by directing perceived salience, interests, and brain specializations along asocial paths. Such effects, which are notably represented by a theory for autism aetiology based on enhanced perceptual functioning [39], may be mediated by overdevelopments of sensory perception and mechanistic, systemizing cognition [33].

Second, increased asocial cognition may itself be a direct result of reduced social cognition, as a compensatory or trade-off-based neurodevelopmental mechanism akin to the overdevelopment of non-visual senses among the blind.

Third, some such asocial cognition and behaviour, and phenotypes such as insistence on sameness and stimulus over-selectivity, may, as noted above, represent defences that aid in coping with challenging symptoms such as increased perceptual sensitivity or avoidance of stress from dealing with inexplicable social cognitive tasks.

Finally, one possible resolution, based on reduced expression of a phenotype virtually unique to humans, is that autism is, in part, underpinned psychologically by underdeveloped imagination, defined as 'the faculty or action of forming new ideas, or images or concepts of external objects not present to the senses'. This hypothesis, originally described by Rutter [40], and Wing and Gould [41], can, in principle, jointly explain social and asocial alterations in autism, including reduced pretend play, reduced social imagination as expressed in theory of mind, restriction of interests and repetition of behaviour, and insistence on sameness.

Determining the degree to which these hypotheses are correct, in general or for any particular individual, is crucially important to autism therapy, especially to prevent enhancements or conditionally adaptive defences of autistic individuals from being treated as deleterious symptoms.

20.2.3 Psychotic Affective Spectrum Conditions

Psychotic affective spectrum conditions include a set of DSM disorders, mainly schizophrenia, bipolar disorder, and depression, which broadly overlap in their symptoms, neurological and psychological correlates, and genetic and environmental risk factors [42] (Fig. 20.1). All of these conditions exhibit substantial genetic components and mediation in part by rare, penetrant risk factors, although most genetic risk appears to be underlain by many alleles each with small effect.

Schizophrenia, as well as other conditions that involve psychosis, can be understood most directly and simply in terms of dysfunction of the human adaptive system for assigning salience (causal meaning) to external, and internally generated, stimuli [30, 43]. Salience assignment, which is underpinned by a dedicated neural system involving the anterior cingulate cortex and insula, is fundamental to cognition, behaviour, and goal-seeking, in that it mediates subjective causal understanding of perceptual inputs. Psychosis thus involves overdeveloped and inappropriate salience, usually in the contexts of social interactions, agency, intentionality, self-other associations, and other aspects of mentalistic (social and mind-related) thought, apparently due to the primacy of social cognition in human goal-directed behaviour [24]. Paradigmatic manifestations of psychosis thus involve paranoia, other social delusions, megalomania, belief that events always refer to the self, alterations to self-other distinctions, and assignment of mind, agency, and intentions to inappropriate subjects and inanimate objects. Such reality distortions are mediated by top-down cognitive processes, and they can be considered as attempts to 'make sense' of the excessively high and inappropriate salience assignment, for external stimuli, that is driven by hyperdopaminergic neurotransmission [30, 44, 45]. Hallucinations, in turn, can be understood as misinterpreted and exaggerated internal perceptions, mediated by overdeveloped salience of internal representations, such that given certain neurophysiological alterations, thought, inner speech, and imagination come to be considered as external percepts. Like delusions, hallucinations are usually expressed as social phenomena, especially auditory hallucinations.

Schizophrenia is predominantly considered as a disorder of cognition, whereby the causal meanings that guide striving become overdeveloped and dysfunctionally overmentalistic. Bipolar disorder and depression, by contrast, represent mainly disorders of emotion, the set of neural and hormonal systems that motivate and modulate striving and goal-seeking across different contexts. Understanding such mood disorders requires consideration of the adaptive significance of condition-dependent variation in human emotions, especially with regard to the social interactions that permeate human thought and behaviour [46]. In this context, considerable evidence indicates that low mood is normally adaptive in situations where individuals benefit by disengaging from unreachable or unprofitable goals, as it facilitates such disengagement and motivates alternative behavioural patterns of goal-seeking that should be more advantageous [47]. High, positive mood, in comparison, represents an emotional mechanism whereby human reward systems motivate continuation of beneficial behaviour, because one's goals are being reached. Depression, then, can be conceptualized and studied as overly low and overly stable mood, a maladaptive extreme of an adaptation, whereby individuals fail to disengage from deleterious thought patterns and striving [3, 47]. Conversely, mania represents an emotional opposite to depression, as the expression of inability to emotionally restrain high mood and intensity of striving, even if and when its consequences become detrimental [48, 49]. Behaviours associated with mania and hypomania can, moreover, be directly interpreted in the context of extreme striving for social dominance, power, and influence, which, if successful, leads to substantial benefits [50, 51]. This evolutionary perspective can explain shifts between mania and depression in bipolar disorder, in that mania is expected to foster pursuit of goals that become more and more risky, unreachable or unsuccessful, eventually prompting the generation of mixed states and descent into depression.

In bipolar disorder, then, cognitive salience systems, and choices of goals, commonly remain functional, but the homoeostatic regulation of the emotions that underlie goal pursuit becomes dysregulated, towards overly low or overly high moods and their sequelae. Moreover, like schizophrenia, mania and depression both centrally involve extremes of social, mentalistic thought and behaviour, here in the context of guilt, shame, embarrassment, perceived social defeat, and social rumination in depression, and social dominance pursuit and pride in mania. Affective psychoses, which comprise psychosis with alterations of mood, may thus be mediated by self-punishment-driven, or reward-driven, overattributions of social salience, in the context of emotionality that becomes sufficiently strong to dysregulate salience. These considerations can help to explain well-documented, otherwise-inexplicable associations of bipolar disorder with high social motivation and achievement [49, 52, 53]. Moreover, bipolar disorder, as well as schizophrenia and schizotypy, have been associated across a wide diversity of studies with increased social imagination, divergent thinking, creativity, and goal attainment, especially in the arts and humanities [54–59]. Imagination can indeed be considered, under Bayesian models of cognition and learning, as directly associated with causal cognition and inference of meaning, such that salience, causal thinking, and imagination should tend to increase, or decrease, in concert with one another [60].

20.2.4 The Relationship Between Autism Spectrum and Psychotic Affective Spectrum Disorders

Bleuler invented the term 'autism' to describe withdrawal from reality and social interactions in schizophrenia, but Kanner was careful to point out that his conceptualization of autism referred to children who had never participated in social life [61]. The relationship between autism and schizophrenia, and psychotic affective disorders more generally, has since been considered in terms of two main hypotheses: (1) partial overlap, with some degree of shared social cognitive deficits and genetic risk factors; (2) a diametric (opposite) relationship, based, at a psychological level, on underdevelopment of social cognition and affect in autism, normality at the centre, and dysfunctional forms of their overdevelopment in psychotic affective conditions [62] (Fig. 20.2). The partial overlap hypothesis is data-driven and motivated primarily by the prominence of social deficits especially in autism and schizophrenia. By contrast, the diametric hypothesis follows directly from evolutionary and neurodevelopmental considerations, under the premises that human evolution has been characterized primarily by elaboration of social cognition (generating increased scope for altered development of specific phenotypes) and that the neurodevelopmental systems that underlie it, like all biological systems, can vary and be perturbed in two opposite directions towards lower or higher expression (Fig. 20.2).

A central prediction of the diametric hypothesis is that autism and psychotic affective conditions (especially schizophrenia, for which most of the relevant data are available) should exhibit opposite phenotypes and genetic risk factors. A suite of such evidence is described in Table 20.1, which provides support for the

Under-development on autism spectrum	Uniquely-human or human-elaborated trait	Over-development on psychotic-affective spectrum
No speech, literal speech	(1) Language	Auditory hallucination
Reduced sense of self	(2) Sense of self	Megalomania, delusions of reference
Low mentalistic skill	(3) Mentalistic skill	Paranoia, other social delusions
Basic emotions	(4) Social emotionality	Shame, guilt, embarassment increased in depression
Mechanical logic	(5) Logical, analytic skill	Thought disorder, loose associations
Reduced goal pursuit	(6) Complex, regulated goal pursuit	Mania, hypomania
Reduced empathizing, empathic abilities	(7) Empathic drive, skills	Enhancements in borderline personality disorder and mild depression
Over-selective attention, restricted interests and repetitive behavior	(8) Drive for causal meaning, salience	Hyper-salience in schizophrenia prodrome, and in psychosis

Fig. 20.2 The autism spectrum and the psychotic affective spectrum can be conceptualized as diametric disorders, with regard to the direction of alterations in uniquely human or human-elaborated phenotypes that comprise their core features

Table 20.1 Diametric genetic risk factors, phenotypes, and correlates of autism spectrum and psychotic affective spectrum conditions

Trait	Autism spectrum	Psychotic affective spectrum	Comments
Copy number variants	Duplications of 22q11.2 increase autism risk [63, 71]	Duplications of 22q11.2 decrease schizophrenia risk; deletions of 22q11.2 greatly increase schizophrenia risk [72]	Deletions of 22q11.2 suggested to increase ASD risk but pattern not found in ASD CNV cohorts [63]
Copy number variants	Duplications of 1q21.1 increase autism risk and increase head size [63, 73]	Deletions of 1q21.1 increase schizophrenia risk and reduce head size [71, 73]	Deletions may increase autism risk, or be false positive [63]
Copy number variants	Deletions of 16p11.2 increase autism risk and increase head size [74]	Duplications of 16p11.2 increase schizophrenia risk and reduce head size [71, 74]	Duplications may increase autism risk, or be false positive [63]
Copy number variants	Duplications of 15q11.2 (BP1-BP2) increase autism risk [75]	Deletions of 15q11.2 (BP1-BP2) increase schizophrenia risk [71]	Deletions and duplications of CYFIP1, a key gene in this CNV region, cause opposite alterations to dendritic spine complexity [76]
Birth size (weight, length)	Smaller size protects against autism; larger size increases autism risk [64]	Larger size protects against schizophrenia; smaller size increases schizophrenia risk [64]	Each of the patterns of risk has been replicated across many other studies
Brain size	Larger brain size in children with autism [77, 78]	Smaller brain size in schizophrenia [79]	Autism involves faster brain growth in early childhood, in particular
Neurological function	Congenital blindness increases risk of autism [80, 81]	Congenital blindness protects against schizophrenia [82, 83]	
Neurological function	Sensory abilities increased in autism [39, 84–90]	Sensory abilities decreased in schizophrenia; sensory deprivation induces features of psychosis [91–98]	Strong, highly consistent pattern in schizophrenia; substantial although somewhat mixed evidence in autism
Neurological function	Prepulse inhibition increased in autism [99, 100]	Prepulse inhibition decreased [101]	Findings highly consistent for schizophrenia, variable for autism
Neurological function	Mismatch negativity increased in autism [102]	Mismatch negativity decreased in schizophrenia [103, 104]	Findings highly consistent for schizophrenia, variable for autism
Neurological function	Mirror neuron system activation decreased in autism [105, 106]	Mirror neuron system activation increased in actively psychotic individuals with schizophrenia [107]	Same protocol used to measure mirror neuron function, in autism and schizophrenia [107]; other studies of schizophrenia usually show reduced activation [108] but do not involve actively psychotic subjects

(continued)

Table 20.1 (continued)

Trait	Autism spectrum	Psychotic affective spectrum	Comments
Neurological function	Default mode system activation reduced in autism, in association with reduced self-referential and imaginative cognition [109–112]	Default system overactivated in schizophrenia, in association with reality distortion and increased imaginative cognition [110]; also less deactivation of this system [113]	Some studies of autism show reduced deactivations of default system that may be associated with reduced activation [110]; Immordino-Yang et al. [114] also contrast autism and schizophrenia as opposite with regard to the default network
Neurological function	Reduced connectivity within default mode in autism [115, 116]	Increased connectivity within default mode in schizophrenia [117–119]	Some mixed results in both autism and schizophrenia, but two reviews support opposite nature of the alterations [120, 121]
Neurological function	Increased local brain connectivity, decreased long-range connectivity, in association with early brain overgrowth [78]	Decreased local brain connectivity, increased long-range connectivity, in association with increased cortical thinning, in childhood-onset schizophrenia [78]	Findings based on the review of neuroimaging findings [78]
Neurological function	Temporal–parietal junction region shows reduced activation in autism and underlies mentalizing reductions [122, 123]	Temporal–parietal junction region shows increased activation in schizophrenia and underlies some psychotic symptoms [124]	
Emotionality and motivation	Reduced social motivation in autism [125]	Increased social motivation in mania, hypomania [49, 51]	Motivation in general decreased in negative symptom schizophrenia, depression
Emotionality and motivation	Cognitive empathic abilities reduced in autism [126]	Some cognitive empathic abilities enhanced in borderline personality disorder and subclinical depression [21, 127]	Cognitive empathic abilities lower in schizophrenia, bipolar disorder, and depression, in association with general cognitive deficits (e.g. [128, 129])
Emotionality and motivation	Reduced social emotion in autism [130]	Increased social emotion expression in bipolar disorder and depression (e.g. guilt, shame, embarrassment, pride) [50, 131]	Reduced general expressed emotionality in negative symptom schizophrenia
Cognitive function	Decreased inattentional blindness in autism [132]	Increased inattentional blindness in schizophrenia [133]	

(continued)

Table 20.1 (continued)

Trait	Autism spectrum	Psychotic affective spectrum	Comments
Cognitive function	Overselective attention [37, 134]	Reductions in selective attention in schizophrenia and positive schizotypy [135, 136]	
Cognitive function	Enhanced Stroop task performance in autism [137]	Decreased Stroop task performance in schizophrenia, by meta-analysis [138]	Results mixed for autism, highly consistent for schizophrenia
Cognitive function	Enhanced Iowa gambling task performance in high-functioning autism [139]	Reduced Iowa gambling task performance in schizophrenia, in most studies [140]	Results mixed for autism, consistent for schizophrenia
Cognitive function	Reduced susceptibility to rubber hand illusion in autism and in healthy high ASD trait individuals [141–143]	Increased susceptibility to rubber hand illusion in schizophrenia [144]	Same general pattern also found for visual illusions, with some inconsistencies [36]
Cognitive function	Literal word interpretation, underinterpretation of social relevance, in autism [145]	Overinterpretation of word meaning and social relevance in schizophrenia [145]	
Cognitive function	Decreased induction of false memories [146, 147]	Increased induction of false memories associated with psychosis phenotypes [148–150]	Results somewhat mixed (some non-significant) for autism
Cognitive function	Semantic memory network states overly rigid in autism [151]	Semantic memory network states chaotic in schizophrenia [151]	
Cognitive function	Working memory deficits in autism [152]; extraordinary working memory enhancements in child prodigies, who score above autism range in attention to detail on autism quotient test and exhibit high rates of autism in their families [153]	Large working memory deficits in schizophrenia; highly consistent finding [154, 155]	Findings of Ruthsatz and Urbach [153] would benefit from replication; areas of excellence in child prodigies notably overlap with those found in savantism in autism [156]
Cognitive function	Hyperlexia found predominantly in autism [85, 157, 158]	Dyslexia associated with schizophrenia and schizotypy [159–161]	Williams and Casanova [162] contrast autism and dyslexia for cortical microstructure
Cognitive function	More deliberative decision-related processing in autism [163]	'Jumping to conclusions' associated with delusions in schizophrenia [164, 165]	

(continued)

314 B.J. Crespi

Table 20.1 (continued)

Trait	Autism spectrum	Psychotic affective spectrum	Comments
Cognitive function	Bias towards hypopriors in Bayesian models of perception and cognition [166, 167]	Bias towards hyperpriors in Bayesian models of perception and cognition [44, 166]	
Cognitive function	Reduced inference of intentions in autism [168]	'Hyperintentionality' in schizophrenia and schizotypy [169, 170]	Bara et al. [171] contrast autism and schizophrenia directly in this regard
Cognitive function	Reduced theory of mind in autism spectrum children by ToM storybooks test [172]	'Hypertheory of mind' in children with more psychotic experiences by ToM storybooks test [173]	
Cognitive function	Theory of mind abilities reduced in autism, using MASC test, due to combination of hypomentalizing, lack of mentalizing, and hypermentalizing [174, 175]	Theory of mind abilities reduced in association with positive symptoms of schizophrenia, using MASC test, due to hypermentalizing [176, 177]; hypermentalizing also found in borderline personality disorder using MASC [178]	
Cognitive function	Reduced salience of social stimuli, and overly specific and inflexible salience of primary perceptual and non-social stimuli [179, 180]	Overdeveloped and arbitrary salience in prodrome and psychosis, mainly involving social phenomena [43, 45, 65]	
Cognitive function	Decreased perception of biological motion, entities, in autism; fail to see humans who are there [181]	Increased and false perception of biological motion, entities, in schizophrenia; see humans in random dots [182]	
Cognitive function	Selectively enhanced visual–spatial abilities in autism [183, 184]	Reduced visual–spatial skills, relative to verbal skills, positively associated with genetic liability to schizophrenia [19, 185]	
Cognitive function	Enhanced embedded figures test performance among healthy individuals with more autistic traits [186]	Reduced embedded figures test performance among healthy individuals with more positively schizotypal traits [186]	
Behaviour	Reduced imagination and creativity in autism [187, 188]	Increased imagination and creativity, in schizophrenia, schizotypy, and bipolar disorder and in relatives [54, 189–192]	The literature relating psychotic affective spectrum phenotypes and conditions to aspects of increased imagination and creativity is large and diverse; reduced imagination has been considered as a diagnostic criterion for autism

(continued)

Table 20.1 (continued)

Trait	Autism spectrum	Psychotic affective spectrum	Comments
Behaviour	Reduced pretend play and social play in autism [193, 194]	Higher levels of dissociation, hallucination, psychotic affective psychopathology associated with the presence of childhood imaginary companions [195–198]	
Social correlates	Autism associated with technical professions in fathers, mothers, and grandfathers [199–201]	Schizophrenia, schizotypy, bipolar disorder, and depression associated with careers and interests in arts, humanities, and literature [202, 203]	
Social correlates	Autism in family associated with technical college majors [204]	Bipolar disorder, depression in family associated with arts and humanities majors [204]	Insufficient data on schizophrenia for analysis, in [204]
Social correlates	Autism associated with higher socioeconomic status [205, 206]	Schizophrenia associated with lower socioeconomic status [207]	

For phenotypes with large sets of evidence, only recent articles or reviews are cited. Crespi and Badcock [24] present additional evidence, from less recent literature diametric hypothesis from diverse and independent sources of data. The partial overlap hypothesis is consistent with the sharing of deficits, especially in social cognition, between autism and schizophrenia, but such deficits can also be considered as deriving from opposite alterations both of which reduce performance on standard tests. Genetic risk factors, such as some genomic copy number variants and some SNPs, have also been associated with both autism spectrum disorders and schizophrenia [62]. Such findings, however, are subject to the caveat that premorbidity to schizophrenia in children and young adolescents, in the form of social deficits and associated developmental problems, can be realistically diagnosed only as autism spectrum since there is not (and never has been) a diagnostic category for schizophrenia premorbidity [63]. This structural limitation in the DSM is expected to lead to a non-negligible incidence of false-positive diagnoses of autism among children who are actually premorbid for schizophrenia, especially among individuals harbouring relatively penetrant genetic risk factors such as copy number variants. Patterns of diagnoses for well-studied CNVs indeed fit with expectations from such false-positive diagnoses [63].

The diametric hypothesis for autism and psychotic affective disorders is novel and controversial and has just begun to be subject to systematic, large-scale testing of its predictions (e.g. [64]). However, to the extent that it is correct, the study of human disorders involving social cognition should be revolutionized and provided its first solid grounding in basic evolutionary principles.

20.3 Implications for Policy and Practice

Risks for human mental disorders have evolved. Evolutionary conceptualizations of autism and psychotic affective disorders lead directly to novel, specific implications for understanding, studying, and treating these conditions.

First, autism, schizophrenia, bipolar disorder, and depression cannot justifiably be considered as 'diseases' under standard medical models of disease, because the neural system adaptations subject to maladaptive alteration in each case remain inadequately understood. Instead, these conditions currently represent broad-scale, heuristic descriptions for suites of related psychological and behavioural problems, none of which has currently specifiable genetic or neurological causes in the same way as do diseases like cancer or diabetes, and all of which grade smoothly in their symptoms into normality. As such, schizophrenia and related psychotic and affective disorders can best be considered as 'syndromes': groups of symptom dimensions that cluster in different combinations across different individuals [65]. Risks and symptoms for these psychiatric conditions have, however, evolved in close conjunction with the evolution of complex human social cognition, affect, and behaviour, which provides the basis for an ultimate understanding, and nosology, of psychiatric maladaptations. In this context, DSM descriptions of autism or a psychotic affective condition should represent starting points for differential diagnosis of their genetic, neurological, social, and environmental causes, for each specific individual. Such causes are expected to involve some combination of effects from deleterious mutations, evolutionarily novel environments, extremes of adaptations, trade-offs, genomic conflicts, and evolved defences.

Second, autism can be considered, from an evolutionary perspective, in terms of underdevelopment of social cognition and affect, centrally involving some combination, and causal conjunction, of reduced social development with increased non-social perception, attention, and cognition. Such social and non-social alterations may have diverse proximate causes, but they appear to commonly converge, psychologically, on reductions in imagination, which can explain both lower levels of sociality and increases in restricted interests and repetitive behaviour. This conceptualization of autism is fully compatible with previously developed psychological models founded on reduced central coherence [66], lower empathizing and higher systemizing [33], and enhanced perceptual function [39].

Third, psychotic affective disorders can be considered as centrally involving dysfunctionally overdeveloped social cognition, affect, and behaviour, expressed as social hypersalience in aspects of psychosis, dysregulated social goal motivation and dominance-seeking in mania, and extremes of negative social emotionality in depression. Each of these disorders, which grade into one another, can best be understood in the individual-level contexts of the developmental causes of negatively valenced and imaginative social salience, and the motivational structure of one's past, current, and future imagined life goals, especially regarding regulation of, and impediments to, success in striving. This framework is fully compatible with current psychological, neurological, cognitive-science-level accounts of psychotic affective conditions (e.g. [30, 43, 49, 51]), but grounds them in evolutionary considerations and in their relationship to the autism spectrum.

Fourth, autism and psychotic affective conditions can be considered and analysed as diametric (opposite) disorders with regard to social development, cognition, affect, and behaviour. This diametric model provides for comprehensive, reciprocal illumination of the diagnoses, causes, and treatments of these disorders, such that insights derived from studying one set of disorders can be applied directly to the other. Most generally, cognitive behavioural treatments for autism should especially focus on enhancing phenotypes that are overdeveloped in psychotic affective conditions, including social imagination, flexible and social salience, and social motivation and goal-seeking. By contrast, treatments for psychotic affective conditions, in addition to focusing more directly on the adaptive, dynamic regulation of social cognitive salience and mood-directed striving, should involve therapies to make perception, cognition, affect, and behaviour relatively 'more autistic'. Similar considerations apply to pharmacological effects: for example, valproate during foetal development represents a well-established human cause, and animal model, of autism [67], but valproate is also used to treat bipolar disorder and schizophrenia [68]; comparably, mGlur5 pathway antagonists are being used to treat fragile X syndrome and autism [69], whereas mGlur5 agonists are being developed to treat schizophrenia [70].

The findings and inferences described here emphasize that evolutionary approaches in medicine, and psychiatry, can offer specific, well-rationalized hypotheses and can help to direct research and treatments along novel and promising paths. Such progress should lead, eventually, to the integration of psychiatry with the standard medical model of disease, as dovetailing evolutionary and proximate approaches to the study of brain development and function uncover the adaptive significance of psychological, cognitive, and affective phenotypes and their neurological and genetic foundations.

References

1. Nesse RM, Stein DJ (2012) Towards a genuinely medical model for psychiatric nosology. BMC Med 10:5. doi:10.1186/1741-7015-10-5
2. Crespi B (2011) One hundred years of insanity: genomic, psychological and evolutionary models of autism in relation to schizophrenia. In: Ritsner M (ed) Handbook of schizophrenia-spectrum disorders, vol I. Springer, Netherlands, pp 163–185
3. Nesse RM (2004) Natural selection and the elusiveness of happiness. Philos Trans R Soc Lond B Biol Sci 359(1449):1333–1347
4. Malhotra D, Sebat J (2012) CNVs: harbingers of a rare variant revolution in psychiatric genetics. Cell 148(6):1223–1241. doi:10.1016/j.cell.2012.02.039
5. Escudero I, Johnstone M (2014) Genetics of schizophrenia. Curr Psychiatry Rep 16(11):502. doi:10.1007/s11920-014-0502-8
6. Plun-Favreau H, Lewis PA, Hardy J et al (2010) Cancer and neurodegeneration: between the devil and the deep blue sea. PLoS Genet 6(12):e1001257. doi:10.1371/journal.pgen.1001257
7. Power RA, Kyaga S, Uher R et al (2013) Fecundity of patients with schizophrenia, autism, bipolar disorder, depression, anorexia nervosa, or substance abuse vs their unaffected siblings. JAMA Psychiatry 70(1):22–30. doi:10.1001/jamapsychiatry.2013.268
8. Singh S, Kumar A, Agarwal S et al (2014) Genetic insight of schizophrenia: past and future perspectives. Gene 535(2):97–100. doi:10.1016/j.gene.2013.09.110
9. Bourque F, van der Ven E, Fusar-Poli P et al (2012) Immigration, social environment and onset of psychotic disorders. Curr Pharm Des 18(4):518–526
10. Stein DJ (2013) What is a mental disorder? A perspective from cognitive-affective science. Can J Psychiatry 58(12):656–662
11. Curtis VA (2014) Infection-avoidance behaviour in humans and other animals. Trends Immunol 35(10):457–464. doi:10.1016/j.it.2014.08.006
12. Trull TJ, Widiger TA (2013) Dimensional models of personality: the five-factor model and the DSM-5. Dialogues Clin Neurosci 15(2):135–146
13. Crespi B (2008) Language unbound: genomic conflict and psychosis in the origin of modern humans. In: Hughes D, D'Ettorre P (eds) Sociobiology of communication: an interdisciplinary perspective. Oxford Universtiy Press, Oxford, pp 225–248
14. Crespi B, Leach E (2015) The evolutionary biology of human neurodevelopment: evo-neuro-devo comes of age. In: Boughner J, Rolian C (eds) Evolutionary developmental anthropology. Wiley, New York (in press)
15. Johnson W, Bouchard TJ Jr (2007) Sex differences in mental abilities: g masks the dimensions on which they lie. Intelligence 35:23–39
16. Nettle D (2007) Empathizing and systemizing: what are they, and what do they contribute to our understanding of psychological sex differences? Br J Psychol 98:237–255
17. Liljenström H (2003) Neural stability and flexibility: a computational approach. Neuropsychopharmacology 28(Suppl 1):S64–S73
18. Jack AI, Dawson AJ, Begany KL et al (2013) fMRI reveals reciprocal inhibition between social and physical cognitive domains. Neuroimage 66C:385–401

19. Kravariti E, Touloupoulou T, Mapua-Filbey F et al (2006) Intellectual asymmetry and genetic liability in first-degree relatives of probands with schizophrenia. Br J Psychiatry 188:186–187

20. Brosnan M, Ashwin C, Walker I et al (2010) Can an 'Extreme Female Brain' be characterised in terms of psychosis? Pers Indiv Diff 49(7):738–742

21. Dinsdale N, Crespi BJ (2013) The borderline empathy paradox: evidence and conceptual models for empathic enhancements in borderline personality disorder. J Pers Disord 27 (2):172–195

22. Crespi B, Foster K, Úbeda F (2014) First principles of Hamiltonian medicine. Philos Trans R Soc Lond B Biol Sci 369:20130366

23. Crespi B (2010) The strategies of the genes: genomic conflicts, attachment theory and development of the social brain. In: Petronis A, Mill J (eds) Brain, behaviour and epigenetics. Springer, Berlin, pp 143–167

24. Crespi B, Badcock C (2008) Psychosis and autism as diametrical disorders of the social brain. Behav Brain Sci 31(3):241–261; discussion 261–320

25. Haig D (2014) Coadaptation and conflict, misconception and muddle, in the evolution of genomic imprinting. Heredity (Edinb) 113(2):96–103

26. Haig D, Ubeda F, Patten MM (2014) Specialists and generalists: the sexual ecology of the genome. Cold Spring Harb Perspect Biol 6(9):a017525. doi:10.1101/cshperspect.a017525, pii:a017525

27. Soni S, Whittington J, Holland AJ et al (2008) The phenomenology and diagnosis of psychiatric illness in people with Prader-Willi syndrome. Psychol Med 38(10):1505–1514

28. Hirstein W, Iversen P, Ramachandran VS (2001) Autonomic responses of autistic children to people and objects. Proc Biol Sci 268(1479):1883–1888

29. Russo DA, Stochl J, Painter M et al (2014) Trauma history characteristics associated with mental states at clinical high risk for psychosis. Psychiatry Res. doi:10.1016/j.psychres.2014. 08.028, pii:S0165-1781(14)00725-2

30. Kapur S (2003) Psychosis as a state of aberrant salience: a framework linking biology, phenomenology, and pharmacology in schizophrenia. Am J Psychiatry 160(1):13–23

31. Nesse RM, Jackson ED (2011) Evolutionary foundations for psychiatric diagnosis: making DSM-V valid. In: De Block A, Adriaens P (eds) Maladapting minds: philosophy, psychiatry, and evolutionary theory. Oxford University Press, Oxford, pp 167–191

32. Brunsdon VE, Happé F (2014) Exploring the 'fractionation' of autism at the cognitive level. Autism 18(1):17–30. doi:10.1177/1362361313499456

33. Baron-Cohen S, Lombardo MV, Auyeung B et al (2011) Why are autism spectrum conditions more prevalent in males? PLoS Biol 9(6):e1001081. doi:10.1371/journal.pbio. 1001081

34. Heil KM, Schaaf CP (2013) The genetics of Autism Spectrum Disorders–a guide for clinicians. Curr Psychiatry Rep 15(1):334. doi:10.1007/s11920-012-0334-3

35. Woodard CR, Van Reet J (2011) Object identification and imagination: an alternative to the meta-representational explanation of autism. J Autism Dev Disord 41(2):213–226

36. Crespi B (2013) Developmental heterochrony and the evolution of autistic perception, cognition and behaviour. BMC Med 11:119

37. Ploog BO (2010) Stimulus overselectivity four decades later: a review of the literature and its implications for current research in autism spectrum disorder. J Autism Dev Disord 40 (11):1332–1349

38. Frith U (2012) Why we need cognitive explanations of autism. Q J Exp Psychol (Hove) 65 (11):2073–2092. doi:10.1080/17470218.2012.697178

39. Mottron L, Dawson M, Soulières I et al (2006) Enhanced perceptual functioning in autism: an update, and eight principles of autistic perception. J Autism Dev Disord 36(1):27–43

40. Rutter M (1972) Childhood schizophrenia reconsidered. J Autism Dev Disord 2:315–337

41. Wing L, Gould J (1979) Severe impairments of social interaction and associated abnormalities in children: epidemiology and classification. J Autism Dev Disord 9:11–29

42. Doherty JL, Owen MJ (2014) Genomic insights into the overlap between psychiatric disorders: implications for research and clinical practice. Genome Med 6(4):29. doi:10.1186/gm546

43. Winton-Brown TT, Fusar-Poli P, Ungless MA et al (2014) Dopaminergic basis of salience dysregulation in psychosis. Trends Neurosci 37(2):85–94. doi:10.1016/j.tins.2013.11.003

44. Cook J, Barbalat G, Blakemore SJ (2012) Top-down modulation of the perception of other people in schizophrenia and autism. Front Hum Neurosci 6:175. doi:10.3389/fnhum.2012.00175

45. Howes OD, Murray RM (2014) Schizophrenia: an integrated sociodevelopmental-cognitive model. Lancet 383(9929):1677–1687. doi:10.1016/S0140-6736(13)62036-X

46. Nesse RM, Ellsworth PC (2009) Evolution, emotions, and emotional disorders. Am Psychol 64(2):129–139

47. Keller MC, Nesse RM (2005) Is low mood an adaptation? Evidence for subtypes with symptoms that match precipitants. J Affect Disord 86(1):27–35

48. Johnson SL (2005) Mania and dysregulation in goal pursuit: a review. Clin Psychol Rev 25 (2):241–262

49. Johnson SL, Fulford D, Carver CS (2012) The double-edged sword of goal engagement: consequences of goal pursuit in bipolar disorder. Clin Psychol Psychother 19(4):352–362

50. Johnson SL, Carver CS (2012) The dominance behavioural system and manic temperament: motivation for dominance, self-perceptions of power, and socially dominant behaviours. J Affect Disord 142(1–3):275–282

51. Johnson SL, Leedom LJ, Muhtadie L (2012) The dominance behavioural system and psychopathology: evidence from self-report, observational, and biological studies. Psychol Bull 138(4):692–743. doi:10.1037/a0027503

52. Coryell W, Endicott J, Keller M et al (1989) Bipolar affective disorder and high achievement: a familial association. Am J Psychiatry 146(8):983–988

53. Higier RG, Jimenez AM, Hultman CM et al (2014) Enhanced neurocognitive functioning and positive temperament in twins discordant for bipolar disorder. Am J Psychiatry. doi:10.1176/appi.ajp.2014.13121683

54. Nettle D (2001) Strong imagination: madness, creativity and human nature. Oxford University Press, Oxford

55. Nettle D (2006) Schizotypy and mental health amongst poets, visual artists, and mathematicians. J Res Pers 40(6):876–890

56. Burns JK (2004) An evolutionary theory of schizophrenia: cortical connectivity, metarepresentation, and the social brain. Behav Brain Sci 27(6):831–855; discussion 855–885

57. Simeonova DI, Chang KD, Strong C et al (2005) Creativity in familial bipolar disorder. J Psychiatr Res 39(6):623–631

58. Carson SH (2011) Creativity and psychopathology: a shared vulnerability model. Can J Psychiatry 56(3):144–153

59. Bilder RM, Knudsen KS (2014) Creative cognition and systems biology on the edge of chaos. Front Psychol 5:1104. doi:10.3389/fpsyg.2014.01104

60. Walker CM, Gopnik A (2013) Causality and imagination. In: Taylor M (ed) The development of imagination. Oxford University Press, New York, pp 342–358

61. Kanner L (1965) Infantile autism and the schizophrenias. Behav Sci 10(4):412–420

62. Crespi B, Stead P, Elliot M (2010) Comparative genomics of autism and schizophrenia. Proc Natl Acad Sci USA 107(Suppl 1):1736–1741

63. Crespi B, Crofts HJ (2012) Association testing of copy number variants in schizophrenia and autism spectrum disorders. J Neurodev Disord 4(1):15. doi:10.1186/1866-1955-4-15

64. Byars SG, Stearns SC, Boomsma JJ (2014) Opposite risk patterns for autism and schizophrenia are associated with normal variation in birth size: phenotypic support for hypothesized diametric gene-dosage effects. Proc Biol Sci 281(1794):20140604

65. van Os J (2009) 'Salience syndrome' replaces 'schizophrenia' in DSM-V and ICD-11: psychiatry's evidence-based entry into the 21st century? Acta Psychiatr Scand 120(5):363–372. doi:10.1111/j.1600-0447.2009.01456.x

66. Happé F, Frith U (2006) The weak coherence account: detail-focused cognitive style in autism spectrum disorders. J Autism Dev Disord 36(1):5–25

67. Markram K, Markram H (2010) The intense world theory—a unifying theory of the neurobiology of autism. Front Hum Neurosci 4:224

68. Haddad PM, Das A, Ashfaq M et al (2009) A review of valproate in psychiatric practice. Expert Opin Drug Metab Toxicol 5(5):539–551. doi:10.1517/17425250902911455

69. Lozano R, Hare EB, Hagerman RJ (2014) Modulation of the GABAergic pathway for the treatment of fragile X syndrome. Neuropsychiatr Dis Treat 10:1769–1779

70. Matosin N, Newell KA (2013) Metabotropic glutamate receptor 5 in the pathology and treatment of schizophrenia. Neurosci Biobehav Rev 37(3):256–268

71. Rees E, Walters JT, Georgieva L et al (2014) Analysis of copy number variations at 15 schizophrenia-associated loci. Br J Psychiatry 204(2):108–114

72. Rees E, Kirov G, Sanders A et al (2014) Evidence that duplications of 22q11.2 protect against schizophrenia. Mol Psychiatry 19(1):37–40

73. Brunetti-Pierri N, Berg JS, Scaglia F et al (2008) Recurrent reciprocal 1q21.1 deletions and duplications associated with microcephaly or macrocephaly and developmental and behavioural abnormalities. Nat Genet 40(12):1466–1471

74. Qureshi AY, Mueller S, Snyder AZ et al (2014) Opposing brain differences in 16p11.2 deletion and duplication carriers. J Neurosci 34(34):11199–11211. doi:10.1523/JNEUROSCI.1366-14.2014

75. Chaste P, Sanders SJ, Mohan KN et al (2014) Modest impact on risk for autism spectrum disorder of rare copy number variants at 15q11.2, specifically breakpoints 1 to 2. Autism Res 7(3):355–362. doi:10.1002/aur.1378

76. Pathania M, Davenport EC, Muir J et al (2014) The autism and schizophrenia associated gene CYFIP1 is critical for the maintenance of dendritic complexity and the stabilization of mature spines. Transl Psychiatry 4:e374. doi:10.1038/tp.2014.16

77. Courchesne E, Mouton PR, Calhoun ME et al (2011) Neuron number and size in prefrontal cortex of children with autism. JAMA 306(18):2001–2010

78. Baribeau DA, Anagnostou E (2013) A comparison of neuroimaging findings in childhood onset schizophrenia and autism spectrum disorder: a review of the literature. Front Psychiatry 4:175. doi:10.3389/fpsyt.2013.00175

79. Haijma SV, Van Haren N, Cahn W et al (2013) Brain volumes in schizophrenia: a meta-analysis in over 18,000 subjects. Schizophr Bull 39(5):1129–1138. doi:10.1093/schbul/sbs118

80. Hobson RP, Bishop M (2003) The pathogenesis of autism: insights from congenital blindness. Philos Trans R Soc Lond B Biol Sci 358(1430):335–344

81. Ek U, Fernell E, Jacobson L (2005) Cognitive and behavioural characteristics in blind children with bilateral optic nerve hypoplasia. Acta Paediatr 94(10):1421–1426

82. Landgraf S, Osterheider M (2013) To see or not to see: that is the question. The "Protection-Against-Schizophrenia" (PaSZ) model: evidence from congenital blindness and visuo-cognitive aberrations. Front Psychol 4:352

83. Silverstein SM, Wang Y, Keane BP (2013) Cognitive and neuroplasticity mechanisms by which congenital or early blindness may confer a protective effect against schizophrenia. Front Psychol 3:624

84. Brown WA, Cammuso K, Sachs H et al (2003) Autism-related language, personality, and cognition in people with absolute pitch: results of a preliminary study. J Autism Dev Disord 33(2):163–167

85. Mottron L, Bouvet L, Bonnel A et al (2013) Veridical mapping in the development of exceptional autistic abilities. Neurosci Biobehav Rev 37(2):209–228

86. Heaton P, Hudry K, Ludlow A et al (2008) Superior discrimination of speech pitch and its relationship to verbal ability in autism spectrum disorders. Cogn Neuropsychol 25.771–782

87. Heaton P, Williams K, Cummins O et al (2008) Autism and pitch processing splinter skills: a group and subgroup analysis. Autism 12(2):203–219

88. Dohn A, Garza-Villarreal EA, Heaton P et al (2012) Do musicians with perfect pitch have more autism traits than musicians without perfect pitch? An empirical study. PloS One 7(5): e37961

89. Falter CM, Braeutigam S, Nathan R et al (2013) Enhanced access to early visual processing of perceptual simultaneity in autism spectrum disorders. J Autism Dev Disord 43(8):1857–1866. doi:10.1007/s10803-012-1735-1

90. Tavassoli T, Miller LJ, Schoen SA et al (2014) Sensory over-responsivity in adults with autism spectrum conditions. Autism 18(4):428–432

91. Bates TC (2005) The panmodal sensory imprecision hypothesis of schizophrenia: reduced auditory precision in schizotypy. Pers Indiv Diff 38(2):437–449

92. Leitman DI, Foxe JJ, Butler PD et al (2005) Sensory contributions to impaired prosodic processing in schizophrenia. Biol Psychiatry 58(1):56–61

93. Leitman DI, Sehatpour P, Higgins BA et al (2010) Sensory deficits and distributed hierarchical dysfunction in schizophrenia. Am J Psychiatry 167(7):818–827

94. Force RB, Venables NC, Sponheim SR (2008) An auditory processing abnormality specific to liability for schizophrenia. Schizophr Res 103(1):298–310

95. Javitt DC (2009) Sensory processing in schizophrenia: neither simple nor intact. Schizophr Bull 35(6):1059–1064

96. Javitt DC (2009) When doors of perception close: bottom-up models of disrupted cognition in schizophrenia. Annu Rev Clin Psychol 5:249–275

97. Mason OJ, Brady F (2009) The psychotomimetic effects of short-term sensory deprivation. J Nerv Ment Dis 197(10):783–785. doi:10.1097/NMD.0b013e3181b9760b

98. Daniel C, Lovatt A, Mason OJ (2014) Psychotic-like experiences and their cognitive appraisal under short-term sensory deprivation. Front Psychiatry 5:106. doi:10.3389/fpsyt.2014.00106

99. Kohl S, Wolters C, Gruendler TO et al (2014) Prepulse inhibition of the acoustic startle reflex in high functioning autism. PLoS One 9(3):e92372. doi:10.1371/journal.pone.0092372

100. Madsen GF, Bilenberg N, Cantio C et al (2014) Increased prepulse inhibition and sensitization of the startle reflex in autistic children. Autism Res 7(1):94–103. doi:10.1002/aur.1337

101. Swerdlow NR, Light GA, Sprock J et al (2014) Deficient prepulse inhibition in schizophrenia detected by the multi-site COGS. Schizophr Res 152(2–3):503–512

102. Orekhova EV, Stroganova TA (2014) Arousal and attention re-orienting in autism spectrum disorders: evidence from auditory event-related potentials. Front Hum Neurosci 8:34

103. Nagai T, Tada M, Kirihara K et al (2013) Mismatch negativity as a "translatable" brain marker toward early intervention for psychosis: a review. Front Psychiatry 4:115

104. Todd J, Harms L, Schall U et al (2013) Mismatch negativity: translating the potential. Front Psychiatry 4:171

105. Oberman LM, Hubbard EM, McCleery JP et al (2005) EEG evidence for mirror neuron dysfunction in autism spectrum disorders. Brain Res Cogn Brain Res 24(2):190–198

106. Kana RK, Wadsworth HM, Travers BG (2011) A systems level analysis of the mirror neuron hypothesis and imitation impairments in autism spectrum disorders. Neurosci Biobehav Rev 35(3):894–902. doi:10.1016/j.neubiorev.2010.10.007

107. McCormick LM, Brumm MC, Beadle JN et al (2012) Mirror neuron function, psychosis, and empathy in schizophrenia. Psychiatry Res 201(3):233–239. doi:10.1016/j.pscychresns.2012.01.004

108. Mehta UM, Thirthalli J, Basavaraju R et al (2014) Reduced mirror neuron activity in schizophrenia and its association with theory of mind deficits: evidence from a transcranial magnetic stimulation study. Schizophr Bull 40(5):1083–1094. doi:10.1093/schbul/sbt155

109. Kennedy DP, Redcay E, Courchesne E (2006) Failing to deactivate: resting functional abnormalities in autism. Proc Natl Acad Sci USA 103:8275–8280

110. Buckner RL, Andrews-Hanna JR, Schacter DL (2008) The brain's default network: anatomy, function, and relevance to disease. In: Kingstone A, Miller MB (eds) The year in cognitive neuroscience. Ann NY Acad Sci, New York, pp 1–38. doi:10.1196/annals.1440.011
111. Iacoboni M (2006) Failure to deactivate in autism: the coconstitution of self and other. Trends Cogn Sci 10:431–433
112. Kennedy DP, Courchesne E (2008) Functional abnormalities of the default network during self- and other-reflection in autism. Soc Cogn Affect Neurosci 3(2):177–190
113. Landin-Romero R, McKenna PJ, Salgado-Pineda P et al (2014) Failure of deactivation in the default mode network: a trait marker for schizophrenia? Psychol Med 21:1–11
114. Immordino-Yang MH, Christodoulou JA, Singh V (2012) Rest is not idleness implications of the brain's default mode for human development and education. Perspect Psychol Sci 7 (4):352–364
115. von dem Hagen EA, Stoyanova RS, Baron-Cohen S et al (2013) Reduced functional connectivity within and between 'social' resting state networks in autism spectrum conditions. Soc Cogn Affect Neurosci 8(6):694–701
116. Jung M, Kosaka H, Saito DN et al (2014) Default mode network in young male adults with autism spectrum disorder: relationship with autism spectrum traits. Mol Autism 5:35
117. Whitfield-Gabrieli S, Thermenos HW, Milanovic S et al (2009) Hyperactivity and hyperconnectivity of the default network in schizophrenia and in first-degree relatives of persons with schizophrenia. Proc Natl Acad Sci USA 106(4):1279–1284
118. Tang J, Liao Y, Song M et al (2013) Aberrant default mode functional connectivity in early onset schizophrenia. PLoS One 8(7):e71061. doi:10.1371/journal.pone.0071061
119. Li M, Deng W, He Z, Wang Q, Huang C, Jiang L, Gong Q, Ziedonis DM, King JA, Ma X, Zhang N, Li T (2015) A splitting brain: Imbalanced neural networks in schizophrenia. Psychiatry Res 232(2):145–153. doi:10.1016/j.pscychresns.2015.03.001
120. Broyd SJ, Demanuele C, Debener S et al (2009) Default-mode brain dysfunction in mental disorders: a systematic review. Neurosci Biobehav Rev 33(3):279–296. doi:10.1016/j.neubiorev.2008.09.002
121. Karbasforoushan H, Woodward ND (2012) Resting-state networks in schizophrenia. Curr Top Med Chem 12(21):2404–2414
122. Lombardo MV, Chakrabarti B, Bullmore ET et al (2011) Specialization of right temporo-parietal junction for mentalizing and its relation to social impairments in autism. Neuroimage 56(3):1832–1838
123. Kana RK, Libero LE, Hu CP et al (2014) Functional brain networks and white matter underlying theory-of-mind in autism. Soc Cogn Affect Neurosci 9(1):98–105
124. Wible CG (2012) Hippocampal temporal-parietal junction interaction in the production of psychotic symptoms: a framework for understanding the schizophrenic syndrome. Front Hum Neurosci 6:180
125. Chevallier C, Kohls G, Troiani V et al (2012) The social motivation theory of autism. Trends Cogn Sci 16(4):231–239
126. Baron-Cohen S (2010) Empathizing, systemizing, and the extreme male brain theory of autism. Prog Brain Res 186:167–175. doi:10.1016/B978-0-444-53630-3.00011-7
127. Harkness KL, Washburn D, Theriault JE et al (2011) Maternal history of depression is associated with enhanced theory of mind in depressed and nondepressed adult women. Psychiatry Res 189(1):91–96
128. Baez S, Herrera E, Villarin L et al (2013) Contextual social cognition impairments in schizophrenia and bipolar disorder. PLoS One 8(3):e57664
129. Konstantakopoulos G, Oulis P, Ploumpidis D et al (2014) Self-rated and performance-based empathy in schizophrenia: the impact of cognitive deficits. Soc Neurosci 9(6):590–600
130. Kasari C, Chamberlain B, Bauminger N (2001) Social emotions and social relationships: can children with autism compensate? In: Burack JA, Charman T, Yirmiya N, Zelazo PR (eds) The development of autism: perspectives from theory and research. Lawrence Erlbaum Associates Publishers, Mahwah NJ USA, pp 309–323

131. Kim S, Thibodeau R, Jorgensen RS (2011) Shame, guilt, and depressive symptoms: a meta-analytic review. Psychol Bull 137(1):68–96. doi:10.1037/a0021466

132. Swettenham J, Remington A, Murphy P et al (2014) Seeing the unseen: autism involves reduced susceptibility to inattentional blindness. Neuropsychology 28(4):563–570. doi:10.1037/neu0000042

133. Hanslmayr S, Backes H, Straub S et al (2012) Enhanced resting-state oscillations in schizophrenia are associated with decreased synchronization during inattentional blindness. Hum Brain Mapp 34(9):2266–2275. doi:10.1002/hbm.22064

134. Reed P, McCarthy J (2012) Cross-modal attention-switching is impaired in autism spectrum disorders. J Autism Dev Disord 42(6):947–953. doi:10.1007/s10803-011-1324-8

135. Morris R, Griffiths O, Le Pelley ME et al (2013) Attention to irrelevant cues is related to positive symptoms in schizophrenia. Schizophr Bull 39(3):575–582. doi:10.1093/schbul/sbr192

136. Granger KT, Prados J, Young AM (2012) Disruption of overshadowing and latent inhibition in high schizotypy individuals. Behav Brain Res 233(1):201–208. doi:10.1016/j.bbr.2012.05.003

137. Adams NC, Jarrold C (2009) Inhibition and the validity of the Stroop task for children with autism. J Autism Dev Disord 39(8):1112–1121

138. Westerhausen R, Kompus K, Hugdahl K (2011) Impaired cognitive inhibition in schizophrenia: a meta-analysis of the Stroop interference effect. Schizophr Res 133(1–3):172–181. doi:10.1016/j.schres.2011.08.025

139. South M, Chamberlain PD, Wigham S, Newton T, Le Couteur A, McConachie H, Gray L, Freeston M, Parr J, Kirwan CB, Rodgers J (2014) Enhanced decision making and risk avoidance in high-functioning autism spectrum disorder. Neuropsychology 28(2):222–228

140. Adida M, Maurel M, Kaladjian A, Fakra E, Lazerges P, Da Fonseca D, Belzeaux R, Cermolacce M, Azorin JM (2011) Decision-making and schizophrenia. Encephale 37(Suppl 2):S110–S116

141. Cascio CJ, Foss-Feig JH, Burnette CP et al (2012) The rubber hand illusion in children with autism spectrum disorders: delayed influence of combined tactile and visual input on proprioception. Autism 16(4):406–419. doi:10.1177/1362361311430404

142. Paton B, Hohwy J, Enticott PG (2012) The rubber hand illusion reveals proprioceptive and sensorimotor differences in autism spectrum disorders. J Autism Dev Disord 42(9):1870–1883

143. Palmer CJ, Paton B, Hohwy J et al (2013) Movement under uncertainty: the effects of the rubber-hand illusion vary along the nonclinical autism spectrum. Neuropsychologia 51 (10):1942–1951

144. Park S, Nasrallah HA (2014) The varieties of anomalous self experiences in schizophrenia: splitting of the mind at a crossroad. Schizophr Res 152(1):1–4. doi:10.1016/j.schres.2013.11.036

145. Chance SA (2014) The cortical microstructural basis of lateralized cognition: a review. Front Psychol 5:820

146. Beversdorf DQ, Smith BW, Crucian GP et al (2000) Increased discrimination of "false memories" in autism spectrum disorder. Proc Natl Acad Sci USA 97(15):8734–8737

147. Hillier A, Campbell H, Keillor J et al (2007) Decreased false memory for visually presented shapes and symbols among adults on the autism spectrum. J Clin Exp Neuropsychol 29 (6):610–616

148. Corlett PR, Simons JS, Pigott JS et al (2009) Illusions and delusions: relating experimentally-induced false memories to anomalous experiences and ideas. Front Behav Neurosci 3:53. doi:10.3389/neuro.08.053.2009

149. Kanemoto M, Asai T, Sugimori E et al (2013) External misattribution of internal thoughts and proneness to auditory hallucinations: the effect of emotional valence in the Deese-Roediger-McDermott paradigm. Front Hum Neurosci 7:351. doi:10.3389/fnhum.2013.00351

150. Grant P, Balser M, Munk AJ et al (2014) A false-positive detection bias as a function of state and trait schizotypy in interaction with intelligence. Front Psychiatry 5:135
151. Faust M, Kenett YN (2014) Rigidity, chaos and integration: hemispheric interaction and individual differences in metaphor comprehension. Front Hum Neurosci 8:511
152. Kercood S, Grskovic JA, Banda D et al (2014) Working memory and autism: a review of the literature. Res Autism Spectrum Dis 8(10):1316–1332
153. Ruthsatz J, Urbach JB (2012) Child prodigy: a novel cognitive profile places elevated general intelligence, exceptional working memory and attention to detail at the root of prodigiousness. Intelligence 40(5):419–426
154. Lee J, Park S (2005) Working memory impairments in schizophrenia: a meta-analysis. J Abnorm Psychol 114(4):599–611
155. Silver H, Feldman P, Bilker W et al (2003) Working memory deficit as a core neuropsychological dysfunction in schizophrenia. Am J Psychiatry 160(10):1809–1816
156. Treffert DA (2009) The savant syndrome: an extraordinary condition. A synopsis: past, present, future. Philos Trans R Soc Lond B Biol Sci 364(1522):1351–1357. doi:10.1098/rstb.2008.0326
157. Cardoso-Martins C, da Silva JR (2010) Cognitive and language correlates of hyperlexia: evidence from children with autism spectrum disorders. Read Writ 23(2):129–145
158. Samson F, Mottron L, Soulières I et al (2012) Enhanced visual functioning in autism: an ALE meta-analysis. Hum Brain Mapp 33(7):1553–1581. doi:10.1002/hbm.21307
159. Revheim N, Butler PD, Schechter I et al (2006) Reading impairment and visual processing deficits in schizophrenia. Schizophr Res 87(1–3):238–245
160. Revheim N, Corcoran CM, Dias E et al (2014) Reading deficits in schizophrenia and individuals at high clinical risk: relationship to sensory function, course of illness, and psychosocial outcome. Am J Psychiatry 171(9):949–959
161. Arnott W, Sali L, Copland D (2011) Impaired reading comprehension in schizophrenia: evidence for underlying phonological processing deficits. Psychiatry Res 187(1–2):6–10
162. Williams EL, Casanova MF (2010) Autism and dyslexia: a spectrum of cognitive styles as defined by minicolumnar morphometry. Med Hypotheses 74(1):59–62
163. Brosnan M, Chapman E, Ashwin C (2014) Adolescents with autism spectrum disorder show a circumspect reasoning bias rather than 'jumping-to-conclusions'. J Autism Dev Disord 44(3):513–520. doi:10.1007/s10803-013-1897-5
164. Speechley WJ, Whitman JC, Woodward TS (2010) The contribution of hypersalience to the "jumping to conclusions" bias associated with delusions in schizophrenia. J Psychiatry Neurosci 35(1):7–17
165. Langdon R, Still M, Connors MH et al (2014) Jumping to delusions in early psychosis. Cogn Neuropsychiatry 19(3):241–256. doi:10.1080/13546805.2013.854198
166. Pellicano E, Burr D (2012) When the world becomes 'too real': a Bayesian explanation of autistic perception. Trends Cogn Sci 16(10):504–510
167. Lawson RP, Rees G, Friston KJ (2014) An aberrant precision account of autism. Front Hum Neurosci 8:302
168. Ciaramidaro A, Bölte S, Schlitt S et al (2014) Schizophrenia and autism as contrasting minds: neural evidence for the hypo-hyper-intentionality hypothesis. Schizophr Bull pii:sbu124. [Epub ahead of print]
169. Backasch B, Straube B, Pyka M et al (2013) Hyperintentionality during automatic perception of naturalistic cooperative behaviour in patients with schizophrenia. Soc Neurosci 8(5):489–504. doi:10.1080/17470919.2013.820666
170. Moore JW, Pope A (2014) The intentionality bias and schizotypy. Q J Exp Psychol (Hove) 67(11):2218–2224
171. Bara BG, Ciaramidaro A, Walter H et al (2011) Intentional minds: a philosophical analysis of intention tested through fMRI experiments involving people with schizophrenia, people with autism, and healthy individuals. Front Hum Neurosci 5:7

172. Blijd-Hoogewys EM, van Geert PL, Serra M et al (2008) Measuring theory of mind in children. Psychometric properties of the ToM Storybooks. J Autism Dev Disord 38 (10):1907–1930
173. Clemmensen L, van Os J, Skovgaard AM et al (2014) Hyper-theory-of-mind in children with psychotic experiences. PLoS One 9(11):e113082
174. Dziobek I, Fleck S, Kalbe E et al (2006) Introducing MASC: a movie for the assessment of social cognition. J Autism Dev Disord 36(5):623–636
175. Lahera G, Boada L, Pousa E et al (2014) Movie for the assessment of social cognition (MASC): Spanish validation. J Autism Dev Disord 44(8):1886–1896. doi:10.1007/s10803-014-2061-6
176. Montag C, Dziobek I, Richter IS et al (2011) Different aspects of theory of mind in paranoid schizophrenia: evidence from a video-based assessment. Psychiatry Res 186(2–3):203–209. doi:10.1016/j.psychres.2010.09.006
177. Fretland RA, Andersson S, Sundet K et al (2015) Theory of mind in schizophrenia: Error types and associations with symptoms. Schizophr Res. doi:10.1016/j.schres.2015.01.024, pii: S0920-9964(15)00028-6
178. Sharp C, Pane H, Ha C et al (2011) Theory of mind and emotion regulation difficulties in adolescents with borderline traits. J Am Acad Child Adolesc Psychiatry 50(6):563–573. doi:10.1016/j.jaac.2011.01.017
179. Bird G, Catmur C, Silani G et al (2006) Attention does not modulate neural responses to social stimuli in autism spectrum disorders. Neuroimage 31(4):1614–1624
180. Sasson NJ, Touchstone EW (2014) Visual attention to competing social and object images by preschool children with autism spectrum disorder. J Autism Dev Disord 44(3):584–592. doi:10.1007/s10803-013-1910-z
181. Blake R, Turner LM, Smoski MJ et al (2003) Visual recognition of biological motion is impaired in children with autism. Psychol Sci 14(2):151–157
182. Kim J, Park S, Blake R (2011) Perception of biological motion in schizophrenia and healthy individuals: a behavioral and FMRI study. PLoS One 6(5):e19971. doi:10.1371/journal.pone. 0019971
183. Almeida RA, Dickinson JE, Maybery MT et al (2013) Visual search targeting either local or global perceptual processes differs as a function of autistic-like traits in the typically developing population. J Autism Dev Disord 43(6):1272–1286
184. Kana RK, Liu Y, Williams DL et al (2013) The local, global, and neural aspects of visuospatial processing in autism spectrum disorders. Neuropsychologia 51(14):2995–3003
185. O'Connor JA, Wiffen BD, Reichenberg A et al (2012) Is deterioration of IQ a feature of first episode psychosis and how can we measure it? Schizophr Res 137(1–3):104–109
186. Russell-Smith SN, Maybery MT, Bayliss DM (2010) Are the autism and positive schizotypy spectra diametrically opposed in local versus global processing? J Autism Dev Disord 40:968–977
187. Craig J, Baron-Cohen S (1999) Creativity and imagination in autism and Asperger syndrome. J Autism Dev Disord 29(4):319–326
188. King D, Dockrell J, Stuart M (2014) Constructing fictional stories: a study of story narratives by children with autistic spectrum disorder. Res Dev Disabil 35(10):2438–2449
189. Jamison KR (1993) Touched with fire: manic-depressive illness and the artistic temperament. Free Press, New York
190. Nelson B, Rawlings D (2010) Relating schizotypy and personality to the phenomenology of creativity. Schizophr Bull 36(2):388–399
191. Claridge G, McDonald A (2009) An investigation into the relationships between convergent and divergent thinking, schizotypy, and autistic traits. Pers Indiv Diff 46(8):794–799
192. Kaufman JC (ed) (2014) Creativity and mental illness. Cambridge University Press, Cambridge
193. Hobson JA, Hobson RP, Malik S et al (2013) The relation between social engagement and pretend play in autism. Br J Dev Psychol 31(1):114–127. doi:10.1111/j.2044-835X.2012. 02083.x

194. Jarrold C (2003) A review of research into pretend play in autism. Autism 7(4):379–390
195. Bonne O, Canetti L, Bachar E et al (1999) Childhood imaginary companionship and mental health in adolescence. Child Psychiatry Hum Dev 29(4):277–286
196. Gleason TR, Jarudi RN, Cheek JM (2003) Imagination, personality, and imaginary companions. Soc Behav Pers 31(7):721–737
197. McLewin LA, Muller RT (2006) Attachment and social support in the prediction of psychopathology among young adults with and without a history of physical maltreatment. Child Abuse Negl 30(2):171–191
198. Fernyhough C, Bland K, Meins E et al (2007) Imaginary companions and young children's responses to ambiguous auditory stimuli: implications for typical and atypical development. J Child Psychol Psychiatry 48(11):1094–1101
199. Wheelwright S, Baron-Cohen S (2001) The link between autism and skills such as engineering, maths, physics and computing: a reply to Jarrold and Routh. Autism 5(2):223–227
200. Spek AA, Velderman E (2013) Examining the relationship between Autism spectrum disorders and technical professions in high functioning adults. Res Autism Spectr Disord 7 (5):606–612
201. Dickerson AS, Pearson DA, Loveland KA et al (2014) Role of parental occupation in autism spectrum disorder diagnosis and severity. Res Autism Spectr Disord 8(9):997–1007
202. Nettle D, Clegg H (2006) Schizotypy, creativity and mating success in humans. Proc Biol Sci 273(1586):611–615
203. Rawlings D, Locarnini A (2008) Dimensional schizotypy, autism, and unusual word associations in artists and scientists. J Res Pers 42(2):465–471
204. Campbell BC, Wang SS-H (2012) Familial linkage between neuropsychiatric disorders and intellectual interests. PLoS One 7(1):e30405
205. Durkin MS, Maenner MJ, Meaney FJ et al (2010) Socioeconomic inequality in the prevalence of autism spectrum disorder: evidence from a U.S. cross-sectional study. PLoS One 5(7):e11551
206. Leonard H, Glasson E, Nassar N et al (2011) Autism and intellectual disability are differentially related to sociodemographic background at birth. PLoS One 6(3):e17875
207. Werner S, Malaspina D, Rabinowitz J (2007) Socioeconomic status at birth is associated with risk of schizophrenia: population-based multilevel study. Schizophr Bull 33(6):1373–1378

Chapter 21
Why Are Humans Vulnerable to Alzheimer's Disease?

Daniel J. Glass, M.A. and Prof. Steven E. Arnold, M.D.

Lay Summary Since Alzheimer's disease (AD) is harmful, it is somewhat of a mystery as to why it hasn't been eliminated by natural selection. Typically, AD has been thought to be "invisible" to natural selection because people have already passed on their genes by the time that AD manifests (usually after age 65). But, this hypothesis may not tell the whole story. One possible explanation is that since the E4 form of the *APOE* gene, a strong risk gene for AD, is very similar to that of our ape-like ancestors but is not currently the most common form in humans, it may have been selected against by natural selection. In addition, as humans take care of their extended family (who shares their genes) even when they are no longer able to have children themselves, the ability to remain cognitively functional in old age may have been selected for. A second possibility is that AD exists in humans because it is the inevitable cost of other features that are beneficial to us—our highly neuroplastic brains, for example. If the benefits outweigh the costs, AD may be maintained despite the severe disadvantages experienced by the elderly. Finally, a third consideration about AD is that it is common due to the mismatch between the environment and lifestyle humans typically sustained during their evolutionary history and the contemporary post-industrialized environment, characterized by high-calorie diets, sedentary lifestyle, and a shortage of pathogens (resulting in autoimmune dysfunction)

   An evolutionary approach has implications for rethinking the biological hallmarks of AD. The brains of AD patients have two characteristic signs—abnormal accumulations of a protein called amyloid-β into "plaques" between

---

D.J. Glass (✉)
Department of Psychology, Suffolk University, Boston, MA 02108, USA
e-mail: djglass@suffolk.edu

S.E. Arnold
Department of Neurology, Massachusetts General Hospital,
Harvard University, Cambridge, USA
e-mail: searnold@mgh.harvard.edu

© Springer International Publishing Switzerland 2016
A. Alvergne et al. (eds.), *Evolutionary Thinking in Medicine*,
Advances in the Evolutionary Analysis of Human Behaviour,
DOI 10.1007/978-3-319-29716-3_21

brain cells and accumulation of abnormal tau protein into "tangles" inside neurons. These lesions are commonly thought to be the cause of neuron death in AD, but there is no direct evidence for this in humans. To date, experimental drugs that have targeted plaques have been ineffective and some have hastened the disease. There is a possibility that plaques and tangles may be harmless by-products of the actual destructive processes, or might even serve an adaptive function, such as trapping free-floating amyloid-β in the brain or protecting against a natural but harmful bodily process called oxidative stress. If this is the case, care should be taken when developing clinical therapies, because interrupting one of the body's beneficial responses may make symptoms worse rather than better.

21.1 Introduction

Alzheimer's disease (AD) is a mysterious disorder for which etiology and treatment remain elusive despite over a century of research. The stakes involved are tremendous. In the USA, AD is currently the sixth leading cause of death. Over five million Americans are estimated to have AD, and this number is expected to rise to 13.8 million by the year 2050 [1]. The global prevalence of AD was 26.6 million in 2006 and is estimated to rise to 106.8 million by 2050 [2]. Although industrialized nations with long-life expectancies are most vulnerable, due to AD usually (but not always) being a disease of advanced age, much of the developing world is struggling with the burden of AD as well [3]. The cost of care per year is over $200 billion in the USA alone and this cost more than doubles if the contributions of unpaid caregivers, including family members, are included [1]. The rise of AD incidence and mortality is a looming public health crisis, and despite many completed and ongoing clinical trials, we await an effective treatment.

An evolutionary perspective may help illuminate a way forward. One of the aims of evolutionary approaches to medicine is to identify the ultimate reasons, that is, the original causes, for illness and disease. Perhaps by understanding the evolutionary origins of AD, future researchers and clinicians will be in a better position to treat or prevent this devastating disease. Toward that end, the current chapter discusses some recent evolutionary perspectives on AD and how they may inform public health policy, research, and medical practice.

The most fundamental question that an evolutionary approach can address with regard to AD is "why are humans vulnerable to the disorder?" Disorders that are "harmful, common, and heritable" [4] should normally be eliminated by natural selection; the fact that AD is such a widespread phenomenon invites the question of how it arose and why it has been maintained in human populations over the millennia. Related to this question is the matter of the major risk gene for AD—*apolipoprotein-E*

(*APOE*) and its isoforms. A number of explanations for the maintenance of AD vulnerability in the human lineage relate to the hypothesis that there are reproductive benefits to the genes responsible for AD, which may outweigh the evolutionary costs of the disorder. Another set of hypotheses is related to the mismatch between the contemporary post-industrial environment and the environmental features humans (and thus AD vulnerability genes) encountered across most of their evolutionary history.

Most of these perspectives have implications for the conceptualization of AD, but of most direct relevance to contemporary research into AD pathology and therapeutic interventions are the hypotheses about the nature of AD pathophysiology, specifically amyloid-β (Aβ) plaques and paired helical filament tau (PHFtau) neurofibrillary tangles. These pathological lesions have generally been seen as causal processes in neurodegeneration, but adopting an evolutionary perspective suggests a rethinking of these biological hallmarks of AD as neutral by-products or even ameliorative adaptations rather than harmful lesions. Additionally, perspectives on mechanisms of pathological transmissibility and phylogenetic analyses of AD-like signs can further illuminate how the medical and research community can move toward therapeutic interventions for the disorder.

21.2   Research Findings

The most straightforward answer to the question of why humans are vulnerable to AD (but one which turns out to be, at best, incomplete) is that AD arose as the result of a genetic mutation but, like other diseases of old age, is invisible to natural selection, since it generally imposes its cost long after people have already reproduced [5, 6]. Genetic disorders that are fatal before puberty are rare because their carriers die without passing on their genes, but AD usually manifests later in life, long after most people would have already had children.

This is a parsimonious explanation that would be satisfactory but for several complicating factors. Firstly, there is evidence that AD-related genes and biological processes may confer some fitness disadvantages early in life, long before the typical manifestation of AD, when the force of natural selection is the strongest. The gene shown to be most reliably associated with the risk of developing AD is *APOE*, which is involved in the production of apolipoprotein E, the major lipid carrier in the brain. *APOE* has three common alleles known as E2, E3, and E4; E4 carriers are at the highest risk of developing AD [7, 8]. However, E4 also confers a higher risk of atherosclerotic cardiovascular disease [9, 10], which does manifest earlier in life and sometimes during the reproductive lifespan. Additionally, there is evidence to suggest that the E3 and especially the E2 alleles may protect against the formation of plaques and tangles that may otherwise occur early in life or following a head trauma [11, 12]. During our evolutionary history, these factors may not have been invisible to natural selection at all, if they meant a differential chance of surviving, reproducing, and successfully raising young. The E4 allele may also be associated with a younger age at menopause [13] (although the evidence is equivocal [14]), which would have presented a salient selection pressure in favor of alternate alleles that allowed for longer reproductive periods.

There is a second potential problem with the notion that AD is invisible to natural selection; phylogenetic analysis reveals that the E4 allele is the ancestral form of the *APOE* gene. In other words, the gene corresponding to *APOE* in all nonhuman primates has the same amino acid structure as the human E4 allele, strongly suggesting that the E2 and E3 forms only arose after the human lineage split from that of chimpanzees and bonobos [15, 16]. The fact that less than 30 % of people presently carry one or two copies of the E4 allele [17] indicates that there has either been selection against this allele (perhaps due to the mechanisms discussed above) or genetic drift or selection favoring the E3 allele, which about 95 % of humans carry [17].

If the E4 allele has been selected against, this indicates by definition that it has a net fitness cost relative to E2 and E3 alleles, and thus is, in fact, visible to natural selection. It is also possible, however, that the prevalence of the newer E3 allele relative to E4 is purely due to random genetic drift and not natural selection. If this is the case, it may have been due to a population bottleneck at some point in our species' history, wherein a sharp reduction in the population randomly happened to leave mostly E3 carriers alive and able to reproduce, and left only a relatively small number of E4 alleles in the population. At any rate, the fact that E4 is the ancestral allele means that this risk allele for AD did not abruptly appear and evade natural selection for millions of years, but rather that it has always been present in humanity and has begun to be replaced by the E3 allele (and, to a lesser extent, E2).

A further challenge to the possibility that natural selection is powerless to remove AD susceptibility genes is the grandmother hypothesis; this is the idea that humans, unique among primates, contribute to their genetic fitness not only by providing parental care to children, but also by helping care for their grandchildren as well [18, 19]. In the harsh environment of pre-industrialized societies, any additional measure of childcare, such as feeding or protection, can mean the difference between life and death. For most of our evolutionary history, we lived in such environments. If post-reproductive individuals can reap fitness benefits by assisting in the care of their grandchildren (which in turn helps those children survive and pass on genes), unimpaired functioning in later life is no longer invisible to natural selection. Thus, AD and other diseases of old age may have been selected against if they prevented elderly individuals from caring for their extended families [18]. On the other hand, there is some controversy surrounding the grandmother hypothesis, in particular over whether the fitness benefit of grandmaternal care outweighs the cost of ceased reproduction. It is possible that menopause (i.e., a period of reproductive senescence) was selected for due not to the benefits of grandmothering, but because offspring borne by older females result in less resources available for offspring borne by younger relatives [20]; this is known as the resource competition hypothesis. Nevertheless, Cant and Johnstone [20] suggest that regardless of why reproductive senescence evolved, older females can definitely contribute to the fitness of their grandchildren, meaning that post-reproductive functioning may be selected for regardless of why it evolved. In other words, it is possible that natural selection could theoretically act on post-reproductive females. It is currently unclear exactly how male longevity and reproductive life history tie into this hypothesis, if at all, since there is little evidence that male care leads to increased fitness of grandchildren [21], although continued ability of longer-lived males to provide for families could possibly play a role.

In sum, there are a number of reasons to think that the continued existence of AD vulnerability in humans is due to more than just the declining power of natural selection as we age. The early-life disadvantages conferred by AD vulnerability genes, particularly *APOE*, is one such reason. Another is the fact that the *APOE*-E4 vulnerability allele was present in our prehuman ancestors and had to be either strongly selected against or encounter significant genetic drift to have such a relatively low frequency now. Finally, there is the possibility that the post-reproductive ability to take care of one's grandchildren has been selected for, meaning that cognitive impairment in old age can be selected against even once a person is beyond his or her childbearing years.

21.2.1 AD and Antagonistic Pleiotropy

If AD is heritable and has been selected against at all, natural selection should have eliminated it, but this has not happened yet. A common explanation for the maintenance of AD vulnerability in humans relates to a concept known as *antagonistic pleiotropy*. In antagonistic pleiotropy, a gene has multiple effects on a phenotype, some of which are beneficial for fitness while others are detrimental. If the net fitness effect of the gene is beneficial, it will be maintained despite its negative effects. Naturally, benefits that occur during peak reproductive years heavily outweigh disadvantages that occur after reproductive years. The concept of antagonistic pleiotropy was originally formulated to explain the general bodily deterioration associated with aging [22] (see also [23]) and has been invoked to explain brain aging in particular [24].

In the last several decades, some evidence has accumulated that AD may be the result of—or *APOE* the subject of—antagonistic pleiotropic processes. There is some evidence that the E4 allele may guard against spontaneous abortion [25], cardiovascular reactions to psychological stress [26], and liver damage in the cases of hepatitis C [27]. It is conceivable that the fitness benefit accrued from these and other potential advantages may outweigh the cardiovascular and AD risks conferred by the E4 allele and lead to its maintenance in the gene pool, although there is currently no direct empirical support for this. Finally, there is a body of evidence showing that the E4 allele may confer cognitive advantages when carriers are young, and the cognitive impairments come only later in life [28].

AD might also be the long-term, accumulated cost of the neuroplasticity responsible for humans' characteristically long period of brain development and profound ability to learn. Human synaptic plasticity undoubtedly contributes to our ability to survive, reproduce, and care for the young. According to Bufill et al. [29], the same characteristics of human neurons that allow them to form new associations well into adulthood (such as high levels of aerobic metabolism and partial myelination) also put them at increased vulnerability to oxidative stress. This oxidative stress is hypothesized by Bufill et al. [29] to precede, and perhaps initiate, the cascade of events leading to Alzheimer's pathophysiology and cell death.

Another perspective, elaborated by Reser [30], is that the neurodegeneration seen in AD may have been selected for because it is an adaptive downregulation of brain metabolism. According to Reser [30], as organisms age, *crystallized intelligence*—i.e., existing knowledge and abilities—becomes more important than *fluid intelligence*—the ability to reason abstractly and problem-solve (see [31]). Since the brain is an energetically costly organ, it may be adaptive for the body to divert crucial metabolic resources away from the less crucial processes as the individual ages; in this way, the neurodegeneration process can be selected for the calories that would otherwise be mobilized for less important brain areas are available for survival and reproduction. Note that this hypothesis refers to the preclinical neurometabolic changes that precede AD, not the debilitating condition of AD itself. In Reser's [30] view, it was very rare that early humans would have lived long enough to see this process culminate in AD as we recognize it. Thus, the hypothesis states that while the metabolic precursors of AD [32] may have been selected for, clinical AD is an unintended result of the extended lifespan made possible by modern public health and medicine (see also Mismatch, below). This hypothesis naturally depends on whether the metabolic downregulation seen in the brains of pre-AD individuals is associated with any significant savings of energy—if these changes are not associated with lowered energy expenditure, they could not have evolved as metabolic reduction mechanisms. Longitudinal fluorodeoxyglucose PET (positron emission tomography) neuroimaging studies that measure resting glucose metabolic rates may be able to provide further evidence for Reser's [30] hypothesis.

Antagonistic pleiotropy is one form of balancing selection, in which multiple alleles are maintained in a gene pool because the level of fitness they confer differs under different circumstances. If the *APOE* E4 allele is not subject to some form of balancing selection (thus far, antagonistic pleiotropy is the only form of balancing selection that has been proposed for AD), then the allele's relatively low prevalence may mean it is, in fact, on its way to extinction. This is true whether it was natural selection or genetic drift that caused the takeover of the E3 allele relative to E4. If E4 confers a net disadvantage, natural selection will continue to select against it until it is eliminated. Even if it does not, Keller and Miller [4] believe that genetic drift will eventually eliminate the E4 allele because over time, drift tends to drive the predominant neutral allele, in this case E3, to fixation (i.e., the rarer alleles get crowded out and the more common allele eventually becomes the only one).

21.2.2 Mismatch Hypotheses

Another set of hypotheses regarding AD views the disorder as an unfortunate by-product of mismatch between the environmental features humans have encountered across evolutionary time and the contemporary post-industrialized environment. AD and the *APOE* E4 allele have been linked to cardiovascular disease, hypertension, obesity, and insulin resistance [33, 34]. Not only do these conditions share risk factors such as lack of exercise, smoking, and poor diet, but insulin resistance in the brain is being investigated as an important mediator of the neurodegeneration found in AD [35, 36] Based on these associations, the explanation behind high rates of AD may be similar to evolutionary perspectives on obesity, cardiovascular disease, and diabetes; the sedentary lifestyle and high fat, high sugar diets of the developed world are not what our bodies were evolved to deal with. Very few hunter-gatherers would have regularly expended as few calories daily as a person who drives to work, sits in front of a computer for ten hours, drives home, and goes to bed. Even in light of the evidence that differences in levels of physical activity may not fully account for differences in obesity between industrialized societies and hunter-gatherers [37], sedentary lifestyles may still contribute to related disorders, including AD. Similarly, our appetitive and metabolic systems were designed to seek out the most energy-rich foods available (those naturally high in fats and sugars) and expend the calories in a thrifty manner, since sustenance was not always plentiful. Today, those same systems drive us toward calorie-rich and nutrient-poor foods designed to take advantage of our fat and sugar preferences, and our metabolisms may be too energy efficient for our low levels of physical activity, leading to obesity and the accompanying maladies.

The result of this environmental mismatch for many humans is the metabolic syndrome. This condition is characterized by abdominal obesity, insulin resistance, dyslipidemia, oxidative stress, inflammation, hypertension, atherosclerosis, and risk of cardiovascular disease. The metabolic syndrome is increasingly recognized as an important risk factor for Alzheimer's disease as well. A central feature of the metabolic syndrome is the resistance to the glucose-lowering effects of insulin and chronically elevated levels of insulin. This often, but not always, leads to diabetes. For the brain, this may result in the cascades of pathophysiological processes seen in AD. Vicious cycles of inflammation; oxidative damage to proteins, lipids, and nucleic acid; resistance to insulin and other trophic factors; calcium dysregulation; caspase activation, mitochondrial, lysosomal, and proteasomal dysfunction; Aβ generation; and tau phosphorylation and fibrillization are set-ups that lead to synaptic loss and neuron death [29].

Fox et al. [38] present intriguing preliminary data that AD may share elements of immune disorders, in that the disorder is more common in regions where environmental pathogens are at low levels. The logic behind this *hygiene hypothesis* is that our immune system is evolved to fight a moderate level of pathogens from our environment; in modern environments where pathogen levels are low, the immune system attacks harmless substances (as in allergies) or other bodily tissues (as in autoimmune disorders)—another case of environmental mismatch (See also Chap. 15 in this volume). Not only does AD present inflammatory features reminiscent of autoimmune disorders, but also it shows a similar pattern as well; rates of AD are strongly negatively correlated with regional parasite stress and other related measures [38]. Future studies controlling for variables such as genetic population markers would provide a powerful test of this hypothesis.

21.2.3 Alzheimer's Pathophysiology

Another focus of evolutionary approaches to medicine involves identifying the nature of disease pathology. The hallmark brain lesions of AD are Aβ plaques and PHFtau neurofibrillary tangles. The traditional view has been that plaques and tangles lead to neuronal death [39], but the nature of this relationship is unclear. The amyloid cascade hypothesis posits that oligomers or plaques of Aβ are toxic to cells and directly cause the sequence of events leading to neurodegeneration. The amyloid cascade hypothesis is supported by early-onset familial AD, which is a Mendelian disorder involving mutations in the *presenilin 1*, *presenilin 2*, and *amyloid precursor protein* (*APP*) genes. These abnormalities result in biochemical processes that create high levels of Aβ, leading to the formation of plaques (as well as tangles) and onset of AD before the age of 65, sometimes much earlier. The causative role of Aβ in the non-familial type of AD, known as sporadic AD, is more circumstantial, based mainly on the correlation, quite modest, between dementia symptoms and plaque load, but the amyloid cascade hypothesis has been at the center of AD research for decades nevertheless.

The amyloid cascade hypothesis has led to a line of therapeutic research aimed at either removing Aβ plaques or preventing their formation. Clinical trials with drugs targeting Aβ have been disappointing so far. For instance, one active immunization trial was ineffective at slowing dementia or reducing PHFtau tangles, vascular injury, or total and soluble concentrations of Aβ, even though it dispersed the plaques themselves [40]. Two other recent large trials of passive immunization directed at Aβ failed to show benefit for their primary outcomes [41, 42]. Another trial using an agent that prevented the formation of Aβ had to be terminated early due to the experimental group's declining at a faster rate than the placebo group [43].

While the failure of any particular drug does not disprove the amyloid hypothesis, cumulatively these results do begin to raise the question as to whether Aβ plaques are a "red herring" in the search for an effective AD treatment. Aβ plaques may be downstream to the primary pathological process(es) of AD rather than the instigator, and may or may not be part of the causal chain of events at all. Accumulating evidence suggests that it may be soluble oligomeric Aβ, rather than Aβ plaques per se, that are neurotoxic [44, 45], and Aβ plaques have even been hypothesized to be protective phenomena that may guard against harmful, free-floating amyloid [46–49]. Perhaps, the molecular structure of Aβ is a design feature to facilitate amyloid aggregation into "sinks" that immobilize oligomeric Aβ to limit neuronal damage [50].

Some evidence suggests that Aβ is functional in small amounts (involved in functions such as synaptic transmission, memory, cholesterol transport, motor activity, and neuroplasticity) and only toxic in higher amounts [51–55]. Soscia and colleagues [48] have shown that Aβ has antimicrobial properties in vitro. Castellani et al. [56] point out that Aβ toxicity has only been demonstrated in vitro but has not been shown to be toxic in vivo. These researchers posit that Aβ is an antioxidant released as a compensatory response to oxidative stress. This is an important perspective in light of the evidence that oxidative stress, rather than Aβ, may be the initiating event in the biochemical cascade that eventually leads to Aβ plaques, PHFtau tangles, and cell death [29, 56]. The hypothesis that Aβ is neutral or even beneficial rather than harmful is consonant with the data that cognitively normal individuals may have heavy plaque load but no brain atrophy [57–60], a finding that is troublesome for the hypothesis that excess Aβ causes neurodegeneration.

Neurofibrillary tangles of PHFtau are the other hallmark sign of AD, but their relationship to Aβ is still a matter of debate. Depending on the nature of the study, tangles appear to either precede [61] or follow [62] plaques in the pathological process. The two pathologies might also interact in a destructive cycle to cause neuronal death [63]. Another possibility, however, is that, similar to Aβ plaques, neurofibrillary tangles act as protective aggregations to mitigate neuron damage from PHFtau oligomers. Lee et al. [64] suggest a mechanism by which phosphorylated tau might work along with Aβ as an antioxidant to protect neurons against oxidative damage.

21.2.4   Transmissibility of AD Pathology

Research into the way that Aβ oligomers and especially PHFtau spread through the brain has revealed that "seeds" of misfolded Aβ and tau proteins can spread by acting as templates that spur further misfolding and thus propagation of abnormal proteins from neuron to neuron across synapses [65, 66]. Whether this transmission is a positive or a negative phenomenon depends on the nature and function of tau (see above), but it should be noted that this transmission method is at least superficially analogous to that of prions, the infectious proteins that cause bovine spongiform encephalopathy, Creutzfeldt-Jakob disease (CJD), and a number of other rapid neurodegenerative disorders in humans and nonhumans. Unlike prion diseases, however, it is important to note that there is no evidence for inter-individual transmission of AD via air, physical contact of any sort, blood or ingestion. Prions are structurally abnormal proteins that spread by inducing normal proteins to misfold as well, resulting in severe and fatal neurodegeneration. While prions are not alive and do not possess DNA, they still spread via a Darwinian process involving mutation and selection [67]. Li and colleagues [67] suggest that this tendency to mutate makes CJD and other prion diseases difficult to treat, as drug-resistant proteins are quick to evolve; to date, no effective treatment has been found for prion diseases. It is conceivable that the comparable transmission mechanism of AD pathology may make the development of protein-targeting drugs difficult for the same reason.

21.2.5 Comparative Perspectives

The incidence of AD-like pathology in nonhuman animals is highly pertinent to this discussion (see also Chap. 19), both because phylogenetic relationships are of interest to evolutionary researchers and because animal models are so popular in the study of AD pathology and treatment. Various combinations of the three hallmark signs of AD—Aβ plaques, PHFtau tangles, and age-related cognitive impairment—have been found in a variety of nonhuman species, including apes, monkeys, dogs, bears, whales, birds, some rodents, and fish [68–75], with amyloid plaques being the most common finding. Crucially, however, all three hallmarks have not been observed in the same nonhuman individual [29], so the term "Alzheimer's disease" has not been readily applied to nonhuman cases. When AD-like patterns of pathology (i.e., age-related cognitive decline accompanied by plaques and/or tangles) are found in nonhumans however, the same characteristic almost always applies: The individual in question is generally an older specimen whose lifespan has been prolonged, via domestication or captivity, beyond what it might achieve in the wild [18, 49].

21.3 Implications for Policy and Practice

While it may be too early for evolutionary approaches to provide specific treatment options for AD, the current research does offer some suggestions on how to conceptualize the disorder clinically and some considerations that may point toward potentially effective interventions. For example, as Ashford [16] points out, the evolutionarily salient discovery that E4 is the ancestral *APOE* allele actually has implications for the development of treatments. Conceptualizing E4 as a harmful allele implies a solution based solely on identifying and blocking those effects of E4 that increase AD risk. On the other hand, conceptualizing E4 as a neutral allele and E2 and E3 as protective would more naturally lead to treatments based on inducing ApoE E2 or E3 protein synthesis, developing ApoE E2 or E3 agonists, or—as recent lines of research have begun to examine—changing the configuration of ApoE E4 to make it behave like ApoE E2 or E3 or using peptides that mimic endogenous ApoE E2 or E3 [16, 76]. Furthermore, the reconceptualization of E4, E3, and E2 as "neutral," "somewhat protective," and "most protective," respectively (rather than the current "bad," "neutral," and "protective," respectively) may also inform how clinicians choose to communicate the results of genetic testing to patients.

The implications for how we should conceptualize Aβ are far from conclusive; yet, perspectives from evolutionary approaches to medicine suggest that we at least proceed with caution where amyloid plaques are concerned. Answering ultimate questions about the origination of traits has not been an emphasis of standard modern medical approaches, at least until the recent emphasis on evolutionary approaches to health [77, 78]. Yet, certain implicit assumptions can be identified in the literature and viewed through an evolutionary lens; in the case of AD, plaques and tangles have traditionally been treated as if they are harmful by-products of the brain's biochemistry—that is, toxic proteins that the human brain happens to be exposed and vulnerable to via an accident of its evolution. The amyloid cascade hypothesis views AD lesions in this way, and thus points inexorably toward a solution based upon removing plaques and tangles. By contrast, a perspective that seeks but does not assume ultimate answers about AD lesions encourages caution and further inquiry into the nature of plaques and tangles, to determine whether they are, in fact, harmful by-products, neutral epiphenomena, or beneficial adaptations.

The disappointing results from recent clinical trials that targeted Aβ plaques do not necessarily suggest that such treatments are ineffective or harmful. Yet, the assumption that plaques are themselves the proper target of clinical intervention may be premature, in light of evolutionary perspectives on their pathophysiology. Researchers developing new AD drugs would be advised to consider the theoretical implications of targeting plaques without knowing whether they are a cause of neurodegeneration, an inert byproduct, or a compensatory response to some other "upstream" process such as metabolic dysfunction, inflammation, or oxidative stress. Clinicians who are discussing opportunities for clinical trials with patients should also be aware of the uncertainty regarding the role of Aβ plaques as they present the risks and benefits of enrolling in particular studies.

If the pathogenic protein propagation process [65] is substantiated as an important mode of AD pathology transmission, as it is in prion diseases, the recommendation of Li et al. [67] that upstream processes are better targets for therapeutic interventions than the abnormal proteins themselves may also hold true for AD. In prion disease, stabilizing or reducing the expression of the normal prion protein is more likely to be effective, according to Li et al. [67], than targeting the abnormal and ever-changing misfolded proteins, and the same may hold true for AD; even if Aβ plaques and PHFtau tangles are determined to be injurious lesions, which has not been confirmed in humans, directly attacking them may not be fruitful if their conformation is variable, thwarting specifically targeted drugs. In such a case, modifying upstream processes such as apoliporotein-E or β- and γ-secretases, enzymes that cleave amyloid precursor protein into Aβ, may be more effective at clearing Aβ. Caution is advised, however, as these enzymes play roles in other functions as well. Indeed, while the γ-secretase inhibitor semagacestat was effective at preventing Aβ formation, it resulted in worsening of dementia and other symptoms [43]. This has been widely presumed to be due to off-target activity of the drug, as γ-secretase plays a role in Notch signaling and other pathways. Still, another consideration is that the prevention of Aβ formation itself may have been deleterious.

Observations of AD-like pathology in nonhuman animals have revealed that such pathology is most likely to be found in elderly specimens who lived beyond the typical natural lifespan for their species. It should be noted that a similar condition applies to humans—in the environment in which we evolved, before modern hygiene and healthcare, life expectancy at birth was around 40 years, and 60 years would have been old age [79]. Our ability to use cultural innovations to extend our lives has resulted in an "unnaturally" long lifespan for humans in developed nations. This perspective is, of course, another mismatch hypothesis; AD can be viewed as a disorder revealed by extended lifespans due to modern advances in health. Individuals who live up to age 85 may have as much as a 50 % risk of developing AD; Terry and Katzman have argued that the difference between people who die with dementia and those who die without it is only a matter of timing, and if we all lived to age 130, everybody would have dementia [80]. This "inevitability" model of AD essentially posits that any human brain, and perhaps those of our close nonhuman cousins as well, will succumb to the disease given enough time.

The argument that AD is an inevitable consequence of aging may or may not be borne out by the evidence; if true, however, it is a practical consideration for physicians, researchers, medical ethicists, and policymakers to take into account. As we continue to develop new ways to prolong the health and longevity of our bodies, we may eventually outstrip the ability of our brains and minds to keep up. This fact should not be viewed as bleak, but rather as a call for increased effort and funding for AD research. Between 2010 and 2013, the National Institutes of Health allocated on average around $476 million of funding per year to AD research. Compare this to the $2 billion, $3 billion, and $5.5 billion that went toward research on cardiovascular disease, HIV/AIDS, and cancer, respectively [81]. We join our colleagues in hoping for successful clinical research and cures to these diseases, but those successes will only underscore the need for greater momentum toward the search for an AD cure.

As investigations into the nature of AD and potential therapeutic interventions continue, it is our belief that evolutionary perspectives will only add to the power of this research, theoretically grounding previous findings as well as revealing new ones. While the applications of evolutionary approaches to medicine, and especially to AD, are still in their infancy, they hold promise toward the goal of a better understanding and treatment for the disease.

Glossary

APOE   A gene involved in the expression of a protein called apolipoprotein-E. This protein is implicated in transporting cholesterol through the bloodstream. There are three variants of the protein caused by different forms of the gene, known as E2, E3, and E4 or sometimes ε2, ε3, and ε4. The E3 allele is the most common. The E4 allele is associated with a higher risk of cardiovascular disease and AD than E3, whereas the E2 allele is associated with a reduced risk of AD relative to E3 [8, 82].

Amyloid-β   (pronounced "amyloid-beta"; a.k.a. beta-amyloid, Aβ) One of the proteins involved in the pathology of AD. It can take one of several forms including oligomers, which are small chainlike molecules, and plaques, which are large insoluble aggregates of Aβ that accumulate in the brains of Alzheimer's patients. Its normal function is not well-understood, although it may be involved in immune response, cellular metabolism, or a number of other processes. Aβ is formed when a protein called amyloid precursor protein (APP) is cut by two other proteins, known as β-secretase (beta-secretase) and γ-secretase (gamma-secretase). Aβ in plaque or oligomeric form has traditionally been thought to be one of the causes of the neurodegeneration seen in Alzheimer's disease.

Tau protein   The other protein whose abnormal processing forms the signature neurofibrillary tangle of Alzheimer's disease. Normally, tau functions to stabilize microtubules, which support the cytoskeleton of neurons in the central nervous system. As part of its normal function, tau changes its shape via phosphorylation, a process wherein phosphate groups bind to tau and allow it to change configurations. In Alzheimer's disease, tau becomes saturated with phosphates, a process called hyperphosphorylation, which promotes its misfolding into filamentous structures called paired helical filaments. These paired helical filaments, in turn, aggregate into neurofibrillary (i.e., "nerve fiber") tangles, which, along with plaques, are a marker of Alzheimer's disease. Tau tangles have been thought to play a role in neuron death.

Oxidative stress   A process caused by by-products of metabolism in which so-called reactive oxygen species or free radicals overwhelm the body's ability to neutralize them, potentially causing damage to DNA and proteins. Reactive oxygen species are molecules or ions of oxygen that are missing one electron, making them highly reactive. Oxidative stress may play key roles in a number of neurodegenerative diseases, including Alzheimer's disease.

References

1. Thies W, Bleiler L (2013) Alzheimer's disease facts and figures. Alzheimer's Dement 9(2):208–245
2. Brookmeyer R, Johnson E, Zeigler-Graham K, Arrighi HM (2007) Forecasting the global burden of Alzheimer's disease. Alzheimer's Dement 3(3):186–191

3. Kalaria RN, Maestre GE, Arizaga R, Friedland RP, Galasko D, Hall K, Luchsinger JA, Ogunniyi A, Perry EK, Potocnik F, Prince M, Stewart R, Wilmo A, Zhang Z-X, Antuono P (2008) Alzheimer's disease and vascular dementia in developing countries: prevalence, management, and risk factors. Lancet 7(9):812–826

4. Keller MC, Miller G (2006) Resolving the paradox of common, harmful, heritable metal disorders: Which evolutionary model works best? Behav Brain Sci 29:385–452

5. Charlesworth B (1996) Evolution of senescence: Alzheimer's disease and evolution. Curr Biol 6(1):20–22. doi:10.1016/S0960-9822(02)00411-6

6. Stearns SC, Nesse RM, Govindaraju DR, Ellison PT (2010) Evolutionary perspectives on health and medicine. Proc Natl Acad Sci 107(suppl 1):1691–1695. doi:10.1073/pnas.0914475107

7. Saunders AM, Strittmatter WJ, Schmechel D, George-Hyslop PHS, Pericak-Vance MA, Joo SH, Rosi BL, Gusella JF, Crapper-MacLachan DR, Alberts MJ et al (1993) Association of apolipoprotein E allele epsilon 4 with late-onset familial and sporadic Alzheimer's disease. Neurology 43:1467–1472

8. Corder EH, Saunders AM, Strittmatter WJ, Schmechel D, Gaskell PC, Small GW, Roses AD, Haines JL, Pericak-Vance MA (1993) Gene dose of apolipoprotein E type 4 allele and the risk of Alzheimer's disease in late onset families. Science 261(5123):921–923

9. Katzel LI, Fleg JL, Paidi M, Ragoobarsingh N, Goldberg AP (1993) ApoE4 polymorphism increases the risk for exercise-induced silent myocardial ischemia in older men. Atertio Thromb Vasc Biol 13:1495–1500

10. Song Y, Stampfer MJ, Liu S (2004) Meta-analysis: apolipoprotein E genotypes and risk for coronary heart disease. Ann Intern Med 141(2):137–147. doi:10.7326/0003-4819-141-2-200407200-00013

11. Ghebremedhin E, Schultz C, Braak E, Braak H (1998) High frequency of apolipoprotein E epsilon 4 allele in young individuals with very mild Alzheimer's disease-related neurofibrillary changes. Exp Neurol 153:152–155

12. Nicoll JA, Roberts GW, Graham DI (1995) Apolipoprotein e epsilon 4 allele is associated with deposition of amloid beta-protein following head injury. Nat Med 1:135–137

13. Koochmeshgi J, Hosseini-Mazinani SM, Seifati SM, Hosein-Pur-Nobari N, Teimoori-Toolabi L (2004) Apolipoprotein E genotype and age at menopause. Ann NY Acad Sci 1019:564–567

14. He L-N, Recker RR, Deng H-W, Dvornyk V (2009) A polymorphism of apolipoprotein E (APOE) gene is associated with age at natural menopause in Caucasian females. Maturitas 62:37–41

15. Hanlon CS, Rubinsztein DC (1995) Arginine residues at codons 112 and 158 in the apolipoprotein E gene correspond to the ancestral state in humans. Atherosclerosis 112(1):85–90

16. Ashford JW (2002) Apo E4: is it the absence of good or the presence of bad? J Alzheimer's Dis 4:141–143

17. Hill JM, Bhattacharjee PS, Neumann DM (2007) Apolipoprotein E alleles can contribute to the pathogenesis of numerous clinical conditions including HSV-1 corneal disease. Exp Eye Res 84:801–811

18. Sapolsky RM, Finch CE (2000) Alzheimer's disease and some speculations about the evolution of its modifiers. Ann NY Acad Sci 924:99–103

19. Lahdenperä M, Russell AF, Tremblay M, Lummaa V (2010) Selection on menopause in two premodern human populations: no evidence for the mother hypothesis. Evolution 65(2):476–489

20. Cant MA, Johnstone RA (2008) Reproductive conflict and the separation of reproductive generations in humans. PNAS 105(14):5332–5336

21. Lahdenperä M, Russell AF, Lummaa V (2007) Selection for long lifespan in men: Benefits of grandfathering? Proc R Soc Lond [Biol] 274(1624):2437–2444

22. Williams GC (1957) Pleiotropy, natural selection, and the evolution of senescence. Evol Dev 11:398–411

23. Rose MR (1991) The evolutionary biology of aging. Oxford University Press, New York

24. Martin GM (2002) Gene action in the aging brain: an evolutionary biological perspective. Neurobiol Aging 23:647–654
25. Zetterberg H, Palmér M, Ricksten A, Poirier J, Palmqvist L, Rymo L, Zafiropoulos A, Arvanitis DA, Spandidos DA, Blennow K (2002) Influence of the apolipoprotein E 14 allele on human embryonic development. Neurosci Lett 324:189–192
26. Ravaja N, Räikkönen K, Lyytinen H, Lehtimäki T, Keltikangas-Järvinen L (1997) Apolipoprotein E phenotypes and cardiovascular responses to experimentally induced mental stress in adolescent boys. J Behav Med 20(6):571–587
27. Wozniak MA, Itzhaki RF, Faragher EB, James MW, Ryder SD, Irving WL (2002) Apolipoprotein E-epsilon 4 protects against severe liver disease caused by hepatitis C virus. Hepatology 36(2):456–463
28. Rusted JM, Evans SL, King SL, Dowell N, Tabet N, Tofts PS (2013) APOE e4 polymorphism in young adults is associated with improved attention and indexed by distinct neural signatures. NeuroImage 65:364–373. doi:10.1016/j.neuroimage.2012.10.010
29. Bufill E, Blesa R, Agusti J (2013) Alzheimer's disease: an evolutionary approach. J Anthropol Sci 91:135–157
30. Reser JE (2009) Alzheimer's disease and natural cognitive aging may represent adaptive metabolism reduction programs. Behav Brain Funct 5:13
31. Horn JL, Cattell RB (1967) Age differences in fluid and crystallized intelligence. Acta Psychol (Amst) 26:107–129. doi:10.1016/0001-6918(67)90011-X
32. Haxby JV, Grady CL, Duara R, Schlageter N, Berg G, Rapoport SI (1986) Neocortical metabolic abnormalities precede nonmemory cognitive defects in early alzheimer's-type dementia. Arch Neurol 43(9):882–885. doi:10.1001/archneur.1986.00520090022010
33. Kuusisto J, Koivisto K, Mykkänen L, Helkala E-L, Vanhanen M, Hänninen T, Kervinen K, Kesäniemi YA, Riekkinen PJ, Laakso M (1997) Association between features of the insulin resistance syndrome and Alzheimer's disease independently of apolipoprotein e4 phenotype: Cross sectional population based study. BMJ 315(7115):1045–1049
34. Martins IJ, Hone E, Foster JK, Sunram-Lea SI, Gnjec A, Fuller SJ, Nolan D, Gandy SE, Martins RN (2006) Apolipoprotein E, cholesterol metabolism, diabetes, and the convergence of risk factors for Alzheimer's disease and cardiovascular disease. Mol Psychiatry 11(8):721–736
35. de la Monte SM (2009) Insulin resistance and Alzheimer's disease. BMB Reports 42(8):475–481
36. Talbot K, Wang HY, Kazi H, Han LY, Bakshi KP, Stucky A, Fuino RL, Kawaguchi KR, Samoyedny AJ, Wilson RS, Arvanitakis Z, Schneider JA, Wolf BA, Bennett DA, Trojanowski JQ, Arnold SE (2012) Demonstrated brain insulin resistance in Alzheimer's disease patients is associated with IGF-1 resistance, IRS-1 dysregulation, and cognitive decline. J Clin Invest 122(4):1316–1338
37. Pontzer H, Raichlen DA, Wood BM, Mabulla AZP, Racette SB, Marlowe FW (2012) Hunter-gatherer energetics and human obesity. PLoS ONE 7(7):e40503
38. Fox M, Knapp LA, Andrews PW, Fincher CL (2013) Hygiene and the world distribution of Alzheimer's disease: epidemiological evidence for a relationship between microbial environment and age-adjusted disease burden. Evol Med Publ Health 1:173–186. doi:10.1093/emph/eot015
39. Karran E, Mercken M, Strooper BD (2011) The amyloid cascade hypothesis for Alzheimer's disease: an appraisal for the development of therapeutics. Nat Rev Drug Discov 10(9):698–712
40. Kokjohn TA, Roher AE (2009) Antibody responses, amyloid-beta peptide remnants and clinical effects of AN-1792 immunization in patients with AD in an interrupted trial. CNS Neurol Disord Drug Targets 8(2):88–97
41. Salloway S, Sperling R, Fox NC, Blennow K, Klunk W, Raskind M, Sabbagh MN, Honig LS, Porsteinsson AP, Ferris S, Reichert M, Ketter N, B. N, Guenzler V, Miloslavsky M, Wang D, Lu Y, Lull J, Tudor IC, Liu E, Grundman M, Yuen E, Black R, Brashear HR, Investigators

BaCT (2014) Two phase 3 trials of bapineuzumab in mild-to-moderate Alzheimer's disease. New Engl J Med 370(4):322–333

42. Doody RS, Thomas RG, Farlow M, Iwatsubo T, Vellas B, Joffe S, Kieburtz K, Raman R, Sun X, Aisen PS, Siemers E, Liu-Seifert H, Mohs R (2014) Phase 3 trials of solanezumab for mild-to-moderate alzheimer's disease. New Engl J Med 370(4):311–321. doi:10.1056/NEJMoa1312889

43. Doody RS, Raman R, Farlow M, Iwatsubo T, Vellas B, Joffe S, Kieburtz K, He F, Sun X, Thomas RG, Aisen PS, Siemers E, Sethuraman G, Mohs R (2013) A phase 3 trial of semagacestat for treatment of alzheimer's disease. New Engl J Med 369(4):341–350. doi:10.1056/NEJMoa1210951

44. Lesné S, Kotilinek L, Ashe KH (2008) Plaque-bearing mice with reduced levels of oligomeric amyloid-β assemblies have intact memory function. Neuroscience 151(3):745–749

45. Hefti F, Goure WF, Jerecic J, Iverson KS, Walicke PA, Krafft GA (2013) The case for soluble Aβ oligomers as a drug target in Alzheimer's disease. Trends Pharmacol Sci 34(5):261–266. doi:10.1016/j.tips.2013.03.002

46. Lee H-G, Castellani RJ, Zhu X, Perry G, Smith MA (2005) Amyloid-β in Alzheimer's disease: the horse or the cart? Pathogenic or protective? Int J Exp Pathol 86(3):133–138

47. Lee H-G, Casadesus G, Zhu X, Takeda A, Perry G, Smith MA (2009) Challenging the amyloid cascade hypothesis: senile plaques and amyloid-β as protective adaptations to Alzheimer disease. Ann NY Acad Sci 1019:1–4

48. Soscia SJ, Kirby JE, Washicosky KJ, Tucker SM, Ingelsson M, Hyman BT, Burton MA, Goldstein LE, Duong S, Tanzi RE, Moir RD (2010) The Alzheimer's disease-associated amyloid ß-protein is an antimicrobial peptide. PloS ONE 5:3

49. Glass DJ, Arnold SE (2012) Some evolutionary perspectives on Alzheimer's disease pathogenesis and pathology. Alzheimer's Dement 8(4):343–351. doi:10.1016/j.jalz.2011.05.2408

50. Ewbank DC, Arnold SE (2009) Cool with plaques and tangles. New Engl J Med 360(22):2357–2359

51. Kamenetz F, Tomita T, Hsieh H, Seabrook GR, Borchelt D, Iwatsubo T, Sisodia SS, Malinow R (2003) APP processing and synaptic function. Neuron 37(6):925–937

52. Yao Z-X, Papadopoulos V (2002) Function of ß-amyloid in cholesterol transport: a lead to neurotoxicity. FASEB J 16:1677–1679

53. Puzzo D, Privitera L, Leznik E, Fa M, Staniszewski A, Palmeri A, Arancio O (2008) Picomolar amyloid-β positively modulates synaptic plasticity and memory in hippocampus. J Neurosci 28(53):14537–14545

54. Garcia-Osta A, Alberini CM (2009) Amyloid beta mediates memory formation. Learn Mem 16:267–272

55. Senechal Y, Kelly PH, Dev KK (2008) Amyloid precursor protein knockout mice show age-dependent deficits in passive avoidance learning. Behav Brain Res 186(1):126–132

56. Castellani RJ, H-g Lee, Nunomura A, Perry G, Smith MA (2006) Neuropathology of Alzheimer disease: pathognomonic but not pathogenic. Acta Neuropathol (Berl) 111(6):503–509

57. Bennett DA, Schneider JA, Arvanitakis Z, Kelly JF, Aggarwal NT, Shah RC, Wilson RS (2006) Neuropathology of older persons without cognitive impairment from two community-based studies. Neurology 66(12):1837–1844

58. White L (2009) Brain lesions at autopsy in older Japanese-American men as related to cognitive impairment and dementia in the final years of life: a summary report from the honolulu-asia aging study. J Alzheimers Dis. doi:10.3233/JAD-2009-1178

59. O'Brien RJ, Resnick SM, Zonderman AB, Ferrucci L, Crain BJ, Pletnikova O, Rudow G, Iacono D, Riudavets MA, Driscoll I, Price DL, Martin LJ, Troncoso JC (2009) Neuropathologic studies of the Baltimore longitudinal study of aging (BLSA). J Alzheimers Dis. doi:10.3233/JAD-2009-1179

60. Iacono D, Markesbery WR, Gross M, Pletnikova O, Rudow G, Zandi P, Troncoso JC (2009) The nun study: clinically silent AD, neuronal hypertrophy, and linguistic skills in early life. Neurology 73(9):665–673. doi:10.1212/WNL.0b013e3181b01077

61. Braak H, Braak E (1997) Frequency of stages of Alzheimer-related lesions in different age categories. Neurobiol Aging 18(4):351–357
62. Hardy JA (2003) The relationship between amyloid and tau. J Mol Neurosci 20(2):203–206
63. Eckert A, Schulz KL, Rhein V, Götz J (2010) Convergence of amyloid-β and tau pathologies on mitochondria in vivo. Mol Neurobiol 41(2–3):107–114
64. Lee HG, Perry G, Moreira PI, Garrett MR, Liu Q, Zhu X, Takeda A, Nunomura A, Smith MA (2005) Tau phosphorylation in Alzheimer's disease: pathogen or protector? Trends Mol Med 11(4):164–169
65. Jucker M, Walker LC (2011) Pathogenic protein seeding in alzheimer disease and other neurodegenerative disorders. Ann Neurol 70(4):532–540. doi:10.1002/ana.22615
66. Guo JL, Lee VMY (2014) Cell-to-cell transmission of pathogenic proteins in neurodegenerative diseases. Nat Med 20:130–138
67. Li J, Browning S, Mahal SP, Oelschlegel AM, Weissmann C (2010) Darwinian evolution of prions in cell culture. Science 327(5967):869–872
68. Finch CE, Sapolsky RM (1999) The evolution of Alzheimer disease, the reproductive schedule, and apoE isoforms. Neurobiol Aging 20:407–428
69. Kimura N, Nakamura S, Goto N, Narushima E, Hara I, Shichiri S, Saitou K, Nose M, Hayashi T, Kawamura S, Yoshikawa Y (2001) Senile plaques in an aged western lowland gorilla. Exp Anim 50(1):77–81
70. Rosen RF, Walker LC, LeVine H (2011) PIB binding in aged primate brain: Enrichment of high-affinity sites in humans with Alzheimer's disease. Neurobiol Aging 32(2):223–234
71. Sarasa M, Gallego C (2006) Alzheimer-like neurodegeneration as a probable cause of cetacean stranding. FENS Forum 3
72. Nakayama H, Katayama K, Ikawa A, Miyawaki K, Shinozuka J, Uetsuka K, Nakamura S, Kimura N, Yoshikawa Y, Doi K (1999) Cerebral amyloid angiopathy in an aged great spotted woodpecker (Picoides major). Neurobiol Aging 20(1):53–56
73. Fahlström A, Yu Q, Ulfhake B (2009) Behavioral changes in aging female C57BL/6 mice. Neurobiol Aging
74. Landsberg G, Araujo JA (2005) Behavior problems in geriatric pets. Vet Clin Small Anim 35:675–698
75. Maldonado TA, Jones RE, Norris DO (2000) Distribution of b-amyloid and amyloid precursor protein in the brain of spawning (senescent) salmon: a natural, brain-aging model. Brain Res 858:237–251
76. Yu J-T, Tan L, Hardy JA (2014) Apolipoprotein E in Alzheimer's disease: an update. Ann Rev Neurosci 37:79–100
77. Nesse RM, Williams GC (1994) Why we get sick: the new science of Darwinian medicine. Times Books, New York
78. Nesse RM, Bergstrom CT, Ellison PT, Flier JS, Gluckman P, Govindaraju DR, Niethammer D, Omenn GS, Perlman RL, Schwartz MD, Thomas MG, Stearns SC, Valle D (2010) Making evolutionary biology a basic science for medicine. Proc Natl Acad Sci 107 (suppl 1):1800–1807. doi:10.1073/pnas.0906224106
79. Finch CE (2010) Evolution of the human lifespan and diseases of aging: roles of infection, inflammation, and nutrition. Proc Natl Acad Sci 107(suppl 1):1718–1724. doi:10.1073/pnas. 0909606106
80. Terry RD, Katzman R (2001) Life span and synapses: will there be a primary senile dementia? Neurobiol Aging 22(3):347–348
81. Health UNIo (2014) Estimates of funding for various research, condition, and disease categories (RCDC). http://report.nih.gov/categorical_spending.aspx. Accessed 25 Aug 2014
82. Corder EH, Saunders AM, Risch NJ, Strittmatter WJ, Schmechel DE, Gaskell PC, Rimmler JB, Locke PA, Conneally PM, Schmader KE, Small GW, Roses AD, Haines JL, Pericak-Vance MA (1994) Protective effect of apolipoprotein E type 2 for late-onset alzheimer's disease. Nature Genet 7:180–184

Chapter 22
Evolutionary Approaches to Depression: Prospects and Limitations

Somogy Varga, Ph.D.

Lay Summary Evolutionary psychiatry has emerged to the status of an important theoretical perspective over the last two decades, and it has generated a sizeable volume of theoretical and empirical studies. It is understandable that many are attracted to the application of evolutionary principles to psychiatric phenomena. Some are attracted by the possibility of providing ultimate explanations for certain mental disorders, while others also think that such an approach can help to counterbalance a naïve understanding of mental disorder. As Nesse and Jackson [1: 194] put it, "campaigns to convince the public and practitioners that depression and anxiety are brain diseases have motivated much useful research and have decreased stigma, but they are biologically naive. An evolutionary approach supports a more medical model in which clinicians recognize many symptoms as defenses shaped by natural selection that are aroused by more primary causes, and others arising from defects in the systems that regulate defenses". Nonetheless, while evolutionary psychiatry is assuming an increasing presence within psychiatric science, the "adaptive turn" has also generated a range of criticisms. Many researchers appreciate the contributions that evolutionary explanations offer for a number of mental disorders, but highlight serious problems that different versions of evolutionary explanations face. The investigation in this chapter was limited to addressing two evolutionary approaches to depression, the mismatch explanation and the persistence explanation. Although both accounts exhibit deficiencies, the conclusion that we should reject applications of evolutionary theory to depression is not warranted. Evolutionary psychiatry should be considered as a potential source of knowledge, and its heuristic value in the development of testable assumptions should not be ignored.

S. Varga (✉)
Philosophy Department, University of Memphis, Memphis, USA
e-mail: svarga@memphis.edu

© Springer International Publishing Switzerland 2016
A. Alvergne et al. (eds.), *Evolutionary Thinking in Medicine*,
Advances in the Evolutionary Analysis of Human Behaviour,
DOI 10.1007/978-3-319-29716-3_22

347

22.1   Introduction

Darwin's theory [1] about the evolution and transmutation of the species has advanced into a pivotal concept in biology, and it has increasingly become a valuable source for explaining structural and behavioural variation between species, groups and individuals [2]. Important works by Hamilton [3] and Wilson [4] have helped to establish evolutionary biology as an independent discipline, but it is only recently that this field of research began to narrow the gap between medicine and evolutionary biology [5]. The prevalence of mental disorders such as depression presents a puzzle from an evolutionary point of view, in particular, because the risk profiles of individuals affected by such incapacitating conditions can partly be explained through different genetic make-up [6]. As Adriaens and De Block [7: 134] put it:

> while ethological psychiatrists are mainly interested in understanding mental disorders by observing psychiatric patients and relating their symptoms to behaviour patterns found in other animal species, the second group of evolutionary psychiatrists considers mental disorders to be evolutionary oddities that need explaining. For why is it, they wonder, that natural selection is so slack in getting rid of mental disorders? Biological psychiatrists invariably assume that there are genes involved in man's vulnerability to mental disorders – how come such genes have managed to escape natural selection?

The prevalence of highly incapacitating (fitness-reducing) mental disorders such as schizophrenia, depression, phobias, anxiety and obsessive-compulsive disorder raises crucial and difficult questions. Depression is both prevalent enough throughout history to make appropriate the investigation of its evolutionary origins, and it is an extremely common debilitating condition [8]. However, beside the psychological distress, cognitive and emotional difficulties, patients with depression are more likely to develop diabetes and cardiovascular disease, to commit suicide and to die childless [9–11].

Partly motivated by what appears to be a deeply puzzling fact (i.e. genes causative of mental disorders have escaped natural selection), researchers have begun to systematically investigate mental disorders within the framework of contemporary evolutionary theory. Bringing this research programme under a name, the term "evolutionary psychiatry" was coined in 1985 by MacLean in an influential editorial published in *Psychological Medicine*. MacLean [12: 219] argued that "evolutionary psychiatry provides counteractive leaven for reductionist views inherent in the molecular approach" and envisaged that the domain of evolutionary psychiatry could encompass both microscopic and macroscopic aspects. Shortly after, Cosmides and Tooby [13] put forward a highly influential account of evolutionary psychology, which quickly became an important reference, together with *Evolutionary Psychiatry: A New Beginning* by Stevens and Price. As Stevens and Price [14: 275] maintain in the second edition of their influential book, their work is committed to the idea that "no theory in psychology or psychiatry could hope to possess any lasting value unless it was securely founded on knowledge of the evolution of our species". Because evolutionary approaches were sometimes criticized for being politically motivated,

Stevens and Price [14: 277] explicitly maintain that "to adopt an evolutionary approach is not to espouse a political cause, nor is it an invitation to submit to 'biological determinism' or an encouragement to abandon a proper concern with ethical or value-oriented premises".

Since the publication of MacLean's editorial, evolutionary psychiatry has emerged as a significant theoretical perspective, and the growing number of studies by researchers in this field is beginning to assume a noticeable presence within psychiatry [6, 14, 15, 16, 17, 18, 19, 20]. One of the attractions of evolutionary psychiatry is its potential to offer explanations that identify the ultimate causes of mental disorders. Traditional psychiatric research has mostly focused on proximate causes, shedding light on the relevant processes or structures in individual organisms. While such explanations seek to understand the mechanisms that underpin a trait or behaviour, ultimate explanations are concerned with evolutionary function [21].

22.2 Research Findings: Evolutionary Psychiatry and Depression

When it comes to the applications of evolutionary theory to psychopathology, several types of explanations can be distinguished [22]. To provide a brief outline of the structure of evolutionary explanations in psychiatry and to introduce some reflections on the benefits and problems associated with such approaches, the investigation in this chapter will be limited to addressing two different evolutionary accounts of depression.

(1) Mismatch explanation [23, 24]: mental disorder is connected to mechanisms that were once adaptive, but that in our present environment are best seen as maladaptive.
(2) Persistence explanation [25–27]: some mental disorders qualify as adaptive even in the present environment.

These accounts proceed in opposing directions, but they share the basic idea that the mechanisms activated in depression evolved to manage certain hostile situations. The chapter will not consider another type of evolutionary explanation that is to a certain degree compatible with the mismatch explanation. According to Nettle [18], depression itself is not selected for. Instead, evolution has produced an optimal reactivity of affect systems, with a normal population distribution around this optimum. But individuals situated in the upper tail of the population distribution are vulnerable to depression.

22.2.1 The Mismatch Explanation

The authors in this camp propose that the human mind consists of hierarchically organized systems of very different evolutionary ages (including a "reptilian", a "palaeo-mammalian" and a "neo-mammalian" system) and mechanisms specialized for language and symbolic processing (see [14], Chap. 2). In this view, our minds can be understood as composite integrated assemblages that consist of a number of functionally specialized adaptations that evolved as solutions to different adaptive problems (e.g. foraging and mating). The core of the proposal is that these systems are still active beneath the threshold of consciousness, albeit sometimes acting in conflicting ways [28, 29]. The idea is that depression evolved as an adaptive response to specific problems that arise in the small, status-oriented social groups of our ancestors [16, 25]. However, in a radically changed environment, those evolved patterns of behaviour now promote unfavourable and maladaptive conditions [30].

Stevens and Price [14] argue that certain forms of depression were constructive human responses to situations in which a desired social goal appeared impossible to achieve. It was speculated that depression assisted the restoration of exhausted resources by forcing the individual to withdraw, helping to maximize pay-offs by resource reallocation [16, 23, 31]. In small groups of hunter-gatherers, depressed states might have induced reflection on weaknesses leading to altered behaviours and ultimately to better reproductive chances. Research on serotonin-level down-regulation in depression appears to offer some support for this theory. For instance, when animals change their place in a power hierarchy, their behavioural changes are accompanied by changes in serotonin levels [19, 32, 33, 34]. The characteristic sense of incapability to fulfil tasks, pessimism, behavioural inactivity and the well-documented exaggerated interpretation of the difficulty of a task restrains the depressed individual from allocating resources in demanding activities with low probability of success [35]. Depression-like states occur in animals and humans who have been defeated and lost rank, and the advantage might be that depressive states help the individual to accept the loss of status and lowered rank [28, 36, 37].

In all, some authors maintain that depression is an adaptive response to the loss of status in small social groups. According to the mismatch explanation, while depression might have been a productive strategy in small groups of hunter-gatherers, in contemporary Western societies it is no longer adaptive [16] and associated with decreased reproductive success [14: 79, 38].

22.2.2 The Persistence Explanation

The persistence hypothesis makes the bold claim that depression and the related genetic material are still adaptive in current Western environments. Researchers in this camp of course recognize that depression causes serious pain and distress, but they argue that depression is the expression of an overall adaptation to changed circumstances [25–27]. It is well-known that whether or not a trait is adaptive is a matter of degree and that some adaptations can be associated with fitness benefits as well as fitness costs. Some mechanisms, such as fever, can be understood as adaptations, while they simultaneously reduce metabolism, sexual and social activity [39–41]. In much the same way, the idea is that while depression clearly interrupts normal functioning, its aversive and disruptive characteristics might actually help us understand its adaptive function. Following the same path of thought, Darwin (1859/2003, 431) himself considered depression as adaptive.

If the characteristic rumination and the down-regulation of positive affect systems that characterize depression incite the depressed individual to re-evaluate and abandon impossible or unmanageable undertakings, then depression might be seen as an adaptive response to social circumstances. This view goes back to earlier work by D.A. Hamburg, who maintained that in a case where the individual estimates that the probability of achieving a goal is very low, "the depressive responses can be viewed as adaptive" [42: 240]. Thus, characteristic features in depression such as rumination, down-regulation of positive affect, diminished responsiveness and the lack of motivation may be seen as fostering disengagement from unachievable goals, which at the end could harm the individual. For instance, Andrews and Thompson [43: 623] argue that depressive rumination harbours a beneficial cognitive effect. The point is that depression as a response mechanism is triggered by analytically difficult problems, and depressive rumination helps people generate and evaluate potential solutions. Thus, the persistence explanation claims that the adaptive aspect is that depressive rumination enables the individual to engage in a profound analysis of the triggering problems. Watson and Andrews [27] are well aware of the costs of such solitary rumination, but conjecture that the benefits are great enough to compensate for the costs.

Focusing on post-partum depression, Hagen [26] maintains that depressed mothers obtain greater care from both their partners and their social network. Watson and Andrews [27] extend this idea to depression generally, and argue that depressive responses revitalize social relationships. In this view, adults' depression conveys a plea for help to others and should be understood as involving a type of communication designed to manipulate others into providing resources. Watson and Andrews [27] review evidence that depression is associated with social problems and suggest that depression plays a crucial role in motivating close social partners to provide help and to make concessions in favour of the depressed. Social partners are aware of the costs imposed on them when a partner is depressed [44], and the costs motivate the members of the depressive's social network to make investments that under normal circumstances they would hesitate to make. In this sense, depression may function like an "instrument" to motivate individuals within the network to overcome their reluctance to help [26]. Overall, despite the fact that depression sometimes causes abandonment and produces social deterioration, the main idea is that depression qualifies as adaptive even in the present environment.

22.2.3 Some Challenges

Having explored the main tenets of the mismatch and persistence explanations, the last part of the chapter will be dedicated to a brief survey of some of the most important challenges that they face (for a more complete evaluation, see [15]).

Let me start with a general concern. In the absence of detailed knowledge about selective events, caution is warranted when it comes to the conclusions we are able to draw. This limitation is aggravated by the fact that evolutionary psychiatry displays a comparative deficiency in explaining both individual and cultural differences in the diagnosis and experience of depression. Further limitations arise from the logic of evolutionary explanations. We have noted in the beginning of this chapter that the prevalence of some mental disorders is usually taken to support the view that they are adaptations. Although the meaning of optimization in biology and medicine is not identical, Ravenscroft [46] notes that this argument contains a problematic and suppressed premise, according to which natural selection is a mechanism that not only optimizes systems, but also eliminates imperfections. However, Dawkins [47] maintains that this is not how selection operates. For instance, he demonstrates how the blind spot of the human eye is a maladaptive property that natural selection has not eliminated. In the same way, Ravenscroft [46: 453] holds that "some mental disorders may be maladaptive features of the human cognitive apparatus which are unlikely be driven from the lineage because the brain is trapped on a local optimum".

Let us now turn our attention to the mismatch explanation. A crucial concern is that the mismatch account fails to address why under identical environmental conditions only certain individuals will be afflicted by the condition. And, given the ubiquity of status competitions and status changes in our contemporary societies, why are not more individuals afflicted by the condition? The persistence explanation fares better on this issue, as it is compatible with the view that the way our modern societies have evolved has created conditions under which higher rates of depression are likely to occur. But a limitation of the persistence explanation is that it is difficult to see how depression can be understood as an adaptation to allow optimal functioning if it significantly increases the risk of another depressive episode, diabetes, cardiovascular disease, sexual dysfunction, physical pain and suicide [8, 9]. Individuals with depression have a shorter life expectancy than those who are not suffering from the condition, in part because of the significantly increased risk of dying by suicide [48].

In addition, the fitness enhancing nature of depressive rumination may be questioned. When analysing typical themes that occur in typical ruminations (such as "Why am I such a bad person?"), we see that they are often of hypothetical nature, and it is in many cases doubtful that they can be understood as problems that require analytical attention or that need to be "solved". Also, depressive rumination may not be triggered by complex social problems. Often, there is no apparent external trigger for a depressive episode that can be identified and ruminated on. But even if this were the case, social dilemmas might not have analytical structure that requires a particular type of analytical approach. Furthermore, depressive rumination may not be fitness enhancing at all, as it actually exacerbates and prolongs distress in depression [49], impairs problem-solving capacities and hinders instrumental behaviour. Depressive ruminations appear to play an important part in the development, maintenance and recurrence of depression [50, 51]. Longitudinal studies demonstrate that people engaging in rumination as a reaction to stress are more vulnerable to develop depressive disorders and to have prolonged periods of depression [52–56]. Thus, depressive rumination not only fails to be solution-oriented, but also directly hampers problem-solving abilities and tends to result in an assessment of problems as over-powering and impossible to solve [57, 58].

22.3   Implications for Policy and Practice

When introducing evolutionary psychiatry in the beginning of this chapter, I have noted that some psychiatrists are concerned that although the conceptual framework of evolutionary psychiatry assists psychiatry's understanding of disorders, it does not directly contribute to the creation of practical applications. Although Stevens and Price [14: 278] close their book by expressing hope that evolutionary psychiatry will eventually be able to help providing effective measures for the prevention and treatment of mental disorders, sceptics may emphasize that it is difficult to claim that progress in this area has been satisfactory. Although effective clinical applications are still rare, some recent developments nevertheless indicate that ideas from evolutionary psychiatry can yield practical benefits in the form of clinical applications. For example, in recent years, a treatment for depression has been developed based on the mismatch theory of depression. Taking seriously the enormous differences between past and modern diet, physical environments and social relations, Ilardi has developed a treatment for depression that simulates ancestral (or just pre-industrial) living conditions, emphasizing exercise, exposure to sunlight, good sleep hygiene and anti-ruminative activity. The therapy had a surprisingly high success rate in reducing the symptoms of depression [59, 60].

Glossary

Rumination   A way of responding to distress that is characterized by a com-
               pulsively focused attention on the symptoms of distress and their
               possible causes and consequences. Usually, the rumination in
               individuals with depression is negative in valence

References

1. Nesse RM, Jackson ED (2011) Evolutionary foundations for psychiatric diagnosis: making DSMV valid, In: Maladapting minds, pp 173–94
2. Darwin C (1859) The origin of species by means of natural selection or the preservation of favoured races in the struggle for life. The Modern Library, Random House, New York
3. Hamilton WD (1964) The genetical evolution of social behaviour II. J Theor Biol 7:17–32

4. Wilson EO (1975) Sociobiology. Belknap Press, Boston
5. Elton S, O'Higgins P (2008) Introduction to medicine and evolution: current applications, future prospects. Taylor & Francis, Boca Raton
6. Brüne M (2008) Textbook of evolutionary psychiatry: the origins of psychopathology. Oxford University Press, Oxford
7. Adriaens PR, De Block, A (2010) The evolutionary turn in psychiatry: a historical overview. Hist Psychiatry 21:132–43
8. Gonzalez HM, Vega WA, Williams DR et al (2010) Depression care in the United States: too little for too few. Arch Gen Psychiatry 67(1):37–46
9. Lonnqvist JK (2009) Suicide. In: Gelder MN, Andreasen, Lopez-Ibor J, Geddes J (eds) New Oxford textbook of psychiatry, 2nd edn. Oxford University Press, Oxford, pp 951–957
10. Knol MJ, Twisk JW, Beekman AT, Heine RJ, Snoek FJ, Pouwer F (2006) Depression as a risk factor for the onset of type 2 diabetes mellitus. A meta-analysis. Diabetologia 49(5):837–845
11. Alboni P, Favaron E, Paparella N, Sciammarella M, Pedaci M (2008) Is there an association between depression and cardiovascular mortality or sudden death? J Cardiovascular Med (Hagerstown, Md.) 9(4):356–362
12. MacLean PD (1985) Evolutionary psychiatry and the triune brain. Psychol Med 15:219–221
13. Cosmides L, Tooby J (1987) From evolution to behavior: evolutionary psychology as the missing link. In: Dupre J (ed) The latest on the best. Essays on evolution and optimality. MIT Press, Cambridge, pp 277–306
14. Stevens A, Price J (2000) Evolutionary psychiatry, 2nd edn. Routledge, New York
15. Burns J (2007) The descent of madness: evolutionary origins of psychosis and the social brain. Routledge, London
16. Nesse RM, Williams GC (1995) Why we get sick. Times Books, New York
17. McGuire MT, Troisi A, Raleigh MM (1997) Depression in evolutionary context. In: Baron-Cohen S (ed) The maladapted mind. Erlbaum, Hillsdale
18. Nettle D (2004) Evolutionary origins of depression: a review and reformulation. J Affect Disord 81:91–102
19. McGuire M, Troisi A (1998) Darwinian psychiatry. Oxford University Press, New York
20. Brüne M et al (2012) The crisis of psychiatry—insights and prospects from evolutionary theory. World Psychiatry 11(1):55–57
21. Scott-Phillips TC, Dickins TE, West SA (2011) Evolutionary theory and the ultimate/proximate distinction in the human behavioural sciences. Perspect Psychol Sci 6 (1):38–47
22. Murphy D (2005) Can evolution explain insanity? Biol Philos 20(4):745–766
23. Nesse RM (2000) Is depression an adaptation? Arch Gen Psychiatry 57:14–20
24. Gilbert P, Allan S (1998) The role of defeat and entrapment (arrested flight) in depression: exploring an evolutionary view. Psychol Med 28:585–598
25. Price J, Sloman L, Gardner R, Gilbert P, Rohde P (1994) The social competition hypothesis of depression. Br J Psychiatry 164:309–15
26. Hagen EH (1999) The functions of postpartum depression. Evol Human Behav 20:325–359
27. Watson PJ, Andrews PW (2002) Towards a revised evolutionary adaptationist analysis of depression: the social navigation hypothesis. J Affect Dis 72:1–14
28. Stevens A, Price J (1996) Evolutionary psychiatry: a new beginning. Routledge, London
29. Cosmides L, Tooby J (1999) Toward an evolutionary taxonomy of treatable conditions. J Abnorm Psychol 108:453–464
30. Baptista T, Aldana E, Angeles F, Beaulieu S (2008) Evolution theory: an overview of its applications in psychiatry. Psychopathology 41:17–27
31. Schmale A (1973) The adaptive role of depression in health and disease. In: Scott JP, Senay FC (eds) Separation and depression. King, Baltimore, MD, p 187
32. Kravitz EA (2000) Serotonin and aggression: insights gained from a lobster model system and speculations on the role of amine neurons in a complex behavior. J Comp Physiol 186:221–238

33. Drummond JM, Issa FA, Song CK, Heberholz J, Yeh S-R, Edwards DH (2002) Neural mechanisms of dominance hierarchies in crayfish. In: Wiese K (ed) The crustacean nervous system. Springer, Berlin, pp 124–135

34. Grant KA, Shively CA, Nader MS, Ehrenkaufer RL, Line SW, Morton TE, Gage HD, Mach RH (1998) Effects of social status on striatal dopamine D2 receptor binding characteristics in cynomologus monkeys assessed with positron emission tomography. Synapse 29:80–83

35. Sloman L, Gilbert P, Hasey G (2003) Evolved mechanisms in depression: the role and interaction of attachment and social rank in depression. J Affect Disord 74(2):107–121

36. Gilbert P (1992) Depression: the evolution of powerlessness. Lawrence Erlbaum Associates, London

37. Gilbert P, Miles JNV (2000) Evolution, genes, development and psychopathology. Clin Psychol Psychother 7:246–255

38. Sharpley CF, Bitsika V (2010) Is depression "Evolutionary" or just "Adaptive"? A comment. Depression Res Treat (631502). doi:10.1155/2010/631502

39. Hasday JD, Fairchild KD, Shanholtz C (2000) The role of fever in the infected host. Microbes Infect 2:1891–1904

40. Cosmides L, Tooby J (2000) Consider the source: the evolution of adaptations for decoupling and metarepresentation. In: Sperber D (ed) Metarepresentations. Oxford University Press, Oxford

41. LeDoux J (1996) The emotional brain: the mysterious underpinnings of emotional life. Simon and Schuster, New York

42. Hamburg DA (1974) Coping behaviour in life-threatening circumstances. Psychother Psychosom 23(13–25):240

43. Andrews PWJ, Thomson A Jr (2009) The bright side of being blue: depression as an adaptation for analyzing complex problems. Psychol Rev 116(3):620–654

44. Segrin C, Dillard JP (1992) The interactional theory of depression: a meta-analysis of the research literature. J Soc Clin Psychol 11:43–70

45. Varga S (2011) Evolutionary psychiatry and depression. testing two hypotheses. Med Healthcare Philos 15(1):41–52

46. Ravenscroft I (2012) What's Darwin got to do with it? The role of evolutionary theory in psychiatry. Biol Philos 27:449–460

47. Dawkins R (1986) The blind watchmaker. W.W. Norton & Company, New York

48. Cassano P, Fava M (2002) Depression and public health: an overview. J Psychosom Res 53 (4):849–857

49. Nolen-Hoeksema S (1991) Responses to depression and their effects on the duration of depressive episodes. J Abnorm Psychol 100:569–582

50. Treynor W, Gonzalez R, Nolen-Hoeksema S (2003) Rumination reconsidered: a psychometric analysis. Cogn Ther Res 27:247–259

51. Wenzlaff RM, Luxton DD (2003) The role of thought suppression in depressive rumination. Cogn Ther Res 27:293–308

52. Just N, Alloy LB (1997) The response styles theory of depression: tests and an extension of the theory. J Abnormal Psychol 106:221

53. Kuehner C, Weber I (1999) Responses to depression in unipolar depressed patients: an investigation of Nolen-Hoeksema's response styles theory. Psychol Med 29:1323–1333

54. Nolan SA, Roberts JE, Gotlib IH (1998) Neuroticism and ruminative response style as predictors of change in depressive symptomatology. Cogn Ther Res 22:445–455

55. Nolen-Hoeksema S (2000) The role of rumination in depressive disorders and mixed anxiety/depressive symptoms. J Abnorm Psychol 109:504–511

56. Nolen-Hoeksema Wisco, Lyubomirsky S (2005) Rethinking rumination. Perspect Psychol Sci 3:400–424

57. Lyubomirsky S, Tucker KL, Caldwell ND, Berg K (1999) Why ruminators are poor problem solvers: clues from the phenomenology of dysphoric rumination. J Pers Soc Psychol 77:1041–1060

58. Donaldson C, Lam D (2004) Rumination, mood and social problem-solving in major depression. Psychol Med 34:1309–1318
59. Ilardi SS, Jacobson et al (2007) Therapeutic lifestyle change for depression: results from a randomized controlled trial. Paper presented at the annual meeting of the Association for Behavioral and Cognitive Therapy, Philadelphia, PA
60. Confer C, Easton J et al (2010) Evolutionary psychology controversies, questions, prospects, and limitations. Am Psychol 65(2):110–126

Chapter 23
The Ups and Downs of Placebos

Pete C. Trimmer, Ph.D. and Prof. Alasdair I. Houston, Ph.D.

> To everyone is given the key to heaven; the same key opens the gates of hell.
>
> Ancient Proverb

Lay Summary Patients are sometimes given dummy tablets or treatments, termed placebos, that sometimes seemingly affect patients' health despite such treatments providing no direct medicinal benefit. The effect is often regarded as somewhat mysterious: why would evolution by natural selection not have resulted in individuals recovering their health as soon as possible, with or without a placebo? Despite the effect being poorly understood, it is often assumed that any 'placebo effect' will be positive. Here, we review the possible effects of placebos from an evolutionary perspective. We identify situations where evolutionary theory predicts that placebos should have positive or negative effects on patients. The outcomes will typically depend on numerous factors relating to the condition and, more specifically, beliefs of the patient. The evolutionary perspective highlights the many trade-offs involved in how the immune system can be affected by placebos, and the potential danger of confusing symptoms with an underlying ailment when considering effects.

P.C. Trimmer (✉) · A.I. Houston
School of Biological Sciences, University of Bristol, Bristol, UK
e-mail: pete.trimmer@gmail.com

A.I. Houston
e-mail: a.i.houston@bristol.ac.uk

© Springer International Publishing Switzerland 2016
A. Alvergne et al. (eds.), *Evolutionary Thinking in Medicine*,
Advances in the Evolutionary Analysis of Human Behaviour,
DOI 10.1007/978-3-319-29716-3_23

23.1 Introduction: The Trade-Off Between Current Health and Other Factors

It is notoriously difficult to define the term 'placebo' satisfactorily [1]. The word is most often used in the context of the 'placebo effect' (frequently regarded as the improvement in health resulting from having taken a pill containing only inert substances), and many definitions of the term assume or imply that any effect of a placebo will be positive. For instance, the on-line Google dictionary defines a placebo as, 'A *harmless* pill, medicine, or procedure prescribed more for the psychological benefit to the patient than for any physiological effect' (our italics). Further such definitions are summarised in Appendix A.

Clearly, it is best not to define a possible treatment by its intended outcomes; the decision of whether to prescribe a placebo would surely then follow from its presupposed effect. Here, we define a placebo simply as, 'a treatment which may have an effect on health only indirectly, through modifying the patient's beliefs'. The beliefs need not be conscious and may be true or false (note that our use contrasts with the anthropological lexicon, where beliefs are generally regarded as conscious [2]).

Many positive effects have been attributed to placebos, such as pain relief. Some of the positive findings are likely to be due to reporting bias. For instance, Okaïs et al. [3] find that, 'vaccinees and healthcare professionals tend to report preferentially the symptoms of the disease against which the nonlive vaccine was administered' and it has occasionally been argued that most findings of placebo effects are simply the result of reporting bias (e.g. [4]). Nevertheless, there is considerable evidence for placebos having positive effects on some conditions: pain, ulcers, etc. For reviews, see [5–9]. The effect of dummy tablets is modulated by many factors, such as the colour of the pills, their size, the regularity with which they are taken and so on. Operations seem to have more effect than tablets, and the manner of the treating physician also modulates the effect. These effects are summarised by Olshansky [8]. However, some conditions (such as cancer or schizophrenia) do not seem to be influenced by placebos.

Evans [5] suggests that placebo effects are related to the modulation of the innate immune system, rather than the acquired immune system. The hypothesis is data-driven: many placebo-responsive conditions (including pain, swelling, stomach ulcers, depression and anxiety) involve the acute phase response of the innate system. Thus, Evans argues, any condition that is not affected by the innate system will not be affected by placebos and, conversely, any condition which is affected by the innate system may be affected by placebos.

This chapter provides an evolutionary perspective on placebos, addressing the question of why an external cue can sometimes prompt individuals to get well, when they already had the capability to do so (since the placebo supplied no biologically active component). We summarise recent work that has shown how the placebo effect could be favoured by natural selection, and identify conditions under which placebos could have deleterious effects. In each case, the effects can be understood as an evolved response to trade-offs in different requirements, as we now discuss.

Biological systems can often be understood in the context of trade-offs (e.g. [10]). From the perspective of natural selection, the fitness of an individual is not just about immediate health, but long-term reproductive success (cf. [11]). The reproductive value of an individual will depend on many factors; age, health, risks of starvation or predation, etc. Fighting off a disease as quickly as possible may reduce the body's ability to deal with other infections or diseases in the near future. Humphrey [12] draws an analogy between the immune system and a hospital administrator who must manage resources when dealing with emergencies (taking into account current resources, expected deliveries and likely future emergencies). Rather than use all the available bandages, blood, etc., it is often better to hold some back in case of future needs. Similarly, it is expected that natural selection has shaped the body to manage its resources when dealing with an illness.

We are interested in the physiological allocation 'decisions' which affect health. Such decisions need not be conscious; evolution by natural selection shapes behaviour (including taking account of the costs and benefits of physiological actions in the immune system), so whether the immune system is adjusted consciously or not does not affect the analysis (cf. [13]).

All else being equal, it is always better to be healthy than ill. But given that someone is ill, the general 'decision' is how much resource (such as energy) to devote to the immune system at the current time (see Fig. 23.1). The key logic is that because the state of the world will affect the optimal level of resource allocation in current health (i.e. effort), perceptions about the world should affect the activity of the immune system.

23.2   Research Findings

There are many ways in which altered beliefs may affect immune systems. We break these down into positive and deleterious effects.

23.2.1   Positive Properties of Placebos

23.2.1.1   Reducing the Perceived Cost of Immune Action

In some situations, a placebo may reduce the perceived cost associated with increased immune effort. For instance, a sick individual could afford to put more effort into fighting an ailment if someone were there to assist (in case food provisions ran low or there was a predation threat). This accords with Houston et al. [14], who

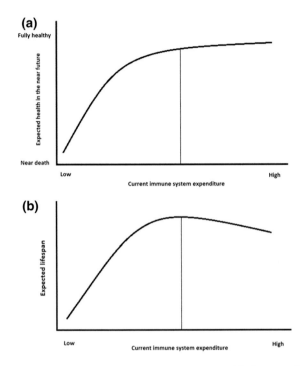

Fig. 23.1 The effect of immune system effort on current health for a particular condition. **a** Provides an example of how expected health in the near future might increase with immune system effort for a given condition. The vertical line shows a hypothetical (evolved) amount of effort typically put in by someone with that condition. To explain why they do not put more effort in (to further improve their expected health in the near future), we need to consider the consequences of increased effort. **b** Shows that for a given set of expectations (regarding current predation risk, prevalence of the disease, food supplies, weather conditions, etc.), some level of immune effort at the current time will maximise expected lifespan. This assumes that some reserves should be held back for potential emergencies in the future, such as other illnesses (this can change over time, depending on current illnesses and expectations about the future). Consequently, the optimal effort level will not maximise health in the near future, allowing beliefs (e.g. through placebos) to have an effect on health

show that the optimal allocation of energy to the immune system increases with energy reserves, when there are risks of death through disease or starvation. In the developed world, food is rarely in such short supply, but the perception of having others there to provide support may still have an effect, because we have not yet evolved to deal with the ecological features of the developed world. In this way, even the presence of someone who appears willing to support a sick patient may produce positive outcomes by increasing the amount of resource allocated to the immune system, much like a dummy tablet.

Trimmer et al. [15] consider a situation where mortality may be caused by disease or other factors. Increased immune activity may prevent the disease from gaining a foothold, or fight it off more readily. Such activity may also reduce energy levels

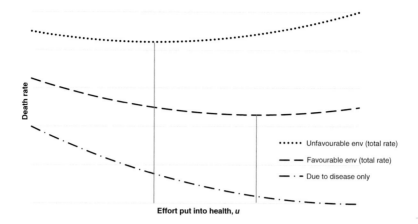

Fig. 23.2 (Reproduced from Fig. 23.1 of [15]): effort put into health should alter with environmental conditions. The death rate due to disease, D, decreases with effort put into health, u. The death rate from other sources (such as starvation or predation), M, decreases with the quality of the environment, a, and increases with u. In an unfavourable environment (where factors such as starvation or predation are more common than in the favourable environment), the optimal effort level, u^*, is low (indicated by the *vertical bar* on the *left*) because there is a high rate of mortality from other sources, so it is best not to increase that rate still further by putting much effort into health. In a favourable environment, where there is a smaller probability of death from other sources, the total mortality rate is minimised by putting more effort into health (indicated by the *right vertical bar*). $D(u) = (1 - u)^2/10$. $M(a,u) = (1 + u)/(20a)$. Unfavourable environment: $a = 0.4$, $u^* = 0.38$. Favourable environment: $a = 0.8$, $u^* = 0.69$

available for escaping predators, avoiding starvation, etc. Consequently, effort put into the immune system reduces the risk of mortality due to disease but increases the risk due to other sources. By assuming that the risk of mortality due to other sources is also dependent on the current conditions, Trimmer et al. show that in good conditions, it is optimal to put more effort into the immune system (Fig. 23.2). Consequently, if a placebo were to give an individual the perception that the world was "safe" (so the main mortality risk was through disease) then more effort should be put into the immune system, resulting in a positive placebo effect.

23.2.1.2 Reducing Perceived Effort of Overcoming Ailment

A placebo may sometimes work by reducing the perceived effort required to overcome an illness. If a dummy tablet is believed to assist recovery, then patients may (whether consciously or subconsciously) respond by increasing their immune activity so as to overcome the ailment now, rather than let it drag on. Consequently, the more powerful a placebo treatment is believed to be (enhanced by being more colourful, larger or prescribed to be taken more regularly), the greater the effect should be, as is often the case [8]. However, we shall see that, in some forms, such a belief may also be detrimental.

23.2.1.3   Reprioritisation Through Increased Urgency

So far, we have considered placebos that increase the perception that the world is currently conducive to improving health: less effort is currently required to fight off an illness, the risks of increasing the immune effort are currently low, or the risks of death through other sources are reduced. Each of these serves to reprioritise (and increase) how much effort to put into the immune system. However, it is also possible to reprioritise effort by regarding the current illness as more (or less) important. So a rather different method of inducing improvements in health would be to emphasise the importance of fighting the illness off as soon as possible. In an extreme form, this might be caused by a doctor saying, 'This is serious, it kills some people—you need to fight it off now!' but (more realistically) being told to take pills at regular intervals may be enough to highlight the importance of dealing with a condition, resulting in more effort being put into improving immediate health.

The extent to which placebos may influence health will, of course, depend strongly on the extent to which beliefs can affect the functioning of the immune system. However, we have seen that there are many ways in which beliefs should, at least in theory, have positive effects on health.

23.2.2   Deleterious Effects of Placebos

We have argued that current investment in the immune system should be traded off against other dangers. If a person's immune system was operating at the optimal level and a placebo altered the amount of effort put into immediate health, the placebo must therefore have increased some risks.

23.2.2.1   The Adjustment of Perception

Pain can be beneficial; it indicates a problem, encourages reduced use of the damaged region (e.g. a broken limb) and reduces activity, thereby giving the healing systems more chance to act. People who do not feel pain (due to a rare genetic mutation) have a low life-expectancy [16]. It is possible that a placebo which acts to reduce pain, without having any direct benefit on the ailment causing the pain, could lead to more danger.

Placebos certainly can have such an effect. By measuring how long people are able to submerge their arms in ice water, it has been shown that a placebo (even one so simple as telling the patients about the positive benefits of ice water immersion) can significantly increase the length of time that subjects are willing/able to expose themselves to such cold (e.g. [17]).

Nevertheless, the use of placebos for pain reduction carries many benefits without the pharmacological side effects of 'real' pain killers; if a placebo will suffice, there is little or no benefit of using active drugs. The downside of placebos in this case also goes for 'real' pain killers; very occasionally, too much pain reduction may be detrimental.

In some cases, placebos seem to improve patients' symptoms (by reducing their perception of pain) while the underlying condition remains unchanged (or even deteriorates). Any placebo that reduces perceived risk may result in time being lost before useful treatment is obtained. This is arguably one of the risks of 'quackery' and alternative (but ineffective) medicines; the more convincing the treatment (i.e. the more effect they have on perception without improving the underlying cause), the more harm may be done.

23.2.2.2 The Perceived Cure-All

Too much belief in a treatment may reduce the effort put into fighting an ailment. If an individual believes that a treatment is sure to fix their ailment, then there is no need for them to put effort in themselves. Because there is nothing medically active in a placebo, and the individual could be misled into putting less effort into their immune system, they may get worse rather than better.

This 'reverse placebo effect' was identified in Trimmer et al. [15] from a theoretical perspective. However, it is difficult to know how such an effect (of strong belief in a placebo treatment leading to deterioration in health) could be shown empirically while remaining ethical.

23.2.2.3 The Dangers of Depleted Resources

Having been influenced by a placebo into putting more effort into health, an individual will have reduced resources such as energy and rare minerals. In the developed world, where food is generally plentiful (all year round), there may be less risk in prescribing placebos to people than those that are at greater risk of starvation (e.g. in low-income countries). However, with depleted resources (e.g. micronutrients such as trace minerals, which are crucial for immune function), they may be more prone to infection and disease.

23.2.2.4 Future Effects

If there is an increased perception that placebos are often prescribed, patients may become more cynical, which may have consequences:

1. Reducing the placebo effect (including that associated with 'real' treatments) in the future, with other illnesses.
2. Reducing faith in doctors—perhaps meaning that people will be less likely to visit doctors to be treated (or delay until symptoms are worse).

23.2.3 Identification of Effects

Placebos are prescribed (and taken) because a once-healthy person has an ailment that needs to be 'fixed'. If taking a placebo results in recovery, then the improvement is (often) attributed to the placebo. But if the ailment gets worse, then the expectation may be that it was going to get worse anyway. Without any expectation that a placebo may have detrimental effects, it is easy for such effects to be missed.

Testing of medicines is most often carried out on new treatments. If a new type of treatment (with both medicinal and placebo components) is having negative effects, the outcomes may be attributed to the medical properties of the treatment rather than the placebo component of the effect. The health of the patients must come first, so it is more difficult to identify deleterious placebo effects (due to discontinued trialling).

Even in 'pure' placebo settings, Mitsikostas et al. [18] note that negative consequences may be more difficult to identify. People experiencing negative outcomes are more likely to drop out of studies and the people who go in for clinical trials may not be a representative selection of the population (cf. [19]).

23.2.4 The Complexity of Causation

One of the complications when studying the effect of placebos arises through the difficulty of distinguishing between an underlying illness (improving or deteriorating) and symptoms (improving or deteriorating). Evolutionary approaches to medicine [20] highlight that there is a potential danger of just treating the symptoms rather than the disease. Symptoms are sometimes caused by the immune defences being active, so a short-term deterioration in symptoms (such as a high temperature) may be associated with a longer-term improvement in health. Repeated stress responses have long-term, deleterious effects on the immune system [21, 22]. Therefore, if a placebo seemingly 'works' by increasing stress, thereby reducing immune activity and reducing symptoms in the short term, the treatment may produce negative outcomes in the longer term.

Table 23.1 summarises the four possible outcomes of a condition that would be *improved with increased immune system activation*, depending on the placebo effects on both the underlying illness and the symptoms.

An additional complication comes through the potential for symptoms to exist without an illness being present. This is because the body's defences are often triggered on a precautionary basis, reacting to risks rather than waiting to know that such a defence is needed. For instance, when eating a group meal, if one person is sick, this can trigger others to vomit. From an evolutionary perspective, this is an understandable precautionary reaction; the benefit of those few calories needs to be

Table 23.1 The correlation between long- and short-term effects of placebos can depend on the condition being treated

Symptoms (short term)		Illness itself (and thus long-term symptoms)	
		Improvement	Deterioration
	Improvement	Symptoms are caused by illness; placebo has increased the immune response	Symptoms are caused by fighting the illness; placebo has reduced the immune response
	Deterioration	Symptoms are caused by fighting the illness; placebo has increased the immune response	Symptoms are caused by illness; placebo has reduced the immune response

weighed against the risk of illness or death [23]. Placebos may reduce the precautionary reactions; whether this is good or not will depend on the situation.

23.3 Implications for Policy and Practice

We have clarified the distinction between the lay-term 'the placebo effect' (often assumed to be positive) and 'the effect of placebos' which may, in some situations, be deleterious. It is important for medical professionals to be aware of the possibility that dummy treatments have adverse effects.

Many people would prefer a world where placebos were not necessary; where any affliction could be dealt with directly, with no need for deception. However, since this is not possible, questions arise over the extent to which it is ethical to use placebos, when placebos will help or hinder, and the associated risks. An evolutionary perspective helps to address these questions; the general effects are shown in Table 23.2.

Present theory suggests that if prescribing placebos, a doctor needs to make a judgement about whether their patient would benefit from increasing or reducing their immune activity, and tailor their message accordingly. For instance, if the

Table 23.2 Summary of the effects of a placebo component of a treatment as a function of the current state of the immune system

Effect of placebo on immunological effort		Increase investment	Decrease investment
Immune system currently	Overactive	Deleterious (e.g. increasing exhaustion, autoimmunity? sepsis?)	Beneficial (e.g. reducing allergies)
	Balanced	Slightly deleterious, but can look beneficial in terms of immediate health	Deleterious; more likely to get ill
	Underactive	Very beneficial (normal 'placebo effect' in public consciousness)	Very deleterious

We ignore the cases where the placebo has no effect on subsequent immunological effort patient has an under-active system (the norm, so the aim is to encourage increased immune function), then the patient should be encouraged to believe that they have something which they need to take seriously, and that the treatment will 'help' to boost their immune system to fight off the ailment.

In terms of policy, Shiv et al. [24] identify that it may be necessary to charge for placebos if they are to have their maximum effect. But charging for something with no active ingredient means that there is a risk of law-suits. However, the finding that more expensive placebo treatments produce larger effects (e.g. for pain, see [25]) may mean that doctors and drug companies are not easily prosecuted. Although this is something which needs to have a current policy, it is also an aspect requiring more research in the future.

23.4   Future Directions

Evolutionary thinking is making progress in understanding the complex and subtle effects of placebos. Rather than prescribe placebos on the assumption that they will cause no harm, we recommend that future studies involving placebos also consider the possibility of deleterious effects.

There is already some empirical evidence that placebos produce negative effects in humans. For instance [26], a meta-analysis of many placebo studies is performed. Most studies involving placebos are carried out to study the effect of 'real' treatments, so they do not include a control case of 'no placebo'. However, by comparing options when different numbers of placebos have been taken, the study identified that those receiving more doses of placebos were more likely to report symptoms such as nausea, headaches and weakness. As Evans [5: 123] concludes, 'more research is needed'.

We regard a placebo as a treatment that can only have an effect on health by modifying expectations, so our analysis is also relevant to the potential for nocebos (typically ascribed the power to do a subject harm through negative expectations; [27]) to have positive effects. From a top-down perspective, there are many ways in which beliefs *should* affect the immune system (positively or negatively). However, more work is required to better understand the proximate mechanisms by which the brain affects the immune system, and how best to discern long-term precautionary reactions from short-term symptoms (especially in relation to [5]).

The key to progress on the placebo effect is in understanding how beliefs can affect physiology. Although there have been relatively few tests of placebos against a control of 'no treatment' for medical problems, there is a strong parallel with sports. While Humphrey talks of a 'health governor' in relation to the placebo effect on health (e.g. [28]), Noakes [29] talks of a 'central governor' in relation to how much effort an individual will put into an activity. The two are arguably closer than analogies; in each case, the brain makes use of information to modulate the body. In the case of physical activity, anticipation affects current effort level, and altering an individual's beliefs (whether consciously or not) may affect how much effort is put into an activity. As Beedie and Foad [29] summarise,

> Findings suggest that psychological variables such as motivation, expectancy and conditioning, and the interaction of these variables with physiological variables, might be significant factors in driving both positive and negative outcomes.

Considerable further work is required to understand how beliefs affect physiology. This research is needed if we are to fully harness the power of the placebo effect to better treat patients in the future.

Acknowledgments This work was supported by the European Research Council (Evomech Advanced Grant 250209 to A.I.H.).

Appendix: Definitions of a Placebo that Presuppose Its Effect Will Be Non-negative

Google online dictionary: A harmless pill, medicine, or procedure prescribed more for the psychological benefit to the patient than for any physiological effect.
Evans [30]: A placebo is a medical treatment that works (i.e. relieves symptoms or cures disease) because the patient believes it works.
www.clarkfamilydental.com/glossary.php: Inert medication or treatment that produces psychological benefit.
en.wiktionary.org/wiki/placebo: A dummy medicine containing no active ingredients; an inert treatment; anything of no real benefit which nevertheless makes people feel better.
http://www.merriam-webster.com/dictionary/placebo: An inert or innocuous substance used especially in controlled experiments testing the efficacy of another substance (as a drug).
Kitsch [1]: A placebo is a chemically inert substance that works by virtue of its presumed psychological effect.
Humphrey [12: 256]: a treatment which, while not being effective through its direct action on the body, works when and because:

- the patient is aware that the treatment is being given;
- the patient has a certain belief in the treatment, based, for example, on prior experience or on the treatment's reputation;
- the patient's belief leads her to expect that, following the treatment, she is likely to get better;
- the expectation influences her capacity for self-cure, so as to hasten the very result that she expects.'

References

1. Kitsch I (1978) The Placebo effect and the cognitive-behavioral revolution. Cogn Therapy Res 2(3):255–264
2. Pelto PJ, Pelto GH (1997) Studying knowledge, culture, and behavior in applied medical anthropology. Med Anthropol Q 11:147–163
3. Okaïs C, Gay C, Seon F, Buchaille L, Chary E, Soubeyrand B (2011) Disease-specific adverse events following nonlive vaccines: a paradoxical placebo effect or a nocebo phenomenon? Vaccine 29(37):6321–6326
4. Hróbjartsson A, Gøtzsche PC (2004) Is the placebo powerless? Update of a systematic review with 52 new randomized trials comparing placebo with no treatment. J Intern Med 256(2):91–100
5. Evans D (2003) Placebo; the belief effect. Harper Collins, London
6. Walach H, Jonas WB (2004) Placebo research: the evidence base for harnessing self-healing capacities. J Altern Complimentary Med 10:S103–S112
7. Vallance AK (2006) Something out of nothing: the placebo effect. Adv Psychiatr Treat 12 (4):287–306
8. Olshansky B (2007) Placebo and nocebo in cardiovascular health. Implications for healthcare, research, and the doctor-patient relationship. J Am Coll Cardiol 49(4):415–421
9. Price DD, Finniss DG, Benedetti F (2008) A comprehensive review of the placebo effect: recent advances and current thought. Annu Rev Psychol 59:565–590
10. Daan S, Tinbergen JM (1997) Adaptation of life histories. In: Krebs JR, Davies NB (eds) Behavioural ecology, 4th edn. Blackwell Science, Oxford, pp 311–333
11. McNamara JM, Buchanan KL (2005) Stress, resource allocation and mortality. Behav Ecol 16 (6):1008–1017
12. Humphrey N (2002) Great expectations: the evolutionary psychology of faith-healing and the placebo effect. In: Humphrey N (2002) The mind made flesh. Oxford University Press, Oxford, pp 255–285
13. Stewart-Williams S, Podd J (2004) The placebo effect: dissolving the expectancy versus conditioning debate. Psychol Bull 130(2):324–340
14. Houston AI, McNamara JM, Barta Z, Klasing KC (2007) The effect of energy reserves and food availability on optimal immune defence. Proc R Soc Lond B 274:2835–2842
15. Trimmer PC, Marshall JAR, Fromhage L, McNamara JM, Houston AI (2013) Understanding the placebo effect from an evolutionary perspective. Evol Human Behav 34:8–15
16. Schulman H, Tsodikow V, Einhorn M, Levy Y, Shorer Z, Hertzanu Y (2001) Congenital insensitivity to pain with anhidrosis (CIPA): the spectrum of radiological findings. Pediatr Radiol 31:701–705
17. Staats P, Hekmat H, Staats A (1998) Suggestion/placebo effects on pain: negative as well as positive. J Pain Symptom Manage 15(4):235–243
18. Mitsikostas DD, Mantonakis LI, Chalarakis NG (2011) Nocebo is the enemy, not placebo. A meta-analysis of reported side effects after placebo treatment in headaches. Cephalalgia 31 (5):550–561
19. Meynen G, Swaab DF (2011) Why medication in involuntary treatment may be less effective: the placebo/nocebo effect. Med Hypotheses 77(6):993–995
20. Nesse RM, Williams GC (1996) Why we get sick: the new science of Darwinian medicine. Vintage Books, NY
21. Shonkoff JP, Garner AS, Siegel BS, Dobbins MI, Earls MF, McGuinn L et al (2012) The lifelong effects of early childhood adversity and toxic stress. Pediatrics 129:e232–e246
22. Ellis BJ, Del Giudice M (2014) Beyond allostatic load: rethinking the role of stress in regulating human development. Dev Psychopathol 26(1):1–20
23. Nesse RM (2001) The smoke detector principle. Natural selection and the regulation of defensive responses. Ann N Y Acad Sci 935:75–85

24. Shiv B, Carmon Z, Ariely D (2005) Placebo effects of marketing actions: consumers may get what they pay for. J Mark Res XLII:383–393
25. Waber RL, Shiv B, Carmon Z, Ariely D (2008) Commercial features of placebo and therapeutic efficacy. J Am Med Assoc 299:1016–1017
26. Rosenzweig P, Brohier S, Zipfel A (1993) The placebo effect in healthy volunteers: influence of experimental conditions on the adverse events profile during phase I studies. Clin Pharmacol Ther 54:578–583
27. Hahn RA (1997) The nocebo phenomenon: concept, evidence, and implications for public health. Preventative Med 26(5):607–611
28. Humphrey N, Skoyles J (2012) The evolutionary psychology of healing: a human success story. Curr Biol 22(17):R1–R4
29. Noakes TD (2011) Time to move beyond a brainless exercise physiology: the evidence for complex regulation of human exercise performance. Appl Physiol Nutr Metab 36(1):23–35
30. Evans D (2004) Placebo. In: GregoryRL (ed) The Oxford companion to the mind, 2nd edn. Oxford University Press, Oxford. ISBN 0-19-866224-6

Index

Note: Page numbers followed by *f* and *t* refer to figures and tables, respectively.